The Circulation of Knowledge Between Britain, India and China

History of Science and Medicine Library

VOLUME 36

Knowledge Infrastructure and Knowledge Economy

Edited by

Karel Davids, VU University, Amsterdam
Larry Stewart, University of Saskatchewan, Saskatoon

VOLUME 3

The titles published in this series are listed at brill.com/hsml

The Circulation of Knowledge Between Britain, India and China

The Early-Modern World to the Twentieth Century

Edited by

Bernard Lightman, Gordon McOuat and Larry Stewart

BRILL

LEIDEN • BOSTON
2013

Cover illustration: The Planetarium in China. Watercolour by William Alexander (1792). Reproduced by permission of the British Library Board. WD 961/42 (122)

Library of Congress Cataloging-in-Publication Data

The circulation of knowledge between Britain, India, and China : the early-modern world to the twentieth century / edited by Bernard Lightman, Gordon McOuat and Larry Stewart.
pages cm. — (History of science and medicine library ; volume 36) (Knowledge infrastructure and knowledge economy ; volume 3)
Includes index.
ISBN 978-90-04-24441-2 (hardback : alk. paper) — ISBN 978-90-04-25141-0 (e-book)
1. Discoveries in science—History. 2. Communication in science—Europe—History. 3. Science—Great Britain—History. 4. Science—India—History. 5. Science—China—History. 6. Knowledge, Sociology of. 7. Enlightenment. I. Lightman, Bernard V., 1950– editor of compilation. II. McOuat, Gordon, editor of compilation. III. Stewart, Larry, 1946–, author, editor of compilation.

Q126.8.C56 2013
303.48'30903—dc23

2013007039

This publication has been typeset in the multilingual "Brill" typeface. With over 5,100 characters covering Latin, IPA, Greek, and Cyrillic, this typeface is especially suitable for use in the humanities. For more information, please see www.brill.com/brill-typeface.

ISSN 1872-0684
ISBN 978-90-04-24441-2 (hardback)
ISBN 978-90-04-25141-0 (e-book)

This book is printed on acid-free paper.

CONTENTS

PART FIVE

CONCLUSION

PREFACE: THE DALHOUSIE UNIVERSITY JAMES DINWIDDIE COLLECTION

Karen Smith

How Eastern and Western scientists of an earlier era exchanged ideas and influenced each other's work is difficult to document as primary source material has been vulnerable to the ravages of time. To help shed light on the intersection of Western and Eastern science in the late eighteenth century, Dalhousie University is pleased to make available the working papers of the noted scientist, Dr. James Dinwiddie (1746–1815), whose career took him from Scotland to India and China and London.

The story of how the Dinwiddie papers arrived at Dalhousie University is a fascinating one that involves a large cast of interesting people. James Dinwiddie was from Dumfries in the Lowlands of Scotland. After Dinwiddie's death in 1815 his papers, library and scientific instruments went to his only child, Ann.[1] Ann had married James Proudfoot, a gunmaker in Dumfries, and they had six children. Two of the six were primarily responsible for preserving their grandfather's papers—notably, the third son William and the youngest, Catherine. The two eldest brothers also played a minor role. Normally the eldest son would inherit all worldly goods but the two older Proudfoot sons were very poor managers, and, according to their bother William, they were worthless alcoholics. They managed to lose the family farm and were forced to vacate the Proudfoot estate, Cottagehead.[2] William, a surveyor in Liverpool, prevailed upon his brothers to allow him to take procession of the James Dinwiddie papers and library. William gathered up the trunk of Dinwiddie papers in October 1846 and added the books in 1848. In a journal entry for Oct. 11, 1846, William vowed to spend his leisure hours writing the memoirs

---

1 William Proudfoot, *Biographical memoir of James Dinwiddie, LL.D.* (Cambridge: Cambridge University Press, 2010: reprint of 1868 edition published by E. Howell, Liverpool, England), 124.

2 Dalhousie University Archives (hereinafter DUA), Halifax, Nova Scotia, Canada. James Dinwiddie fonds, MS-2-726, File I.2. *Journal of William Proudfoot, Oct. 1840–1852*, entry for April 3, 1848.

of Dr. Dinwiddie.[3] Twenty years later he published, *Biographical memoir of James Dinwiddie, LL.D., astronomer in the British Embassy to China, 1792,'3,4': Afterwards professor of Natural Philosophy in the College of Fort William, Bengal: Embracing some account of his travels in China and residence in India* (Liverpool: Edward Howell, 1868). Fortunately this volume has been recently reprinted in the Cambridge Library Collection Travel and Exploration series and is readily available.

After he retired, William relocated to Tunbridge Wells, Kent and lived with his sister, Catherine Bulman. Catherine's son, James Dinwiddie Bulman, immigrated to Canada in the 1870s and ended up in Sweetsburg, Quebec,[4] where he served as Captain and Adjutant in the 79th Battalion. James also worked as a registrar of patents and as a practical watchmaker and jeweller. Not surprisingly he was good with tools and specialized in chronometer, rack level and watch repair.[5] The trunk of James Dinwiddie papers was passed down to James Bulman and made the trip across the Atlantic Ocean.

The Dinwiddie trunk remained in the Bulman family home in Sweetsburg until the 1970s. Ann Dinwiddie Bulman Ashness-Wells, James Bulman's daughter, and her niece May Bulman shared the Bulman family home. Between 1970 and Ann's death in 1974, many family furnishings and items were dispersed to interested family members and others sold. A few letters, two notebooks and two journals were given to a family member in Alberta and are now reintegrated back into the collection. A trunk with most of the Dinwiddie papers still intact was acquired by Dr. Gilles Bisson, a Montreal surgeon, who lived in nearby Cowansville, Quebec.[6] He switched specialities and accepted a position as pathologist at the Victoria General Hospital in Halifax in 1972. The Dinwiddie trunk was packed up with all the Bisson furnishings and transported to Halifax. In the late 1990s, a Bisson grand-daughter took a history course at Dalhousie University and mentioned to her professor that her grand-mother was

[3] DUA, MS 2-726. *Journal of William Proudfoot, Oct. 1840–1852*, entry for October 11, 1846.

[4] DUA, MS 2-726, *Journal of William Proudfoot, June 1875–Oct. 1879*, entry for May 21, 1878.

[5] "Sweetsburg, Quebec" in *County of Missisquoi and Town of St. Johns Directory for 1879, 1880 and 1881* (Montreal: O. L. Fuller, 1879), 113; "Dominion of Canada Militia List" in *Canadian Almanac and Repository of Useful Knowledge for the year 1893* (Toronto: Copp, Clark Co., 1892), 66.

[6] "Gilles E. Bisson", in *Canadian Medical Directory* (Don Mills, Ont.: Seccombe House, 1970).

moving and wanted to find a good home for some interesting papers in an old trunk. In January 1998 the Dinwiddie papers were relocated to the Dalhousie University Archives and they were officially donated to Dalhousie University by Mrs. Barbara Bisson in 1999. After the papers were received, Dalhousie Archives assistant, Hansel Cook, mounted an online exhibit of collection highlights. The online presence attracted the notice of Dinwiddie family, friends and researchers from all around the world. As a consequence more Dinwiddie material has come to light and is contributing to a fuller understanding of James Dinwiddie's career and life.

The Dalhousie James Dinwiddie collection is remarkable for the variety of material present, the range of topics explored, the length of the time period covered and for the international scope of his work. Historians working with the Dalhousie James Dinwiddie Collection will have available to them the following:

Correspondence—In the collection are letters to Dr. James Dinwiddie from 151 colleagues, friends, and family members dating from 1770 to 1815. Also present are a number of letters Dinwiddie wrote to family members and colleagues.

Personal Journals (83 items)—There are 83 personal journals with almost daily entries dating from Jan. 5, 1776 to Jan. 17, 1812.

Scientific Journals (38 items)—Dinwiddie kept numerous scientific journals of his experiments and observations. In the collection are 22 volumes of his *Queries and Hints* series and 16 volumes on specific topics such as optics, galvanism and astronomy. The earliest such journal present is dated 1779 and the latest is 1805 just before his return to London.

Notebooks (20 items)—Dinwiddie kept separate notebooks of notes he took at Royal Institution meetings and of other information relevant to his work. Among the most interesting notebooks in this series are the Meteorological Journal (on voyage from London to China 1792–1794) and Chronometer Journal (on voyage from London to China 1793).

Lecture Notes (79 sets) There are 79 sets of lecture notes compiled by Dinwiddie on topics ranging from ballistics, fine arts, metaphysics, motion and force, to naval architecture. The lecture notes date from 1781 to 1811.

Experiment Reports (3 reports)—In the 1770s Dinwiddie conducted a series of experiments and recorded all aspects of the process. The experiments are also outlined in his scientific journals.

Royal Institution (15 items)—When living in London after 1807, Dinwiddie was deeply involved in the administrative business of the Royal Institution. Meeting notices, minutes, ballots for executive positions and annual reports were carefully retained.

Family Journals and Portraits (9 items)—The personal journals of Dinwiddie's grandson and biographer, William Proudfoot and the short biography he wrote about his father, James Proudfoot, are present. William also copied oil paintings of his parents, James and Ann and they are part of the collection.

As in all collections there is also a miscellaneous section. In the Dinwiddie collection this section has 13 items that include a translation of a Malay law, an essay on Roman legions and Dinwiddie's University of Edinburgh degree.

The entire collection is available for onsite consultation in the Dalhousie University Archives on the 5th floor of the Killam Library, Dalhousie University or available as a digitized collection. The digitized Dinwiddie collection is available online via two routes. Go to http://www.library.dal.ca/DUASC/Digital-Collections/Dinwiddie/ or go to DalSpace (http://dalspace.library.dal.ca) and select the Dalhousie University Libraries option. The Dinwiddie material is within the Dalhousie University Archives and Special Collections section.

While Dr. James Dinwiddie was not associated with Nova Scotia and Dalhousie University in any way it is fitting that his papers are now in Halifax. Both Dalhousie University's founding patron, Lord Dalhousie, and the first president of Dalhousie University, Dr. Thomas McCulloch, were graduates of the University of Edinburgh in the same time period that Dinwiddie was there. They all grew up amid the intellectual ferment of the Scottish enlightenment and they all valued scientific and philosophical innovation. They each left an important legacy for future scientists and for history of science historians.

ACKNOWLEDGEMENTS

The editors would like to thank Michael Moosberger, Chief Archivist and Karen Smith, Archivist at the Dalhousie University Library, and Kathryn Harvey, Archivist at Guelph University. Barbara Bisson donated the bulk of the Dinwiddie papers to Dalhousie University. Margaret C. Jacob provided the initial funding for the digitization of the Dinwiddie papers. Sundar Sarukkai of Manipal University has been a great facilitator of this project. Funds were provided by the Social Sciences and Humanities Research Council of Canada International Opportunity Fund and the "Situating Science Strategic Knowledge Cluster." We would also wish to thank the external reviewers for their informed and useful comments along with our Brill editors, Rosanna Woensdregt, Sabine Steenbeek and Michiel Thijssen.

LIST OF CONTRIBUTORS

ARUN BALA is Senior Research Fellow with the Asia Research Institute at the National University of Singapore. He is the author of the book *The Dialogue of Civilizations in the Birth of Modern Science* (Palgrave Macmillan, 2006) which shows how Asian ideas—in particular Chinese mechanical inventions and cosmological views, Indian computational techniques and atomic hypotheses, and Arabic planetary and optical theories—combined with ancient Greek ideas to generate modern science. He is also the editor of *Asia, Europe and the Emergence of Modern Science: Knowledge Crossing Boundaries* (Palgrave Macmillan 2012).

FA-TI FAN is the author of *British Naturalists in Qing China: Science, Empire, and Cultural Encounter* (Harvard University Press, 2004) and two dozen articles on a variety of topics, including science in modern East Asia and the global history of science. He is writing a book on science and mass politics in communist China. He teaches at Binghamton University, State University of New York.

JAN GOLINSKI is Professor of History and Humanities at the University of New Hampshire, and has recently held fellowships at the Huntington Library and at the Chemical Heritage Foundation. He is the author of *Science as Public Culture: Chemistry and Enlightenment in Britain, 1760–1820* (Cambridge University Press, 1992); *Making Natural Knowledge: Constructivism and the History of Science* (new edition, University of Chicago Press, 2005); and *British Weather and the Climate of Enlightenment* (University of Chicago Press, 2007). His current work is on Humphry Davy's self-fashioning as a man of science, and on the sciences of weather and climate in the long eighteenth century.

BERNARD LIGHTMAN is Professor of Humanities at York University, where he is Director of the Institute for Science and Technology Studies. He is also the Editor of the History of Science Society's flagship journal, *Isis*. Lightman's most recent publications include *Victorian Popularizers of Science*, *Evolutionary Naturalism in Victorian Britain*, and *Science in the Marketplace* (co-edited with Aileen Fyfe). Lightman is also general editor of a monograph series titled "Science and Culture in the Nineteenth

Century," published by Pickering and Chatto. He is currently working on a biography of John Tyndall and is one of the editors of the John Tyndall Correspondence Project.

GORDON MCOUAT teaches history and philosophy of science at the University of King's College and Dalhousie Universities. He is also the Director of the Canadian national strategic research cluster, "Situating Science: Cluster for the Humanities and Social Studies of Science". His own research concerns the concepts and history of classification, natural history and logic.

KHYATI NAGAR is a PhD candidate in the Graduate Humanities Program at York University, Toronto. Her dissertation focuses on art and science in botanical publications of the East India Company in the 19th century. The chapter in this book is her first publication. Khyati's major areas of expertise are: the circulation of science between Asia and Europe from 1750–1900, the visual cultures of science and the social and cultural history of India.

SAVITHRI PREETHA NAIR received her doctorate from the School of Oriental and African Studies, University of London, in 2003 for her dissertation on the museum and the shaping of the sciences in colonial India. Nair's research interests include modernity and enlightenment at the turn of the 19th c, the history and politics of collecting for science, and women and science in colonial and post-colonial India. Amongst her most recent publications is the co-authored *Science and the Changing Environment in India: A Guide to Sources in the India Office Records 1780–1920* (British Library, 2010) and *Raja Serfoji II: Science, Medicine and Enlightenment in Tanjore* (Routledge, 2012). As New India Foundation Fellow, she is at present engaged in writing the biography of Indian woman cytobotanist E. K. Janaki Ammal.

JAHNAVI PHALKEY is Lecturer in History of Science and Technology at the King's College, London. She has studied politics and history of science at the University of Bombay, the School of Oriental and African Studies, London, and Georgia Institute of Technology, Atlanta. Her research is focused on contemporary history of science at its intersection with transformations on the Indian subcontinent during the twentieth century, and after, as seen through laboratory histories and histories of scientific practice.

VEENA RAO is an obstetrician and gynaecologist by profession. She worked at the department of human genetics at the Institute for Research in Reproduction in Mumbai and at the Indian Institute of Science, Bangalore, for a short period. A visit to the Haldane Collections at the Centre for Cellular and Molecular Biology in Hyderabad, India, stimulated her to take up a project titled, "J. B. S. Haldane's Indian Period" in 2003. She has published a manuscript titled "J. B. S. Haldane, Ernst Mayr and the Beanbag Genetics Dispute" in the *Journal of the History of Biology* in 2010 and an article titled "Haldane Archives" in *Wellcome History* issue 24 in October 2003. Currently she is an Honorary Research Associate at the National Institute of Advanced Studies, IISc Campus, Bangalore 560012.

SUNDAR SARUKKAI is trained in physics and philosophy, and has a PhD from Purdue University, USA. His research interests include philosophy of science and mathematics, phenomenology and philosophy of language and art, drawing on both Indian and Western philosophical traditions. He is the author of the following books: *Translating the World: Science and Language* (University Press of America, 2002), *Philosophy of Symmetry* (Indian Institute of Advanced Studies, 2004), *Indian Philosophy and Philosophy of Science* (CSC/Motilal Banarsidass, 2005), *What is Science?* (National Book Trust, 2012) and *The Cracked Mirror: An Indian Debate on Experience and Theory* (Oxford University Press, 2012, co-authored with Gopal Guru). Presently, he is the Director of the Manipal Centre for Philosophy & Humanities, Manipal University, India.

GRACE YEN SHEN is Assistant Professor of Chinese History at Fordham University, in New York. She works on the history of modern science in China and is interested in questions of national identity, scientific internationalization, spirituality, and empiricism. She is in the midst of writing on the development of fuel sciences in China and beginning a project on China's participation in the formation of the United Nations. Her book, *Unearthing the Nation: Modern Geology and Nationalism in Republican China*, is forthcoming from University of Chicago Press.

KAREN SMITH has been an academic librarian in the Dalhousie University Library system since 1979. Currently she is the Special Collections Librarian responsible for all aspects of collection development and reference service for the Dalhousie rare book and special collections. She serves on the Editorial Board of the *Journal of the Royal Nova Scotia Historical Society*.

Research interests include Canadian literary small press development; nineteenth century Nova Scotia publishing and early libraries in Nova Scotia.

LARRY STEWART is Professor of History and director of the Situating Science node at the University of Saskatchewan, Canada. He is the author *The Rise of Public Science* (1992) and, with Margaret Jacob, *Practical Matter* (2004), as well as various essays on the dissemination of scientific knowledge since Newton. He is currently writing a study of experiment during the Enlightenment and the first industrial revolution, and is editing, with Erica Dyck, a collection of essays on the uses of humans in experiment.

JONATHAN R. TOPHAM is a Senior Lecturer in History of Science at the University of Leeds, UK. His research relates mainly to the history of printed communication in science, especially in late Georgian Britain. Among his co-publications are *Science in the Nineteenth-Century Periodical: Reading the Magazine of Nature* (Cambridge University Press, 2004), *Culture and Science in the Nineteenth-Century Media* (Ashgate, 2004), and *Science in the Nineteenth-Century Periodical: An Electronic Index* (HRI Online, 2005). He is currently completing a monograph about the Bridgewater Treatises entitled *Reading the Book of Nature: Science, Religion and the Culture of Print in the Age of Reform*. He is also preparing a book-length study of science and print culture in Britain, 1789–1832.

YANG HAIYAN is an associate professor in the Department of Medical Humanities at Peking University and was a visiting scholar in 2008–09 under the supervision of Professor James Secord at the University of Cambridge. With an educational background in biology and in the history and philosophy of science (PhD, Peking University, 2003), her research focuses on history of evolutionism, communication of Darwinism in China, and medical humanities. Dr. Yang works with the "Darwin Now Network" established between the Cambridge HPS Department and the Philosophy Department at Peking University to investigate the global spread of Darwinism. She also works on "The Academic Tree of Chinese Geneticists" and "Birth Control: Key Technologies and Its Social Implications", which are funded by the China Association for Science and Technology.

LIST OF ILLUSTRATIONS

Veena Rao

INTRODUCTION

Bernard Lightman, Gordon McOuat, and Larry Stewart

A Failure to Circulate

In the spring of 1793, at the end of a sea voyage of many months out of Portsmouth, an attempt to breach the resistance of the Chinese floundered amid the politics of disappointment and diplomacy. The first serious attempt by the British to engage with the Chinese Emperor, through the embassy of the experienced and well-connected Lord George Macartney, was candidly accepted by the ambassador as a venture confused by misunderstanding and ritual. Even so, the initial venture did offer future success.[1] No matter how well-conceived, such an encounter failed to survive the first plan laden as it then was with many commercial objectives. Macartney's entourage, under direction from King George III and the Board of Trade, successfully made the ocean voyage and the inland trek to Pekin, but there it faltered. There were too many obstacles for the embassy to carry the weight assigned to it. The expedition has been deemed a diplomatic failure—perhaps symbolized by the delicate negotiations needed to address Viscount Macartney's refusal to conduct the kow-tow, to prostrate himself before his Imperial Majesty, or the status of gifts from King George III.[2] There was to be no immediate opening of China to British trade beyond the foreign factory otherwise indulged at Canton. As he put it, Macartney was thus "obliged to give up on any signal success."[3] Such, at least, was the immediate assessment. Yet the story was rather more complex and the consequences more interesting beyond Peking—from Canton, to Calcutta, to London. As we shall see in the following essays, many an encounter and many an expedition often produced problematic consequences.

---

1 James L. Hevia, *Cherishing Men from Afar. Quing Guest Ritual and the Macartney Embassy of 1793* (Durham and London: Duke University Press, 1995), 218.

2 Hevia, *Cherishing Men*, 77, 104, 180, 220.

3 J. L. Cranmer-Byng, ed., *An Embassy to China. Being a journal kept by Lord Macartney during his embassy to the Emperor Ch'ien-lung 1793–1794* (Hamden: Archon Books, 1963), 206.

Fig. 1 The Planetarium in China. Watercolour by William Alexander (1792). Reproduced by permission of the British Library Board. WD 961/42 (122).

In search of a grand impression and Enlightenment display, Macartney had also intended his embassy as a scientific and manufacturing exchange—symbolic ultimately of Western superiority. Much, for example, can be made of the extravagant and much repaired planetarium taken by the embassy.[4] But the real key was in the supercargo. Macartney intended to

[4] British Library, IOR G/12/20, fols. 596–618. Articles bought for Macartney expedition January–August, 1792. The Planetarium, by P. M. Hahn and Baron de Mylius in Bavaria, was repaired at great expense by Vulliamy and Sons, Pall Mall, London. See Henry C. King with John R. Millburn, *Geared to the Stars. The Evolution of Planetariums, Orreries, and Astronomical Clocks* (Toronto and Buffalo: University of Toronto Press, 1978), 237–238. The plan to send scientific apparatus originated with the ill-fated Cathcart Embassy of 1787 and consisted, in part, of articles returned and repaired. IOR G/12/20, fols. 638–641. The

take along painters, musicians, a machinist, and "Gentlemen of Science capable of making Philosophical Experiments."[5] And here was the impecunious Scottish schoolmaster and itinerant lecturer James Dinwiddie (1748–1816), accompanying the diplomatic mission nominally as its "mathematician" and machinist. His perspective on the expedition, rather less heavily encumbered by the strategic cares of state, was equally reflective. Dinwiddie's private, unpublished, notes have only recently emerged as the best first-hand account of philosophical and experimental encounter with the Far East. These were first discovered in the basement of a house in the Eastern Townships of Quebec, Canada, before being deposited at Dalhousie University, Halifax, in most instances half a world away from where they were first penned. We thus have the considerations of an insider which provides a richness unequalled by the published semi-official reports of those in charge. Indeed, one can surely argue that it is not the diplomats but the practitioners who dealt with the day to day who tell us much about the nature of scientific encounter.[6] Dinwiddie evidently then took a dim, if cautious, view of imperial presumptions. He had by 1793, in a career that spanned half the globe, learned a lot from disappointments. But his reflections on the appearance of failure in China also reveal much about the meaning that lies behind scientific exchange. It was Dinwiddie's manuscripts which occasioned the international conference in Halifax, on the circulation of scientific knowledge, out of which this volume has arisen.

When Dinwiddie had signed on to the Macartney expedition in 1792 his duties, founded upon his reputation as a public showman, or demonstrator, were precisely tied to a raft of scientific instruments intended to impress.[7] The general trade in lecturing had prepared him well. He

cargo intended as gifts included a vast array of instruments, including some of the latest chronometers; see fol. 608; cf. Hevia, *Cherishing Men*, 77; J. L. Cranmer-Byng and Trevor H. Levere, "A Case Study in Cultural Collision: Scientific Apparatus in the Macartney Embassy to China, 1793," *Annals of Science* 38 (1981): 503–525; Maxine Berg, "Macartney's Things: Were They Useful? Knowledge and the Trade to China in the Eighteenth Century," (http://webfirstlive.lse.ac.uk/economicHistory/Research/GEHN/GEHNPDF/Conf4_MBerg.pdf)

[5] British Library, IOR G/12/91/ Part I. Papers relating to Lord Macartney's Embassy to China, 1792–93. Vol. I, fol. 35. 'Tableau or Sketch of an Embassy from His Majesty to the Emperor of China'.

[6] For a recent exploration of such issues see Simon Schaffer, Lissa Roberts, Kapil Raj, and James Delbourgo, eds., *The Brokered World. Go-Betweens and Global Intelligence, 1770–1820* (Sagamore Beach: Science History Publications, 2009).

[7] It seems likely that Dinwiddie was actually a second thought for those who carefully considered the best way of impressing the Chinese court. While a recommendation may actually have come through Matthew Boulton in Birmingham, it seems that the inventor,

was to be engaged in "erecting and regulating the planetarium; the constructing, filling and ascending in the balloon; descending in the diving bell; together with experiments on air, electricity, mechanics, and other branches of experimental philosophy; astronomical and other calculations."[8] What might have amused his audiences in Britain, in the numerous sites of scientific culture, was surely to amaze the Emperor's Court. The result was entirely the opposite. From the point of view of the officers of the embassy, whatever Dinwiddie accomplished was not what was expected. But no fault would attach to him. According to Macartney, neither the Emperor "nor those about him appeared to have any curiosity in these matters.... It was observed that most of the great men who came to see the globes, the planetarium, the barometers, the pendulums put up at Yuan-ming Yuan affected to view them with careless indifference, as if such things were quite common and familiar to them, and the use of them well understood."[9] British treasures seemed like trifles.[10] Experimental philosophy, as demonstrated by Dinwiddie, seemed at first not to travel well.

Despite reflections on the apparently obstinate refusal of the Chinese Court to admire British skill in instrumental design, Dinwiddie clearly believed the obstacles were rather more subtle, cultural and complex than first presented in public memoirs of the encounter. Macartney acknowledged "We are quite ignorant of their language," although he assumed that,

---

chemist, and balloonist James Sadler was the first choice. Sadler was assistant to the chemist and radical physician Thomas Beddoes. He was also a well-known balloon adventurer and inventor of engines. Although he seemingly purchased six hundred yards of silk for a balloon, it is probable that one of the reasons he actually failed to sail with Macartney was his promotion in 1792 of an engine which he was convinced would be more powerful than that of Boulton and Watt but at half the price. See Cornwall Record Office, Davies Gilbert Correspondence, DG 41/17, Thomas Beddoes to D. Giddy, nd; DG 41/21, James Sadler to Revd. Stacey, August 10, 1792; also J. L. Cranmer-Byng and Trevor H. Levere, "A Case Study in Cultural Collision: Scientific Apparatus in the Macartney Embassy to China, 1793," *Annals of Science* 38 (1981): 503–525, 505 n.2.

[8] Cranmer-Byng, ed., *An Embassy to China*, 310. See William Jardine Proudfoot, *Biographical Memoir of James Dinwiddie, LL.D., Astronomer in the British Embassy to China, 1792, '3, '4: Afterwards Professor of Natural Philosophy in the College of Fort William, Bengal: Embracing Some Account of His Travels in China and Residence in India.* Compiled from his Notes and Correspondence by his Grandson, William Jardine Proudfoot (Liverpool: Edward Howell, 1868), 27; R. H. Phillimore, *Historical Records of the Survey of India, Volume II (1800 to 1815)* (Dehra Dun, India: Offices of the Survey of India, 1950), 395–396. The best assessment of Dinwiddie's role and his responsibilities remains J. L. Cranmer-Byng and Trevor H. Levere, "A Case Study in Cultural Collision," op. cit., note 3.

[9] Cranmer-Byng, ed., *An Embassy to China*, 266.

[10] Hevia, *Cherishing Men*, 77.

since the young son of his secretary Sir George Staunton readily picked it up, "it cannot be a very difficult one."[11] Dinwiddie was far less sanguine. Indeed, Dinwiddie seems to have understood the trading zone, where encounters and exchange took place, to be a complex space.[12] Certainly, given the duty of explaining apparatus and experimental principles to the mandarins, as well as attempting to deduce the mechanical capacities of the Chinese, Dinwiddie was frustrated by the language barrier. "What information," he lamented, "could we derive respecting the arts and sciences in a country where we could not converse with the inhabitants?"[13] His irritation was first revealed many years later, in a biography published in the nineteenth century. With obvious access to the original documents, his grandson, William Jardine Proudfoot, recounted the efforts of Dinwiddie and his contemporaries to set up the often elaborate devices at the Emperor's court. We must understand, in the light of the picture of Chinese reluctance, that the hurdles were not entirely the elusive ones of protocol or of cultural convention.

Dinwiddie was left by the physician Dr. Hugh Gillan and the mathematician John Barrow to fend for himself in setting up the remarkable planetarium brought as a gift for the Emperor. Dinwiddie's account, by someone adept at instrumental displays, reveals a level of unaccustomed frustration:

> At the moment we turned round the lens-house, and while I stood at the mirror had a full view of his Majesty. He was very near us, and expressed nothing observable in his countenance. His address was much plainer, but in the same manner as the first order of mandarines [sic]. He looked at the lenses not more than two minutes, and retired. When viewing the air-pump, &c., he said, *These things are not good enough to amuse children.*"
>
> The air-pump and mechanical powers were selected for experiments, a variety of which were performed in presence of the chief minister and a number of mandarines [sic], who seemed little pleased even with the most entertaining. When the lenses were ready for experiments, the audience again attending, but the result only proved that a Chinaman will entertain ideas truly provoking a European philosopher. Wood set on fire, even the Chinese cash melted by the power of this apparatus, seemed to excite no other feelings in the prime minister than lighting his pipe at the focus in

---

[11] Cranmer-Byng, ed., *An Embassy to China*, 210.

[12] For the concept of 'trading zone' as a "social, material, and intellectual mortar binding together the disunified traditions of experimenting, theorizing, and instrument building," see Peter Galison, *Image and Logic: A Material Culture of Microphysics* (Chicago: University of Chicago Press, 1997), 798, 803.

[13] Cranmer-Byng, ed., *An Embassy to China*, 53, 347.

> derision of its usefulness. A little mirth was occasioned by a eunuch who had more curiosity to approach the focus with his finder, and felt more than he had desired for.[14]

Worse was to come. Even with all the effort taken to impress, notably with the spectacular burning lens purchased personally by Macartney in Macao, and made by William Parker of Fleet Street—the next day after its display the lens housing had already been dismantled. Dinwiddie complained about the end of the finest lens brought from Europe: " 'That lens–of which there is not an equal in the world—is consigned to everlasting oblivion.' A great mistake appears to have been committed in furnishing the Emperor with a list of the presents, which he was extremely solicitous to have delivered up in full. Access to the instruments occasioned further tension. "The two gentlemen were asked to perform some experiments on the air-pump, which they refused, and only walked about till noon, when they returned to town, meditating on the storm that was brewing."[15] Display was tangled in the web between gifts and tribute. Even given the opportunity to witness the latest experiments of ballooning which had been planned even before Dinwiddie had been recruited, and attempts with a diving bell and novel fireworks displays, the mandarins were evidently not to be moved.

The tale of encounter with the court was, at first appearance, a litany of polite obstruction of the wishes of the British. This has been an easy explanation. It was all too obvious that British objectives to secure greater access to the China trade required Chinese authorities acquiesce in revising their own, long-considered, astute sense of security. The Emperor would have none of it. Reasonably suspicious, surely the Chinese had enough experience with European traders at Canton who would have loved nothing better than to expand the ports of trade. Yet, on the other hand, if the Court was not amused by scientific machines it was a different matter in the international entrepot of Canton. It was here that Dinwiddie was very much on display and seemingly more appreciated. Macartney, in his reflections on the Chinese and the sciences, reported that

> Dr. Dinwiddie gave a few lectures and exhibited some experiments at Canton to the English factory, which were constantly attended by the principal Chinese merchants, who seemed highly delighted with them, and showed the strongest desire for further instruction. Had Dinwiddie remained at

14 Proudfoot, *Biographical Memoir of James Dinwiddie*, 53–54.
15 Ibid.

> Canton and continued his courses, I dare say he might have soon realized a very considerable sum of money from his Chinese pupils alone.[16]

Macartney's vision of the relative openness of the merchants was a stark contrast to the mandarins. The discovery of Dinwiddie's own manuscript notes of his Canton lectures tells us much more about the how natural and experimental philosophy could travel just as well as Macartney's ship's store of instruments. Many of the devices appear to have been retained by Dinwiddie when he left China. Even with the destruction of some instruments, others were returned and, evidently, "passed over to Dinwiddie as part payment for his services..."[17] Dinwiddie did not miss the next opportunity. He soon lectured in Canton, in October, 1793, on attraction and repulsion, on gravity, the division of matter, and the acceleration of falling bodies. His subjects were well within a Newtonian framework which, before the expedition, he had long delivered to audiences throughout Britain. If the patronage of Macartney had once saved him from the vagaries of the market for public lectures, Macartney was correct that Dinwiddie could have generated much wealth from cultivating an interested audience. That audience, however, would prove to be not in Canton but, on a long extended sojourn, in Calcutta. As Sir John Barrow sniffed, "it attended in India." But Barrow's impression was a bit misleading. If anything India was something of a beginning—ultimately, at the very least, of Dinwiddie's "little fortune."[18] This trajectory is the subject of the first essays in this volume.

James Dinwiddie's reputation as an experimental philosopher has largely been viewed as incidental to the expedition to China. But, as this volume will show, he was also a superb example of the careers of the many itinerant lecturers and scientific showmen with which the eighteenth century abounds. Moreover, he was himself a purveyor and creator of experimental apparatus. These were primarily to be used for the amusement of his auditors in Scotland, England, Ireland, as well as in China and India. His career was, especially, a profoundly important reflection of the role of instruments, not merely as presents to impress potential clients, but of the significance of devices as much as lectures that attest to the circulation of scientific knowledge. The instruments in which he was heavily invested were essential to the transport of philosophical ideas in

16 Cranmer-Byng, ed., *An Embassy to China*, 266.
17 Phillimore, *Historical Records*, 251.
18 Phillimore, *Historical Records, Biographical Notes*, vol. II, 181.

the various trading zones between instrument makers, experimentalists, and in spectacles across the spaces of Enlightenment in Europe, China and India. In this regard, the transport of instruments carried with it the latest philosophical knowledge. This is evident from his early days in Scotland when, as a schoolteacher, he first sought the purchase of an orrery to teach boys the rudiments of astronomy. That apparatus was the beginning of Dinwiddie's addiction to devices which drove him to debt and the edges of bankruptcy. But it was also the secret to his securing a place in the early public theatre of scientific circulation.

Circulating Knowledge

The "transit" of knowledge and its material culture has come under sharp scholarly focus as of late. Replacing one-sided stories of centre-to-periphery "dissemination", cultural and political historians and historians of science are now unpacking the multifarious and multidirectional nature of global exchange and encounter, by following things, theories and practices rather closely.[19] Earlier "dissemination" models of encounter rehearsed self-described notions of unequal development, where centres (*inter alia*, Europe) provided the scene and grounds of development, which in turn disseminated knowledge to the "periphery".[20] It is remarkable how much pro and anti-colonial discourse resonated with the same ethos.

In 2004, the historian of science James Secord advised us (at a conference on "circulating knowledge" held in Halifax) to follow things around in circulation, to examine how science moves from local, or even individual,

---

[19] This time around, we can point to the germinal work of Bruno Latour, *Science in Action* (Cambridge, Mass.: Harvard University Press, 1987). Francis Bacon had already encouraged the study of the circulation of things and their non-European origin at the very birth of the "Modern". Francis Bacon, *Franciscus de Verulamio Summi Angliae Cancellaris Instauratio magna* [*The Great Instauration*], 1620.

[20] The classic statement, in economics, is to be found in Samir Amin, *Unequal Development: An Essay on the Social Formations of Peripheral Capitalism* (New York: Monthly Review Press, 1976), and in science: George Basalla, "The Spread of Western Science," *Science* 156 (1967): 611–22. An impassioned view of the unequal development of scientific knowledge is to be found in the writings of the Sri Lankan scholar, Susantha Goonatilake, *Aborted Discovery: Science and Creativity in the Third World* (London: Zed Press London, 1984). It is surprising how the resistance to dominant knowledge of the "centre" rehearses the dissemination model. E.g., see Lewis Pyenson, *Civilizing Mission: Exact Sciences and French Overseas Expansion, 1830–1940* (Baltimore: Johns Hopkins University Press, 1993). For an early vociferous debate over the place of science in imperialism, see Paulo Pallidino and Michael Warboys, "Science and Imperialism" *Isis* 84 (1993): 91–102, Lewis Pyenson, "Cultural Imperialism and the Exact Sciences Revisited," *Isis* 84 (1993): 103–8.

significance to the "taken-for-granted." In this he was summing up years of scholarship beginning with the disintegration of the so-called "positivist school" of historiography, through the discombobulation of the Kuhnian Revolution to the short-lived tempest of social-constructivism.[21] The turn to practice, and then the turn to "things", marked a new materialism and particularism in the study of the history and meaning of science.[22] Secord suggests that we burrow down—a worm's eye view—to the way that science is "communicated" as a kind of communicative action across borders.[23] Science, East and West, is just such a border. Dinwiddie himself moved from itinerant Newtonian circulation, to Orientalist magi and bringer of knowledge, to Enlightenment accumulator of things and knowledge. This volume follows that legacy by following knowledge, things and people as they move within the East and the West and especially between. It resists taking seriously Eurocentrism as an abstract category. It starts by looking at the ways and means that knowledge circulated, first in Europe, but then beyond. It engages the knowledge and encounters of those Europeans as they moved over well-trodden territory, in search of the new and new markets. It also explores places where this new and old encounter could be resisted, refashioned, or transmuted.

Macartney's mission to China was centred on tricks and tokens of trade and science, accompanied by mathematicians and technicians to set up and maintain the displays. All was meant to impress but even the most careful negotiation encountered difficulties in different nuances attached to the reciprocity of "the gift".[24] The overarching goal, of course, was mercantile, and reflected the problematic relationship between the gift and

[21] James Secord, "Knowledge as Transit," *Isis* 95 (2004): 654–72.

[22] The turn from practice to things can be marked in the work of Susan Star & James Griesemer, "Institutional Ecology, 'Translations' and Boundary Objects: Amateurs and Professionals in Berkeley's Museum of Vertebrate Zoology, 1907–39," *Social Studies of Science* 19 (1989): 387–420, and perhaps Peter Galison, *Image and Logic.* The new movement has yet to finalise its relationship to "the theory of the thing". See Raiford Guins, ed. *The Object Reader* (London: Routledge, 2009).

[23] Secord's notion of building communication from the ground up is markedly different from the Kantian theory of reason and communicative action proposed by Jürgen Habermas, *Theory of Communicative Action.* (New York: Beacon, 1987).

[24] The scientific instruments of the Macartney mission are ably discussed in J. L. Cranmer-Byng and Trevor H. Levere, op. cit. (note iii), and Simon Schaffer "Instruments as cargo in the China trade," *History of Science* 44 (2006): 1–30, Hevia, *Cherishing Men*, 104ff.; and the more generalised discussion in Maxine Berg "Britain, Industry and Perceptions of China: Matthew Boulton, 'Useful Knowledge' and the Macartney Embassy to China 1792–4," *Journal of Global History* 1 (2006): 269–288.

the commodity (and especially of science) in the modern.[25] The Chinese too had their "gifts", and duty of protocol also aiming at the display of wonder. In such a display of *scientific* wonder, to achieve the true effect the construction must be concealed. But this was a lesson Dinwiddie had learned in Britain. Audiences admired the show as much as scientific substance. Dinwiddie knew that

> An ignorant people should always be taken by surprise. When a grand machine is shown all at once, and the principles of motion concealed, it seldom fails in its effect. But when it is shown piecemeal; when the various detached parts are exhibited and put together in their sight—in short, when the machine is built from its foundation before them, the time seems long; they impute every trifling mistake or hesitation in the workman about placing this or that part, which can be known most frequently only by trial, to a want of knowledge in the machine, and want of ability in the profession.[26]

In an attempt to impress the East, following the failure of Cathcart, the Macartney mission brought with it a particular display of manufactured goods, fruits of the new factory process of an industrialising Britain.[27] Here too, the very source of the wonder—the manufacturing method—could not be displayed (ironically, because of distance), and so the goods stood on quality alone. It was the process of manufacture that made them all so special. But the Chinese demurred. They had seen Western instruments and toys before. Compared to the fine craftsmanship of the Chinese goods, and indeed to Western devices already collected at the Emperor's court, these products of new process would readily be deemed derivative.

The *scientific* apparatus, on the other hand, were emblems of the new wares and ways of the Enlightenment Europe. They were meant by Macartney to impress on the Chinese the shock of innovation. We have learned to suspect, through the monumental work of Joseph Needham

[25] The classic study of "The Gift" in reciprocal exchange is Marcel Mauss, *The Gift* (London: Routledge, 2001). An interdisciplinary expansion is to be found in Mark Osteen, ed., *The Question of the Gift: Essays Across Disciplines.* (London: Routledge, 2002).

[26] Dinwiddie's journal, as quoted in William Jardine Proudfoot, *Biographical Memoir of James Dinwiddie* (Liverpool: Edward Howell, 1868), 47. It would perhaps be extremely profitable to compare this account of display and construction with that of the account of natural religion and artisanal construction in David Hume, *Dialogues on Natural* (London: 1776), Part 5, when our marvel at the ingenuity of the workman (and thus, by analogy, of the Prime Mover) is diminished we discover the actual method of artifice, achieved through "multiplied trials, mistakes, corrections, deliberations, and controversies."

[27] For the difference between the Chinese and British manufacturing process, and the striking achievements of Chinese porcelain production, see M. Finlay, "The Pilgrim Arts: The Cultures of Porcelain in World History," *Journal of World History* 9 (1998): 148, 156.

and his students, that the East was far from "carte blanc" with respect to science and technology, and even with the new ways of European goods and practices.[28] Contact with the "West" was already extensive and interactive—the Europeans had long borrowed and traded (and emulated even more) the processes and cargo of the "East", and, in turn, the "East" had already its fair share of European clocks and planetariums, transported by the Jesuits and other early circulators. But, that said, there was a marked "incommensurability" amongst the meaning of gifts. The Chinese Emperor's response is most telling. The "Middle Kingdom" had no use for such trade. According to the much cited letter to George III, the emperor made it clear that "[W]e have never valued ingenious articles, nor do we have the slightest need of your Country's manufactures."[29] The machineries of European Enlightenment wonder were pale reflections of the achievements of the East. And so the meanings were lost, and the Chinese were rather ostentatiously unimpressed.[30]

Settling Knowledge

While Dinwiddie's temporary successes at Canton were much greater than in Peking, his real chance to settle and shine came with an appointment at the College of Fort William at Calcutta. This was at the centre of a burgeoning and bustling empire carved out of the East by an unchallenged post-Napoleonic Britain. Established under the auspices of the East India Company in 1800 to impart general education and oriental languages to budding young British civil servants, the College has been described as "one of the most significant and interesting institutions of nineteenth-century India," even though its life was rather short-lived.[31] Acute tension between mercantile and administrative aspects of the East India Company played

[28] Joseph Needham, ed. *Science and Civilisation in China.* (Cambridge: Cambridge University Press, 1954+). The work spurred by Needham's original thesis is far too large to list here. For a recent encounter, see Morris Low, ed. *Beyond Joseph Needham: Science, Technology and Medicine in East and South East Asia. Osiris* 13 (Chicago: University of Chicago Press, 1998).

[29] Hevia, *Cherishing Men*, 238.

[30] A fine collection of original documents relating to the Chinese perspective of the Macartney mission can be found in Alain Peyrefitte, ed., *Un choc de cultures. La vision des Chinois.* (Paris: Fayard, 1992).

[31] A thoroughly interesting, well-researched and balanced account of the origins, and decline, of the College at Fort William is to be found in Sisir Kumar Das, *Sahibs and Munshis: An Account of the College of Fort William* (Calcutta: Rupa, 1978).

heavily on the institutional function of the College. The original plan for the College curriculum was wide ranging and liberal and was to include instruction in Law, Jurisprudence, History, Political Economy, Chemistry, Botany, Greek and Latin, Experimental Philosophy and Indian and European Languages—that is, more than a centre for "Orientalism".[32] Through his connections at the India Company, solidified by his performance in China, Dinwiddie was employed as the College's first Professor of mathematics and experimental natural philosophy. Here, finally, he could find a stable home for his style of experimental and mathematical display.

Again, contact and exchange between indigenous, mostly Brahmin, Indian and the Sahib's of the College found its limits on the unequal playing field of colonial domination, with Indians always playing a subordinate role in the College. But even there, on this uneven playing ground, we experience a landscape of exchange. Brahmin's occasionally visited Dinwiddie's lecture room (one occasion being an eclipse of the moon),[33] and Dinwiddie himself actively sought out the achievements and meanings of local knowledge. The accumulation was moderate. Yet, Dinwiddie's stay at Fort William was rather short, punctuated by a dispute within the Company regarding the purpose of a general and scientific education. He started for home in September 1806, having amassed a small fortune, and much machinery of the new science.

Circulating

James Dinwiddie circulated in three spheres: First and always as an itinerant lecturer of Enlightenment Newtonian science. Second as the mathematician, technician, and experimental showman of the first Embassy to China. And then, finally, as a professor of mathematics and natural philosophy at the College of Fort William. Around him circulated the robust theories of the "new science" and the apparatus of exchange and practices

---

[32] Das, *Sahibs and Munshils*, 8. While opening up the academic subfield of Post-colonial studies, Edward Said's *Orientalism* did much to construct the notion Western "Orientalism" as a decidedly one-sided enterprise, originating in the (European) centre, and written hegemonically upon the dominated subaltern. Ironically, in its attempt at pricking "essentialism", Said essentalised the exchange and, as a result, prevented the "subaltern" from speaking. For a thorough engagement and critique of Said's legacy, see Daniel Martin Varisco, *Reading Orientalism: Said and the Unsaid* (Washington: University of Washington Press, 2007).

[33] Proudfoot, *Biographical Memoir*, 104–5.

of European Enlightenment encounters with nature.[34] Here, through these exchanges, in the early-modern intellectual, instrumental and commercial trading zones, we follow the making of the Modern and the sciences' place within them. We note that this wasn't always a one-way street. Yet, the stories returned as accounts of ascendency and domination, the building of an imperial centre and scientific "universality". In that, Dinwiddie failed in reciprocity. Europe, while reaching out to the Orient, was self-assured in its ascendancy. Embedded in its trading networks was the mission of enlightening the natives. The direct site of exchange was one of comparison, translation, exchange and resistance. And those conditions remained even in the post-colonial period.

This collection of papers is about the history and meaning of that circulation. It participates in the attempt to open up more nuanced and balanced trajectories of colonial and post-colonial encounters. It also dovetails into recent general studies of science which eschew older stories of the "triumph of reason and evidence" (and logical skill) over localised indigenous knowledge. Dinwiddie certainly sought out indigenous techniques on behalf of the Company. Thus, exchange, translation, resistance—all are localised, or at least multi-central. In this we recognise a hope at giving voice and agency to the many purportedly "bit-players" and things originally relegated to the margins in the development of late modern science.

In these different centres, all opening up to contact and exchange, we find different sorts of things—not just the theories, laws and fact of received science, but science in the making, and the making of lots of material *things* too. The mercantalists who spurred Macartney's mission knew this long ago, and it was reflected in the cargo of the mission. The circulation of knowledge, where Dinwiddie was so exemplary, was also the circulation of things and people. Macartney brought lenses, clocks, porcelain and planetaria, along with those to attend to them. These cargo, or kinds of instruments, moved (or sometimes did not move) not only from one place to another, but from one cultural ecology to another, being restructured in their meanings and trajectories in those new ecologies. Their success as "science" was the correlate of this success at moving around. Sometimes, it should be noted, success came at the barrel of a gun, as the Chinese would soon discover. But sometimes it came in a more

34 For European enlightenment style, see William Clark, Jan Golinski and Simon Schaffer, eds. *The Sciences in Enlightened Europe* (Chicago: University of Chicago Press, 1999).

subtle, transformative, manner, bringing with it a new ecology of practices and ways of being with the world. The success of "European science" was a result of this. As the following essays argue, becoming universal, in science and in culture, was intimately related to this circulation.

The comparison of Chinese and Indian experiences of colonial and post-colonial encounters with "science" says much about the way science and its landscape gets circulated, and transcends the local. How science got transformed during and after this strict form of mercantile imperialism is both a story of globalisation and encirclement. The following studies form a foundation for those comparisons.

The Volume

The volume is divided into five sections. The first focuses on James Dinwiddie as the embodiment of the great age of the go-between at the turn of the eighteenth and nineteenth centuries. The three papers in this section draw extensively on the rich collection of Dinwiddie papers located in large measure at Dalhousie University to illuminate Dinwiddie's astonishing career over the course of half a century. Stewart's examination of Dinwiddie's instruments demonstrates their crucial role in his self-fashioning as a public showman in Britain, India, and China. His telescopes, air pumps, prisms, and electrical machines attested to the universality of natural phenomena as they provided the means of duplicating experiments in sites around the globe. Through them, Stewart argues, Dinwiddie conveyed his notions of physical powers and of natural theology to enthusiastic audiences. In so doing, he reflected the increasing reach of the itinerant scientific lecturers of the late Enlightenment—beyond Europe, to colonies in the Americas of course, and incidentally to China and India.[35] Nair centers on Dinwiddie's activities in Calcutta from 1794–1806. She asserts that Dinwiddie established a new culture of rational-recreation in Calcutta based on public lectures and spectacular experimentation that led to the emergence of public science in colonial India. Her fine-grained analysis of the traders, East India Company officials, artists, women, and members of the Indian elite who were members of his audience reveals hitherto unexplored dimensions of colonial science in this period. In his study of Dinwiddie's use of galvanism in Calcutta and London, Golinski

[35] Cf. James Delbourgo, *A Most Amazing Scene of Wonders. Electricity and Enlightenment in Early America* (Cambridge, Mass. and London: Harvard University Press, 2006).

turns the conventional story about circulation upside down. While teaching at the new College of Fort William in Calcutta Dinwiddie deployed the voltaic pile, a European discovery, to enhance his authority among colonials and natives. But when he returned to London he presented galvanism as an exotic curiosity from the East. Dinwiddie is clearly a key to understanding the complex patterns of circulation in the late eighteenth and early nineteenth century.

In section two, "Circulation Beyond Dinwiddie" three papers examine features of circulating knowledge both before and during the time of Dinwiddie's travels. Bala explores the possibility that key scientific ideas, developed by the Indian Kerala School of mathematical astronomy, were transmitted to Europe during the seventeenth century. But they were not transmitted through Jesuit missionaries. Rather, Indian craftsmen, who had learned them from their native scholars, conveyed them as a system of computing techniques to European craftsman aboard visiting Portuguese ships. Topham investigates the establishment of Britain's earliest scientific journals between 1793 and 1806. He argues that these journals were conceived of as international anthologies of scientific discovery, which were economically viable because they drew together materials from a wide range of printed sources from inside and outside Britain. These were the very volumes to which Dinwiddie, in India, gained access, purchasing directly from London. These new journals were a significant component of a new communication network by which Continental science came to circulate in Britain. Nagar brings us back to India and the complicated circulation of knowledge between Calcutta and London. The usual flow of knowledge broke down when locally produced botanical catalogues by William Carey and William Roxburgh were published in Calcutta without approval from Joseph Banks at Kew. Nagar asserts that this is compelling proof that colonial science practiced by European botanists in India was not merely an extension of London-based imperial science.

The chapters in section three explore how modern China later became an integral part of the global circulation of knowledge. When Dinwiddie and the Macartney expedition failed to persuade the Chinese emperor to open up trade routes in China to the British, it was clear that China still had the option in the late eighteenth century to refuse to participate in the expanding global exchange of knowledge. However the growing force of western imperialism over the course of the nineteenth century, fuelled in part by scientific contributions to new technologies of empire, made it more and more difficult for Chinese rulers to sequester themselves from the rest of the world. If the early modern period was the age of go-betweens like Dinwiddie, the late nineteenth ushered in a period when the circulation of knowledge was mediated by international organizations and by publishing institutions.

In her chapter Yang Haiyan examines how evolutionary theories circulated widely in China for the first time at the end of the nineteenth century. They circulated in the periodical press as a result of the first boom of Chinese journalism. Journalists became interested in evolutionary theory because it provided a means for coming to terms with China's humiliating defeat in the Sino-Japanese War of 1894–1895. By no means a passive recipient of western evolutionary ideas, Chinese intellectuals reworked them for their own purposes to build up their sense of cultural and national identity. Fan's chapter shifts the discussion to the circulation of materials objects, such as antiques and fossils, in twentieth-century China. Early twentieth century Chinese intellectuals reacted with hostility to the way that foreign scientific expeditions to Chinese Central Asia carried away numerous archaeological objects. The international controversy that resulted demonstrates that circulation is rarely a smooth process. The measures set up by China to regulate, monitor, and control the flow of antiques and fossils reveals that circulation involves a series of stops and starts, negotiations, and struggles. Shen's chapter draws our attention to large-scale modes of circulation, such as international conferences. She analyses Chinese participation in various international geological congresses in the early twentieth century against the backdrop of the formation of the discipline of geology in China. Increasingly in this period the Chinese found that it was necessary to participate in these meetings, since it was here that the movement of objects, ideas, and individuals were regularized. Shen's study illustrates how circulation patterns defined by international conferences could both marginalize and stabilize new scientific communities.

In section four, on modern India, Phalkey and Rao explore the circulation of science between the west and postcolonial India. Phalkey argues that the period between 1945 and 1950 is crucial for understanding the role of science in the imperatives of state formation in India. During this period the strategy of the India National Congress was to strengthen scientific industrialism as an overarching purpose of the state. This strategy was intended to make India productive and autonomous, but it also prevented the resurgence of religious and other nationalisms. This is how science and technology became central to the self-understanding and reinvention of India's place in the current world order. Rao's study of the Indian period of the Marxist J. B. S. Haldane provides an example of cultural exchange between East and West in the early years of post-colonial India. She examines Haldane's attempt to develop modern biology in India. Rather than push Indian science towards imitating the West, he tried to establish a strong base for a distinct Indian science. He encouraged young researchers to work on animals and plants in India, especially those of economic importance to the future growth of the nation.

In a final concluding chapter Sarukkai tackles the broad issue of translation and the history of science—a topic that cuts across many of the papers in the volume. Sarukkai argues that historians of science now need to integrate translation studies into their work, especially when they attempt to understand the transmission and circulation of knowledge. Rarely, he asserts, is a scientific idea transmitted in its structural entirety because it is supported by various metaphysical presumptions and belief structures unique to the culture in which it is originally found. Sarukkai suggests that historians of science consider how translation studies can help us understand the generation of new meaning in scientific discourses across cultures.

What began with us as an exercise in the exchange of early-modern science—with the traces of the peripatetic Dinwiddie as exemplar—was transformed by deliberation into a consideration of the wider ramifications of translation and trade. From the eighteenth century, it would appear that considerations of state were not only important—as Francis Bacon had once proposed would and should be the case—but the state and its institutions had important implications for the translation of science across intellectual and cultural boundaries and over many borders. We hope the collection of essays which follow will contribute to the instauration of that long debate.

PART ONE

CIRCULATING KNOWLEDGE: JAMES DINWIDDIE IN CHINA, INDIA, AND BRITAIN

THE SPECTACLE OF EXPERIMENT: INSTRUMENTS OF CIRCULATION, FROM DUMFRIES TO CALCUTTA AND BACK

Larry Stewart

The attractions of enlightenment science often rested as much in spectacle and entertainment as in the advance of theory and mathematics. Since the early eighteenth century, performance and experimental philosophy were not exclusive domains. Numerous public displays of Newtonian science increasingly followed often obscure, itinerant lecturers who wound their way from coffee house to tavern to lecture hall, from industrial cities to spa towns, in search of subscribers. This was the case throughout much of Europe, and particularly amidst what has recently been called Britain's knowledge economy.[1] Where available, the many published versions of their public lectures were obviously designed to give the impression of a well-polished performance. Print, however, is far from telling us much about the actual shows or their reception. Neither the place nor devices of display were clearly reflected in print and were very rarely revealed in manuscript. Consequently, the skill of many a lecturer and often his contrivances remain obscure. Performance has unfortunately left little more than the ephemera of the theatre. Even so, demonstrations were not simply entertainment. They were much more than an end in themselves.

In the age of enlightened economy, an especially good example of the lecturers and of their audience may be seen through the traces of James Dinwiddie. As he rarely published, among public performers Dinwiddie remains much less well-known than ought to be the case. Like others, he wandered through Scotland south to the Midlands, to London, Dublin and Paris, and ultimately, in his case, to the edges of Empire. When he is

[1] Cf. Joel Mokyr, *The Enlightened Economy. An Economic History of Britain 1700–1850* (New Haven and London: Yale University Press, 2009), esp. 40–62; Peter M. Jones, *Industrial Enlightenment. Science, technology and culture in Birmingham and the West Midlands 1760–1820* (Manchester and New York: Manchester University Press, 2008), esp. 70–109. Among the recent studies of intinerant lecturers see Roger Neil Bruton, "An Examination of the Extent to which Scientific Lecturers were contributing to the Dissemination of Knowledge in the mid-Eighteenth Century West Midlands or were largely a source of Entertainment" (M.A. diss., University of Birmingham, 2009); and Paul A. Elliott, *The Derby Philosophers. Science and Culture in British Urban Society, 1700–1850* (Manchester: Manchester University Press, 2009), esp. 41–46.

familiar at all, it is as 'astronomer' and mathematician to Lord Macartney's expedition to China in 1792. Dinwiddie first emerged as a mathematics teacher in the Scottish market town of Dumfries, likely matriculating at the University of Edinburgh in 1778. This was the beginning of a grand trajectory. In the astonishing survival, given their travels, of his many unpublished and frequently amended lecture notes and diaries, we can follow in great detail a career over almost half a century and half the globe. Dinwiddie was perpetually in between, a creature of opportunity and the culture of connection. His traces moreover take us much closer to daily performances in physics than the finely-crafted publications of many better-known lecturers. Dinwiddie printed little—merely a few surviving course outlines, and calculations made for almanacs in 1778 and 1787.[2] Yet on tours of Ireland, or trailing throughout many centres in Britain, to scientific demonstrations at the limit of the second British Empire, in Peking, Calcutta and Bombay, we can see not only an extensive expansion of experimental spaces but also gain useful hints about interactions with his audiences. Dinwiddie was far more important than his short, if apparently quixotic, experience at the Emperor's court.

Dinwiddie was constantly in between spaces. In the late 18th century, Dinwiddie's peripatetic encounters included audiences and instruments, theatres, lecture halls, schools and commercial networks—on behalf of Crown and East India Company—amid ship's company, as scholar in Calcutta, and as producer of apparatus. Here were his zones of demonstration and philosophical exchange. We might usefully appreciate the exchanges in what the philosopher of science Peter Galison has described as trading zones.[3] Dinwiddie was forever looking for patrons and places, with his wagon train of apparatus much newly-constructed to meet the circumstances of entertainment and experiment. But his devices were also go-betweens. They were objects with their own trajectory. Instruments were essential because they readily attested to the universality of

---

[2] *The town and country almanack for the year MDCCLXXVIII.... With new lists, and an improved kalendar... The calculations by Mr. Dinwiddie* (Edinburgh, 1778); see also 1791; *The Glasgow almanack, for the year MDCCLXXXVII. Being the third after leap year.... The calculations by James Dinwiddie, A.M.* (Glasgow, 1787); James Dinwiddie, *Syllabus for a Course of Lectures on Experimental Philosophy* (Dublin: D. Graisberry, 1782). I wish to thank the National Library of Scotland for providing a copy of this document; see RB.s. 2639. Much more extensive is the 65 page outline in James Dinwiddie, LL.D., *Analysis of a Course of Lectures in Experimental Philosophy* (Calcutta, 1801) [British Library, Tracts 44].

[3] Peter Galison, *Image and Logic: A Material Culture of Microphysics* (Chicago: University of Chicago Press, 1997).

natural phenomena. Even with relatively simple devices like the telescope or the barometer, observations were endlessly duplicated by observers, collectors and experimenters from one site to another and, in the process, emphasizing an empire of the senses. Even political empire promoted the evidence of universals—in natural phenomena discovered far beyond Europe, as much as in the more conventional regions of economic regulation or imperial power. So, in the 18th century, the same natural forces might be measured and revealed in Cape Town, on the tides of Tonkin or amid the transit of Venus whether observed at Greenwich, on Black Heath or in Tahiti.[4]

From early days in Dumfries, Dinwiddie demonstrated ability in mathematical and philosophical subjects. Like his contemporary, the well-known lecturer and instrument-maker Benjamin Martin, he had no independent means and the search for a career led him to teaching mathematics and natural philosophy. Dinwiddie was quite unremarkable in that regard. Moreover, his initial forays into Newtonian physics put him squarely within an orthodox Christian, natural philosophical, tradition. Dinwiddie, in 1767, argued "I am perswaded [sic] that knowledge is sufficient to recommend itself.... Can any thing afford more suitable pleasure to the human mind... than to be admitted (If I may be allowed the expression) into the secret councils of providence?" He further suggested, "... tho' all the affairs of Mankind some way or other depend upon Mechanical principles and are consequently subjects of Natural Philosophy yet the dignity of human nature eminently point out something further than what is mechanical and seems to be adapted for the entertainment of more refined knowledge than the mere concerns of the groveling [sic] business of life."[5]

It was nonetheless the grovelling business of survival that demanded Dinwiddie's attention. Like his predecessors, practical mathematics provided opportunity. By 1776, he had become a significant player in the Dumfries Academy and in its Philosophical Society. He was already then

---

[4] Cf. Lorraine Daston, "Preternatural Philosophy," in *Biographies of Scientific Objects*, ed. Lorraine Daston (Chicago and London: University of Chicago Press, 2000), 1–41, esp. 40–41; Simon Schaffer, "The Asiatic Enlightenments of British Astronomy," in *The Brokered World. Go-Betweens and Global Intelligence*, edited by Simon Schaffer, Lissa Roberts, Kapil Raj, James Delbourgo (Sagamore Beach: Science History Publications, 2009), 49–104, esp. 72–74.

[5] Dalhousie University Archives (DUA), James Dinwiddie Collection, MS 2-726, K 1. *Statement Concerning a Course of Natural Philosophy* (taught by Dinwiddie), dated March, 1767.

giving lectures on electricity and magnetism, soon relying on his reading of Stephen Gray, Charles Du Fay, and Benjamin Franklin.[6] There is some evidence that Dinwiddie briefly sought devices in London in 1774. His notes on early experiments indicate an intent to visit the Observatory and getting apparatus from 'Mr. Martin', while buying also Joseph Priestley on air and James Ferguson on astronomy.[7] If Benjamin Martin, instrument maker and publisher of Fleet Street, was an influence alongside the lecturer Ferguson then Dinwiddie had effective contemporary models in the public promotion of science. Certainly Dinwiddie became convinced of the value of demonstrations, in astronomy and navigation, for the boys in the Dumfries Academy and to that end a subscription was taken up for the purchase of an orrery.[8] This was the first indication of the knowledge of instruments that would later recommend him to the Crown.

Orreries figure prominently in the trade of philosophical lecturers.[9] The Dumfries' orrery turned into more than a demonstration device. It was the beginning of Dinwiddie's addiction to apparatus. Indeed, the direction of Dinwiddie's career not only reflected the necessity of instruments, but it also cast light on the market for the latest show pieces which he collected for his own experiments. His life-long attraction to philosophical gadgets caused him considerable grief. Indeed, according to his biographer and grandson, Dinwiddie got himself into such debt that he secured a leave of absence from his school in 1778. He employed his most portable instruments as an itinerant lecturer on a tour of Ireland and Scotland. This instrumental turn was a critical venture.[10] He had left considerable

6 DUA, MS 2-726, D 28. 'Magnetism and Electricity', April, 1776; F 2. Early Experiments, f. 1v, 1777.

7 DUA, MS 2-726, F 1. Early Experiments, ca. 1774. September 29. By November, Dinwiddie was back teaching experimental philosophy in his mathematical classes. On Ferguson see Henry C. King with John R. Millburn, *Geared to the Stars. The Evolution of Planetariums, Orreries, and Astronomical Clocks* (Toronto and Buffalo: University of Toronto Press, 1978), 178ff.

8 DUA, MS 2-726, K 2, Dumfries Philosophical Society, Rules and Minutes, 1776–1777; MS 2-726, A 32, List of the Gentlemen Subscribers for purchasing a Philosophical Apparatus, for the use of the School of Dumfries... [poss. 1774]. I thank Mr. Chris Green for providing a copy of what appears to be the original subscription for the orrery, dated July 16, 1765; see below.

9 On British Orreries see King and Millburn, *Geared to the Stars*, 150ff.

10 See William Jardine Proudfoot, *Biographical Memoir of James Dinwiddie, LL.D., Astronomer in the British Embassy to China, 1792, '3, '4: Afterwards Professor of Natural Philosophy in the College of Fort William, Bengal: Embracing Some Account of His Travels in China and Residence in India.* Compiled from his Notes and Correspondence by his Grandson, William Jardine Proudfoot. (Liverpool: Edward Howell, 1868), 4ff; and Linde Lunney, "The Celebrated Mr. Dinwiddie: an Eighteenth-Century Scientist in Ireland," *Eighteenth-Century*

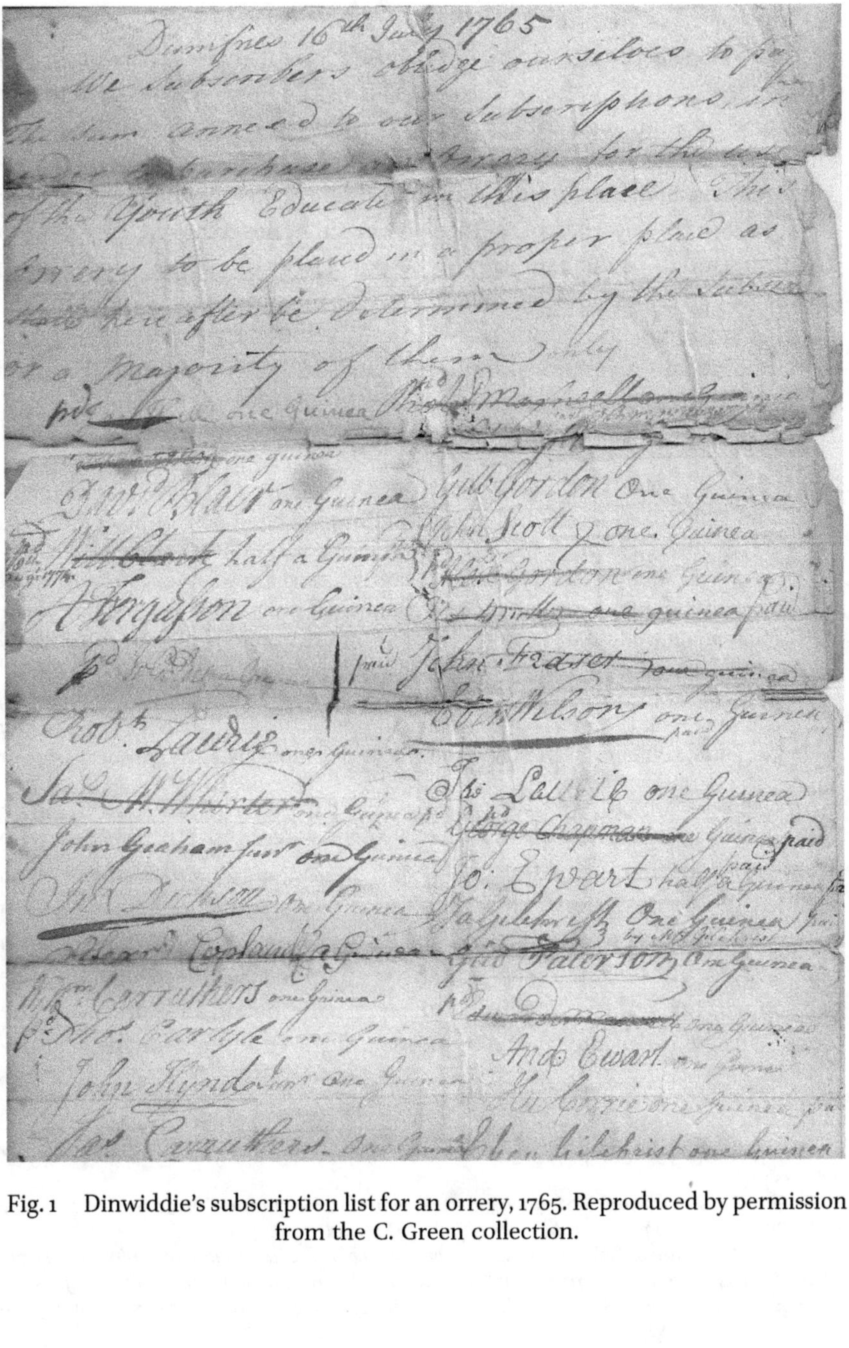
Dumfries 16th Jany 1765
We Subscribers oblige ourselves to pay
the Sum annexed to our Subscriptions for
purchase an Orrery for the use
of the Youth Educate in this place The
Orrery to be placed in a proper place as
shall hereafter be determined by the Subscribers
or a Majority of them only

Fig. 1 Dinwiddie's subscription list for an orrery, 1765. Reproduced by permission from the C. Green collection.

apparatus and apparently much of his library at Dumfries. But in his portable devices he had found a potentially lucrative venture, rather more exciting perhaps than teaching mathematics to school boys.

The shows in physics were the foundation of many a career in the eighteenth century. We can, through Dinwiddie's everyday experience in the zones of philosophical exchange, witness the priorities of the public showman. He wished to be taken more seriously than the numerous marketplace charlatans and conjurors who made many an outlandish claim to conquer the powers of nature. Dinwiddie nevertheless understood the importance of display. Apparatus enabled explorations in which an audience could vicariously engage. This was one critical version of instruments and universalism. But Dinwiddie was not content with the drama. His approach to mechanical force revealed his general method. He tried to set himself apart from those then once described, with distain, as "the common herd of philosophers about town."[11] He stated,

> I appeal to fact. It is only in those parts of science which have been mathematically considered, that natural philosophy can boast of having carried on her investigations with certainty, success, and utility.... In short, gentlemen, without a moderate share of mathematical knowledge, you can expect only a schoolboy's acquaintance with natural philosophy, resembling those religionists who take up their authority on the opinions of their priests, and neither can give a reason for what they believe, nor apply it to any good purpose in life. Without any moderate share of mathematical knowledge, you may read the flimsy and indeterminate writings of a Nollet or a Priestly [sic], but you cannot profit by the truths delivered by a De Lambert [sic] or a Newton.[12]

Like it or not, it was the abbé Nollet and the Reverend Mr. Priestley who were most influential in the promotion of the theatre of science. And, despite his protestations, their experience was more of a model than Dinwiddie wished to admit. Notes of his early experiments, dating from the decade after 1774, reveal extensive demonstration pieces such as telescopes, air pumps, prisms, and electrical machines—precisely the range of devices of any credible showman.[13] We know that by 1777 he

Ireland 3 (1988): 69–83, esp. 70–73. I thank Dr. Lunney for supplying me with a copy of her essay.

[11] British Library, Add. MSS. 30094, f. 41. Smeaton to Benjamin Wilson, August 24, 1747.

[12] Proudfoot, *Biographical Memoir of James Dinwiddie*, pp. 5–6, 10–11.

[13] Note M. l'abbé Nollet, *L'art des expériences, ou avis aux amateurs de la physique, sur le choix, la constitution et l'usage des instrumens; sur la préparation et l'emploi des drogues qui servent aux expériences.* Seconde Edition. I (Paris: Chez P.E.G. Durand, Neveu, Libraire rue

was especially concerned with electrical effects employing static charges, by spirits ignited and gun powder fired, with demonstrations of thunder houses, church steeples and lightning rods, and simulated aurora borealis. His list was extensive and informed by the very reading we might have expected of Grey, Dufay, Franklin and Joseph Priestley on both electricity and air. By 1783 it seems he had cylindrical glass tubes, electrical machines, dancing figures put in motion by static charges, and jars from which he could discharge sparks in support of the electrical doctrine of Ben Franklin.[14]

Dating Dinwiddie's experiments with certainty is problematic as his notebooks were repeatedly amended over several decades. His notes on mechanical apparatus, much in the earlier tradition of J. T. Desaguliers, covered many devices from pendulums, to innovations for reducing fiction, and maximizing the power of pulleys and pumps.[15] Moreover, he paid a great deal of attention to the latest philosophical news. He knew of the numerous reports of balloon experiments in the 1780s involving, among others, Birmingham industrialists like Matthew Boulton, following upon James Watt's description of the expansive power of steam and, moreover, described a vast array of apparatus. This included, inter alia, Benjamin Martin's device for demonstrating the transit of Venus in 1769, along with pumps and thermometers.[16] His visit to the philosophical sites of Paris in 1788 led him to witness the latest of balloon contrivances.[17] He encountered numerous electrical machines, all constructed of

S. Jacques, a la Sagesse, 1770). See Lewis Peyenson and Jean-Francois Gauvin, eds., *The Art of Teaching Physics. The Eighteenth-Century Demonstration Apparatus of Jean Antoine Nollet* (Sillery: Septentrion, 2002). I would especially like to thank Jean-Francois Gauvin and Ms. Sylvie Dauvin, of the Stewart Museum, Montreal, for their assistance.

[14] DUA, MS 2-726, C 5. Lecture 7, Electry [sic]; F 1, Early Experiments, ca. 1774; F 2, Early Experiments 1777, fols 1v, 3v, 6r, 7r, 8v, 9 r/v, 10v, 12v. Cf. John Neale, *Directions for gentlemen who have electrical machines* (London, 1747). In 1781 he was in Dublin in correspondence with William Gleghorn from London who reported on an exhibition of electrical charges by Nairne. MS 2-726, A 19. Cleghorn to Dinwiddie, January 2, 1781. Both Cleghorn and Nairne were members of the Chapter Coffee House Society. See Trevor Levere and Gerard L'E Turner, eds., *Discussing Chemistry and Steam. The Minutes of a Coffee House Philosophical Society 1780–1787* (Oxford: Oxford University Press, 2002), *passim.*

[15] DUA, MS 2-726, C 11. See, for example, Mechanics 1, nos. 15–18, 21–25, 30, 42–43; Mechanics 2, nos. 45–47.

[16] DUA, MS 2-726, C. 60, fols. 29–55. Lectures notes: The effects of heat and mixture on bodies.

[17] Cf. Marie Thebaud-Sorger, *L'Aérostation au temps des Lumières* (Rennes: Presses Universitaires de Rennes, 2009).

the very latest cylindrical plates.[18] While he witnessed many early indications of social disturbance and was extremely critical of the "Bufoonaries on the Boulevards", he attended a meeting of the Academy and saw the philosophical laboratory of the Lyceum with its large plate electrical machine and the demonstration models of the abbé Nollet painted black and decorated with silver.[19] In Jacques Vaucanson's surviving collections he saw dividing engines, boring drills, machines for making chains, pile engines, and various textile apparatus. While he made many notes of new devices, such as Watt's copying machine and several designs for thermometers, it cannot be at all certain that he did any more than note their uses. He could hardly have collected every piece of apparatus that attracted his attention. However, in many instances, it is readily apparent that Dinwiddie was especially interested in experiments to be performed by electrical or chemical devices.[20] In the gap between performance and audience participation, instruments transformed many observers into experimentalists and occasionally into demonstrators in their own right. Dinwiddie was one case of a career born in such trading zones.

Dinwiddie's wide-ranging public mission of magnetism, electricity and chemistry reflected the newest fashions in physics. By the 1770s, after the shocking displays of Nollet, and especially in the wake of the promotions of Joseph Priestley, there was a particular passion for electrical experimentation. Like others, Dinwiddie exploited this fully. From 1776, Dinwiddie admitted that magnetism and electricity had not then proved susceptible to "mathematical demonstration." There were many replications of magnetic and electrical attractions for which no general law had yet been agreed. And the distinction between repulsion and attraction was impossible then to explain, although the phenomena seemed to suggest some universal inverse square law.[21] With so much uncertain, it is not surprising that Dinwiddie, like many others, repeated dramatic electrical phenomena for the curious. Empiricism reigned among the instruments.

---

[18] On the transformation of these devices, see W. D. Hackmann, *John and Jonathan Cuthbertson. The invention and development of the eighteenth century plate electrical machine.* Communication nr. 142 (Leyden: Rijksmuseum, 1973).

[19] DUA, MS 2-726, B 12. Journals, Paris 1788, pp. 2, 14, 18, 26.

[20] DUA, MS 2-726, C 11. Lecture Notes; Instruments, 1, nos. 24, 30, 37, 38, 44–45; Experiments, 1, no. 72, 74.

[21] DUA, MS 2-726, D 28. Scientific Journal, Magnetism and Electricity, April, 1776. Cf. Particia Fara, "'A treasure of hidden vertues': the attraction of magnetic marketing," *British Journal for the History of Science* 28 (1995): 5–35; On magnetic theories see Patricia Fara, *Sympathetic Attractions. Magnetic Practices, Beliefs, and Symbolism in Eighteenth-Century England* (Princeton: Princeton University Press, 1996), 175–185.

Fig. 2 Demonstration device: A version of the Thunder House, ca. 1740. By permission of the Science Museum, London.

This also provided opportunity to test natural phenomena without concern for a dominant over-riding theory. Certainly, by the early 1780s, he had the latest apparatus in his arsenal, including an electrical machine (although its type is unclear), an orrery made to turn by static charge, a "French thunder house the bottle with the Inflammable air placed with the mouth down," and demonstrations of the identity of electricity and lightning, wherein he opted for pointed conductors rather than blunt ones, then an issue of much contention in the Royal Society.[22] He clearly used many of the same showpieces as did other lecturers, such as thunder houses, the electro-phosphorous, and an electrified fountain.[23] While

22 On the orrery driven by static electricity, several of which Ferguson gave away to friends, see King and Millburn, *Geared to the Stars*, 191–193.

23 DUA, MS 2-726, C 5, Lecture Notes, Lecture 7 (1783?), Electricity; C 7, Lecture Notes, Lecture 8 (1783?). See Trent A. Mitchell, "The Politics of Experiment in the Eighteenth Century: The Pursuit of Audience and the Manipulation of Consensus in the Debate over Lightning Rods," *Eighteenth-Century Studies* 31 (1998): 307–331; Jessica Riskin, *Science in the Age of Sensibility. The Sentimental Empiricists of the French Enlightenment* (Chicago and

these all became part of his extensive cabinet of apparatus, little did he know that within a decade his reach would literally become imperial. He not only displayed some of the more elaborate contrivances then available, from the best of manufacturers, but he would be engaged to do so as far afield as Peking, Calcutta and Bengal.

Imperial Experiments

Dinwiddie navigated the world of the lecturer by cultivating connection. We know that during the 1770s he had consulted the Manchester physician Thomas Percival who, in turn, recommended him to William Cullen in Glasgow. While our philosopher is known to have made tours of the Midlands, it would be no surprise that he had made significant industrial connections.[24] It was likely Matthew Boulton in Birmingham who recommended Dinwiddie as philosophical demonstrator to the Macartney expedition. As a means of gaining access to the China trade, the embassy naively intended to impress the emperor's court with British manufactures and scientific instruments. There was much confusion over

London: University of Chicago Press, 2002), 69–103; James Delbourgo, *A Most Amazing Scene of Wonders. Electricity and Enlightenment in Early America* (Cambridge Mass.: Harvard University Press, 2006), chapter 4; R. W. Home, "Points or Knobs. Lightning Rods and the Basis of Decision Making in Late Eighteenth Century British Science," in *Playing with Fire. Histories of the Lightning Rod*, edited by Peter Heering, Oliver Hochadel and David Rhees (Philadelphia: American Philosophical Society, 2009), 97–120. The thunder house was sold for generations as a demonstration device to show the power of lightning, or even possibly to reveal the necessity of a lightning conductor, whether of balled or pointed ends, depending on the state of the controversy in the 1770s. Often made of brass and mahogany, the discharge of electricity could make the part of the walls collapse.

[24] In 1770 Percival told Cullen that he, along with doctors Dobson, Bostock and Haygarth had agreed to "meet for our mutual improvement every three months at Warrington, wch is in the center between Liverpool, Chester, & Manchester..." perhaps thus being the precursor of the Manchester Literary and Philosophical Society. University of Glasgow Library, Special Collections, Cullen MSS 99 (1), Percival to Cullen, November 11, 1770; A. E. Musson and Eric Robinson, *Science and Technology in the Industrial Revolution* (Toronto: University of Toronto Press, 1969), 91–92. It is possible, although by no means likely, that he may have been the Dinwiddie who was a member of the Manchester Literary and Philosophical Society, although this almost certainly confuses him with the firm of Dinwiddie, Kennedy and Dinwiddie, cotton manufacturers. Cf. William Jardine Proudfoot, *Biographical Memoir of James Dinwiddie, LL.D., Astronomer in the British Embassy to China, 1792, '3, '4: Afterwards Professor of Natural Philosophy in the College of Fort William, Bengal: Embracing Some Account of His Travels in China and Residence in India. Compiled from his Notes and Correspondence by his Grandson, William Jardine Proudfoot* (Liverpool: Edward Howell, 1868), 17.

Chinese resistance or even the status of gifts or presents to the Emperor.[25] The Chinese refused to be suitably moved even by the most elaborate instruments.[26] Much of the considerable apparatus seemingly fell into the hands of Dinwiddie who, upon leaving China, settled amidst the East India Company operatives in Calcutta, ultimately lecturing at Fort William.

At the edge of empire, Dinwiddie suddenly found useful employment in Calcutta, both in reviving his experimental shows and as advisor on potential manufactures. In 1793 his lectures turned to achievements in British engineering and manufactures.[27] This could not have but helped his reputation amongst his more practical clients in the East India Company. By 1795 the Board of Trade had been ordered to "consult Dr. D on the several points of Chemistry Mechanics and Natural Philosophy which have relation to the business of their department."[28] Although he had been preceded as scientific lecturer in 1787 by the Revd. Dr. Andrew Bell, for a Scottish natural philosopher India may have seemed lucrative territory, especially amongst the officers and agents of the Company and the Crown.[29] When Dinwiddie initiated his lectures in Calcutta, he returned to the fundamentals of matter and motion. In good Newtonian fashion, as many lecturers before him, he sought to define for his auditors the basic terms of physics. He focussed on time and duration, and tried to determine what he believed to be the "essential properties of Matter" and its

---

25 British Library, India Office Records, G/12/20, fols. 596–618. Articles bought for Macartney expedition January-August, 1792. Cf. James L. Hevia, *Cherishing Men from Afar. Quing Guest Ritual and the Macartney Embassy of 1793* (Durham and London: Duke University Press, 1995), 178–180; J. L. Cranmery-Byng and Trevor H. Levere, "A case study in cultural collision: Scientific apparatus in the Macartney embassy to China, 1793," *Annals of Science* 38 (September, 1981): 503–525; Maxine Berg, "Britain, industry and perceptions of China: Matthew Boulton, 'useful knowledge' and the Macartney Embassy to China 1792–94, *Journal of Global History* 1 (July, 2006): 269–288; Simon Schaffer, "Instruments as Cargo in the China Trade," *History of Science* 44 (June, 2006): 217–46.

26 Proudfoot, *Biographical Memoir of James Dinwiddie*, 53–54.

27 DUA, MS 2-726, C 21, Manufactures & Engineering, ca. 1793.

28 DUA, MS 2-726, Journals, B 21, 13 July, 1795, July 23, 1795; University of Guelph, James Dinwiddie Correspondence, XS1 MS A164, no. 47. W. Emmerston? to Dinwiddie, January 31, 1797.

29 Cf. H. J. C. Larwood, "Science and Education in India Before the Mutiny," *Annals of Science* 17 (July, 1961): 81–96, esp. 87–89; Kapil Raj, "Colonial Encounters and the Forging of New Knowledge and National Identities: Great Britain and India, 1760–1850," *Osiris*, 2nd ser., 15 (2001): 119–134, esp. 125–127, Kapil Raj, *Relocating modern science. Circulation and Construction of Scientific Knowledge in South Asia and Europe. Seventeenth to Nineteenth Centuries* (Delhi: Permanent Black, 2006), 153–170; Simon Schaffer, "The Asiatic Enlightenments of British Astronomy," *The Brokered World. Go-Betweens and Global Intelligence, 1770–1820*, edited by Simon Schaffer, Lissa Roberts, Kapil Raj and James Delbourgo (Sagamore Beach: Science History Publications, 2009), 61–62, 88.

divisibility, of the meaning of space and mobility. His lectures on dynamics thus covered much the same ground as did most contemporary lecturers interested in physics and machines.[30] Such principles were repeated in his 'Notes on a Course of Lectures on Natural Philosophy' penned in 1792 and 1793. Here, as he had earlier done in 1767, he demonstrated Newtonian propositions of matter and extension, along with the laws of repulsion, gravitation, and the acceleration of falling bodies.[31] His instrumental theatre revealed universals.

Dinwiddie in Calcutta essentially extended the experience he had gained in Britain over the previous two decades. From Dumfries to Dublin, from Cork to Peking and Calcutta, his notions of physical powers and of natural theology were translated by sensational demonstrations, seducing his subscribers by empiricism and enthusiasm. Before his auditors in Calcutta, Dinwiddie defined his purpose strictly in terms of sensation. He declared, "Natural philosophy is that Study of the Sensible motions and actions of bodies, with a view to discover the general powers of Nature and explain the Phenomena congruent on their exertion." The purpose of the examination of nature was not limited to natural theology. Dinwiddie's method was explicitly experimental and instrumental, to show "by what means the various changes are affected what are the causes produced of Natural Phenom., [and] we then view Nature as a subject of Philosophy and endeavour to investigate its laws." Most importantly, by the use of experiment "we comport[?] nature to reveal his secrets."[32] The difficulty, however, was limits to the senses. The best that might be determined were facts drawn by induction from experiment. Thus, he told his listeners, "such a course may enable you to read the popular productions of Ferguson or a Nollet–but not to understand all[?] the Sublime Investigations of Dalembert [sic] or a Newton."[33] This was a doctrine he had held since at least the 1770s.

---

30 DUA, MS 2-726, C 35. 'On the General Affections of Motion' (1793?), and 'Of Dynamics', fols. 2, 4–5. See, for example, James Ferguson, *Lectures on Select Subjects in Mechanics, Pneumatics, Hydrostatics, and Optics. With the Uses of Globes, The Art of Dialing. And the Calculation of the Mean Times of New and Full Moons and Eclipses* (London: A. Millar, 1764), Lecture III, 'Of the mechanical powers', 30ff.

31 DUA, MS 2-726, K 1, 'Statement Concerning a Course of Natural Philosophy', March, 1767; C 16, October 14, 1793 and C 17, October 18, 1793.

32 DUA, MS 2-726, C 14, Miscellaneous Subjects, 1792?, fols. 6–7.

33 DUA, MS 2-726, C 14, Miscellaneous Subjects, 1792?, fol. 7v.

Performance

The interface between entertainment and engagement defined both the role of philosophical showmen and that of their audience. While lecturers sought to display basic physical principles through considerable apparatus, public demonstrators sometimes had greater aims in mind. As Dinwiddie's unpublished notes reveal, some lecturers clearly expanded the boundaries of experimental practice. New devices revealed new sensations. Lecturers may even have induced some of their audience to become likewise employed. But it is obvious that Dinwiddie did not stop with merely designing elaborate and amusing performances. Like many a lecturer, he had other roles to play.

During the 1780s in Britain, Dinwiddie had explored the difference between dynamic forces and inter-particulate reactions. Thus, he asserted, "the motions and actions of the great mass of matter . . . is properly the object of Expt. Philosophy." In attractions between gross bodies there was no modification in matter itself; unlike in chemical attractions where bodies undergo a "total change" of state or of composition. For example, the idea that bodies might chemically unite "only in certain proportions" could be demonstrated by simple experiments easily repeated before his subscribers.[34] Given the philosophical preoccupations of the 1780s, notably following the concerns of many experimentalists in Britain and France on the significance of 'caloric', Dinwiddie turned his attention to the "affects [sic] of heat and mixture of bodies." In an elaborate draft of a course, he turned to an historical account of various notions of heat derived in Britain from Bacon, Boyle and Newton, and from Pieter van Musschenbroek and René-Antoine Ferchault de Réaumur. Instruments measured, and determined, the physical distinctions. He was particularly interested in the application of the thermometer in chemical experiments. Here he followed Benjamin Martin in the comparison of thermometric scales.[35] He likewise assessed the distinctions of fluidity and frost, noting especially the measurable temperature changes in experiments on alkalis and salts for which he produced an extensive table of reactions.[36] The important

[34] DUA, MS 2-726, C 3, Lectures on Chemistry, 1st, 1783.

[35] Similar effects of dissemination followed the development of the barometer. Cf. Jan Golinski, "Barometers of Change: Meteorological Instruments as Machines of Enlightenment," in *The Sciences in Enlightened Europe*, edited by William Clark, Jan Golinski, and Simon Schaffer (Chicago: University of Chicago Press, 1999), 83–84.

[36] DUA, MS 2-726, C 60, Lecture Notes, The effects of heat and mixture on bodies, nd., fols. 15, 17, 21, 23, 25, 37, 49, 91; new pamphlet, fols. 1–12.

point is not that Dinwiddie was demonstrating the work of others. He proceeded further with his own explorations of a subject then very much of interest to chemists in Europe. Many of his further speculations are revealed in amendments to notes made in the Far East.

Luckily, for us, much of Dinwiddie's manuscripts and correspondence survived the perils of imperial navigation. He obviously focussed on practice and performance. Despite the lack of published lectures, some of his many notes are in a careful hand and suggest that he may have considered the possibility of having them printed. That is the case with his extensive chemical speculations. The emergence of the rage for pneumatic or gas chemistry, especially following methods promoted by Joseph Priestley, Henry Cavendish and James Watt, may be seen to have culminated in the revolution induced by Guyton de Morveau and Antoine Lavoisier.[37] Dinwiddie's lectures from 1787, while then still in Britain, reveal he was adept at laboratory practice and, like many, especially interested in experiments on the chemical effects of breathing.[38] Just as the French were then constructing tables of chemical elements, Dinwiddie, like Lavoisier, also undertook experimental demonstrations of the effects of airs, in respiration and combustion but, in the spring of 1787, he was clearly still influenced by the older notion of phlogiston.[39] It was precisely this which later led to his carefully crafted manuscript 'Experiments on Factitious Airs'. Here Dinwiddie relied, in part, on the experimental production of airs traced from the Newtonian Francis Hauksbee through to Joseph Black's later work on lime. Dinwiddie recounted the production of so-called inflammable air and fixed air, created through the application of various acids. He notably compared the properties of gases by chemical tests, by specific gravity and absorption in water, and not simply by sensation. This chemical essay appears very much have been drafted with publication in mind, although there is no evidence that he undertook any further steps to do so.[40] But it was a remarkable account of the state of British chemistry on the eve of Lavoisier's revolution. Even after the Macartney voyage, he remained strikingly well informed of philosophical news from Europe. Pneumatic chemistry was then much in vogue. It is worth noting that we also find him amongst the correspondence networks in India and in Britain on the

---

[37] See David Phillip Miller, *Discovering Water. James Watt, Henry Cavendish, and the Nineteenth-Century Water Controversy* (Aldershot and Burlington: Ashgate, 2004), 48–80.

[38] DUA, MS 2-726, C 12, Pneumatics.

[39] DUA, MS 2-726, C 12, Pneumatics, April 14, 1787?, fols. 1–5.

[40] DUA, MS 2-726, C 62, Experiments on Factitious Airs, nd.

production of nitric acid used in the treatment of disease—a matter of considerable public speculation in the 1790s.[41]

Dinwiddie's interest in the virtues of new airs evolved within the contemporary craze in balloons.[42] Ballooning was a particularly dramatic instrumental turn. Not only was it stunningly theatrical and highly unpredictable, but the numbers of witnesses to such demonstrations could often number in the tens of thousands. Surely this was not the solitude of the private philosopher. Experiment was indeed private only if it went unnoticed, as was occasionally the case in the Tuileries, already the site of the analysis of airs in the pit of the theatre by Lavoisier.[43] If the audience at plays was then unsuspecting of airs captured in vials, the opposite was true of witnesses to the balloons rising over Paris and London drawn thereby to the flame of spectacle and the prospect of disaster. Among the practitioners was Dinwiddie in December, 1783 at Burlington Gate, London where he inflated a hot air balloon. Dinwiddie's balloon was exhibited over a number of towns in Britain including Bath and Bristol and, ultimately, in Dublin.[44] What greater show might there be than of men aloft, risking their lives and of those on the ground should the balloon fail

41 University of Guelph Library, Archives and Special Collections, XS1 MS A 164, James Dinwiddie Correspondence, Helenus Scott to Dinwiddie, August 23, 1796. Scott in Bombay was a promoter of nitric acid in the treatment of venereal disease and his letters on the subject were published by Thomas Beddoes. See Beddoes, *Communications respecting the external and internal use of Nitrous Acid; Demonstrating its Efficacy in Every Form of Venereal Disease, and Extending its Use to Other Complaints: with Original Facts, and a Preliminary Discourse, by the Editor, Thomas Beddoes, M.D.*. (London: J. Johnson, 1800), xxxi; Thomas Beddoes, M.D., A *Collection of Testimonies Respecting the Treatment of the Venereal Disease By Nitrous Acid* (London: J. Johnson, 1800). On Scott see Mark Harrison, "Medical experimentation in British India. The case of Dr. Helenus Scott," in *The Development of Modern Medicine in Non-Western Countries*, edited by Hormoz Evrahimnejad (London and New York: Routledge, 2009), 23–41.

42 For details of balloon construction and methods of filling them from 1783 see DUA, Dinwiddie Collection, MS 2-726, C 45, Aerostation, 1793?; F 3, Early Experiments, nd.; and recently, Marie Thebaud-Sorger, *L'Aérostation au temps des Lumières*; and Michael R. Lynn, *The Sublime Invention: Ballooning in Europe, 1783–1820* (London: Pickering & Chatto, 2010).

43 Larry Stewart, "His Majesty's Subjects: From Laboratory to Human Experiment in Pneumatic Chemistry," *Notes and Records of the Royal Society* 63 (September, 2009): 231–245, esp. 232.

44 Proudfoot, *Biographical Memoir of James Dinwiddie*, 17–18; John Penny, "Ballooning in the Bristol region, 1784 to 1786," *Bristol Past. Fishponds Local History Society* (2007): 1–11, esp. 2–3; http://fishponds.org.uk/balloon.html; Linde Lunney, "The Celebrated Mr. Dinwiddie: an Eighteenth-Century Scientist in Ireland," *Eighteenth Century Ireland* 3 (1988): 69–83, esp. 75ff. I wish to thank Dr. Lunney for providing a copy of this article.

to properly inflate?[45] This, indeed, is precisely what happened all too often in several public spaces throughout Europe as in the riot in Lord Foley's gardens, in Portland Place, when John Sheldon's balloon burst into flames. It is not, perhaps, so remarkable that among the devices Dinwiddie was intended to show in Peking was a hot air balloon.

The Lure of Electricity

With the potential spectacle of punctured balloons to the sharp blasts of electrical discharges, Dinwiddie drew a considerable company of witnesses. In India, he was both constructing devices as well as purchasing new apparatus from British sources with a possible view to resale. In his early lecture notes, revised in Calcutta in 1801, he referred to building a Voltaic pile and to electrical machines in need of repair. These are interspersed with a long list of electrical experiments, with sources ranging from Franklin in America to Martinus Van Marum in Haarlem.[46] As with many a lecturer and maker, in electricity Dinwiddie discovered a following.

Dinwiddie was obviously well versed in the electrical experiments then underway throughout Europe. For decades he had been interested in the bodily effects of electricity of the body—taking notes in Cork in 1780 from a copy of the *Philosophical Transactions* on a cure of muscular contraction.[47] At the turn of the century, he undertook an unpublished Journal of Galvanism on an extensive range of experiments, with details on construction of batteries and increasingly powerful Galvanic apparatus. As Jan Golinski reveals in this volume, Dinwiddie's devices also had practical uses and were much in demand, so much so that he made many notes on the efficacy of electricity in medical disorders. It was hardly unusual in the eighteenth century for both chemical and electrical experimentalists to be seduced

---

45 For the most complete account of ballooning see Marie Thebaud-Sorger, *L'Aérostation au temps des Lumières.* I wish to thank Dr. Thebaud-Sorger for allowing me to examine a copy of her work while this essay was being written. See also Michael R. Lynn, "Consumerism and the Rise of Balloons in Europe at the End of the Eighteenth Century," *Science in Context* 21 (2008): 73–98.

46 DUA, MS 2-726, F 2, Early Experiments, 1777; See also W. D. Hackmann, *Electricity from Glass: The History of the Frictional Electrical Machine 1600–1850* (Alphen aan den Rijn: Sijthoff and Noordhoff, 1978), 104ff.

47 DUA, MS 2-726, E 2, Notes from the 'Philosophical Transactions', Cork, September 17, 1780; see Miles Partington, "A Cure of a Muscular Contraction by Electricity," *Philosophical Transactions* 68 (1778): 97–101.

by medical promise. But, by 1804, "The increased number of patients" so overwhelmed Dinwiddie that he declared his intention to "give up Medical Galvanism." In July he had sold four sets of Galvanic apparatus, one to the Medical Board for use in the General Hospital in Calcutta. The lure of electricity proved too much. Within a month, he announced he had already constructed a new "Galvanic Apparatus for Medical and Chemical purposes, consisting of 100 series in two boxes, with glass tubes Gold and Silver wires &c for Chemical Ex.... and have already sold 5 sets, with many orders on hand."[48] Following a pattern that had been obvious much earlier in Scotland, his new apparatus led to new courses of lectures. Even during his final voyage to Britain in 1807, on a stop in Cape Town, Dinwiddie set up his machine to meet the demand of the local ladies to be galvanized.[49] Some of his instruments long survived Dinwiddie's return and were still "an occasional amusement" for his grandson, James Proudfoot, fifty years later. By then, Proudfoot reported, electricity was in such a vogue in Liverpool that in the evenings "several Galvanic machines or batteries are to be met on the streets... Any person may feel the shock, or get galvanized for one half-penny."[50]

Instruments

During early days in Dumfries, Dinwiddie had set upon the purchase of apparatus for the Academy.[51] That first subscription had set him on the track of many shows in physics. His instruments followed philosophical fashions. Such was the power of the marketplace. We have seen that Dinwiddie in his later experience, as many other lecturers, became immensely concerned with electrical phenomena. This was especially true when there were, at the turn of the century, so many remarkable developments in theories of animal electricity and in design of demonstration devices that owed a great deal to the experiments of Volta and Galvani. Electrical apparatus became a staple of the more important London makers who

48 University of Guelph, James Dinwiddie Correspondence, XS1 MS A164, no. 40, Indenture from Medical Board, July 3, 1804; DUA, MS 2-726, D 22. Journal of Galvanism, 1801–1805, esp. June 12, July 10, July 22, 1804.

49 DUA, MS 2-726, Journal B 73, January 4, 1807.

50 DUA, MS 2-726, D 22. Journal of Galvanism, 1801–1805; William Proudfoot, October 25, 1858.

51 Chris Green archive, Subscription list, Dumfries, July 16, 1765; see also DUA, MS 2-726, A 32, List of the Gentlemen Subscribers for purchasing a Philosophical Apparatus, for the use of the School of Dumfries... [poss. 1774].

catered to the philosophical market—and even to medical therapies. Thus, for example, Edward Nairne had published in 1783 The Description and Use of Nairne's Patent Electrical Machine; With the addition of some Philosophical Experiments and Medical Observations. As Paola Bertucci has shown, Nairne chose to promote the device, and to seek a patent, explicitly on the parallel markets in physics and medicine: "Though this machine and its apparatus are constructed with a particular view to the purposes of medicine, yet it will be found equally applicable to philosophical uses."[52] Certainly, it became quite evident that electrical machines might have dual applications.

Like Macartney in China who tied instruments to impression, many a lecturer's credibility and reputation required dramatic demonstration pieces. It is no surprise therefore, that from India, even with some of the Macartney apparatus in tow, Dinwiddie would attempt to secure the latest apparatus from Britain. The instrument trade consumed him. In 1793, during the Macartney expedition, Dinwiddie drafted lectures on electricity in which he sketched some of the instruments he had seen in Paris in 1788. By these a lecturer could reveal the power of electricity to ignite gun powder, to generate varied colours when sent through different metallic filaments, and to simulate lightning.[53]

At least as early as the spring of 1795, he was communicating from India with the London instrument maker William Jones who then offered Dinwiddie a planetarium. Surely this was not as grandiose as the elaborate gift to the Emperor. Jones promoted a device "chiefly after the plan of the late Mr. Benjamin Martin, and in my opinion the most complete and elegant of any for the price that is made: I venture to say that it is more complete than any other that I have seen an imitation of from the shop of a gentleman you know in Fleet Street." This he offered to Dinwiddie at a discounted, but still substantial, price of 32 Guineas. Jones also included in his proposal to Dinwiddie copies of his latest catalogues. He also remarked on the increase in his business as "we are recommended by the first people of science now existing." This was not hyperbole. Jones successfully

[52] [Edward Nairne], *The Description and Use of Nairne's Patent Electrical Machine; With the addition of some Philosphical Experiments and Medical Observations* (London: Printed for Nairne and Blunt, 1783). British Library, shelf mark B730/12, belonged to Sir Joseph Banks. See National Register of Archives, Chancery Petty Bag Office. IND 1/6941, 41v. 20 May, 1782; Patent Rolls, C274/45. On Nairne, see Paola Bertucci, "A philosophical business: Edward Nairne and the Patent Medical Electrical Machine (1782)," *History of Technology* 23 (2001): 41–58.

[53] DUA, MS 2-726, C 39, Electricity (no. 1), 36, 48.

adopted the publication and marketing strategy the over-extended Martin had once attempted. At least since 1784, Jones had been selling a 'new portable orrery' ideal for itinerant philosophers. He also offered a range of larger orreries, planetariums and cometariums which had been built both by the lecturer James Ferguson and by Martin.[54] Not one to miss an opportunity, Jones had already heard that in Calcutta Dinwiddie had "met with some considerable degree of encouragement and patronage." This was the case even before the arrival in India of Dinwiddie's one-time Irish auditor, Viscount Wellesley. Assuming wealth, Jones proposed another option, one only then available to the well-heeled. Jones told Dinwiddie, his "old acquaintance," that "You will see by our catalogue that we have Lord Bute's Grand Orrery. For a public seminary, or a popular lecture I think it is the best instrument ever offered to the public at the price."[55]

At the end of the century, new ventures in electricity and chemistry were all the rage among the lecturers. Instruments as well as chemicals were essential to the showman. Dinwiddie's initial connection with Jones remains hidden. It would hardly be surprising that, as an itinerant lecturer in need of devices, or while in London, the two would not have had some transactions. Dinwiddie similarly had an account with Joseph Hume, chemist in Long Acre, with whom he remained in touch from India, possibly with regard to materials essential to his chemical demonstrations.[56] Itinerant lecturers had continuous need for apparatus, a necessity that was filled not only by the development of new devices but also by a resale market that Jones and his contemporaries vigorously sought to exploit. Consequently, the audience for lectures would also have had plenty of opportunity to come into possession of apparatus through various dealers,

---

54 University of Guelph, James Dinwiddie Correspondence, XS1 MS A164, no. 66. William Jones to Dinwiddie, March 26, 1795; William Jones, *The Description and Use of a New Portable Orrery*. Second edition, Enlarged (London: John Jones, 1784), esp. p. 31, note. See also John R. Millburn, "Benjamin Martin and the Development of the Orrery," *British Journal for the History of Science* 6 (December, 1973): 378–399.

55 University of Guelph, Dinwiddie Correspondence, XS1 MS A164, no. 66. Jones to Dinwiddie, March 26, 1795. On Bute see A. D. Morrison-Low, *Making Scientific Instruments in the Industrial Revolution* (Aldershot and Burlington: Ashgate, 2007), 255–256.

56 Hume, at his shop in Long Acre, was a noted chemist and associate of James Stodart of the nearby Strand. See W. A. Smeaton, "Early Methods of Cladding Base Metals with Platinum," *Platinum Metals Review* 22 (1978): 61–67, esp. 65; *Kent's Directory for the Year 1794. Cities of London and Westminster, & Borough of Southwark*; Chris Green archive, Dinwiddie to John Anderson, November 14, 1795. Also *Philosophical Magazine* 47 (Jan.–June, 1816): 264–265. Joseph Hume to Alexander Tilloch, 'Mr. Hume On Emitic Tartar', April 17, 1816; Proudfoot, *Biographical Memoir of James Dinwiddie*, 133–134. Dinwiddie to Hume, February 27, 1796.

Fig. 3 The Grand Orrery, ca. 1733. By permission of the Science Museum, London.

and auction sales, even if they could not readily frequent the shops of London or Paris.[57] Here was yet another measure of audience and engagement that Dinwiddie sought to exploit.

Science by Subscription

Dinwiddie was only one in an immense range of players in the experimental enterprise. This included public showmen desperately seeking subscribers, perhaps also professors who needed apparatus, to those privately intent on pursuing further experiments, to the curious who sought out entertainment pieces merely to duplicate the dramatic in drawing rooms, to the immensely wealthy whose collections reflected both the status and the cultivation of enlightened practice. For example, a study by Gerard Turner of the instrument collection of John Stuart, third Earl of Bute, was made possible through a surviving sale catalogue of February, 1793. Bute, after the Peace of Paris in 1763 and his disengagement from the ministry of George III, had set up his own laboratory at Luton House, Bedfordshire. While Bute's wealth enabled him to purchase large, and often very elaborate, devices—like the large silver microscope made by George Adams senior in 1761. Yet, much of his apparatus was rather more mundane and, therefore, more useful. One of the most interesting pieces was the so-called "philosophical table" designed to perform various mechanical demonstrations explicitly "according to THE MATHEMATICAL ELEMENTS OF NATURAL PHILOSOPHY of W. J. Gravesande, Desagulier's [sic] edition…" This was later bought at auction by Francis Wollaston, recently appointed in 1792 as Professor of Natural and Experimental Philosophy at Cambridge. Besides mathematical and optical instruments, Bute's laboratory contained many glass bottles and jars "for experiments with airs", hydrostatical balances, electrical machines, furnaces and retorts and the mysterious "machine useful to apply steam to the legs, or any other part of the human body." Numerous instrument makers, like William Jones and Edward Nairne, bought many of the lots at auction in 1793.[58] These purchases were not made for collecting, but for

57 James A. Bennett, "Shopping for Instruments in Paris and London," in *Merchants & Marvels. Commerce, Science, and Art in the Early Modern World*, edited by Pamela H. Smith & Paula Findlen (London and New York: Routledge, 2002), 370–395.

58 G. L'E. Turner, "The auction sale of the earl of Bute's instruments, 1793," *Annals of Science* 23 (1967): 213–242, esp. 239.

re-sale. Similarly, Jones subsequently offered a copy of George Atwood's friction machine. This had been designed at Cambridge in the 1770s and became a standard demonstration device manufactured by George Adams senior. William and Samuel Jones, London instrument makers, purchased most of Adams' instruments in 1795.[59]

The Grand Orrery offered to Dinwiddie in 1795, with a glass pane "that opens as a door, to admit the hands of the lecturer or pupil", had come from the sale of Lord Bute's collection. This great device had been bought new for over £400 but listed in their catalogue at half that. This had been sold to Jones at auction for the price of £68:5:0, thus providing the potential for a tidy profit.[60] There is no indication Dinwiddie was inclined to take up the offer. Ironically, while Jones and Edward Nairne had bought much of George Adams' inventory in 1795, at the same time his successor Dudley Adams had arranged to send to Dinwiddie in Bengal various camera obscura and optical machines useful for dramatic purposes.[61]

Audience ultimately decided the fate of apparatus. Dinwiddie understood this in Calcutta, where he faced incessant demands for machines and even for treatments. Likewise, instrument makers like William and Samuel Jones made an effort to reach across the entire range of the market. Thus, they could offer 'Instruments of Recreation and Amusement' like 'the cube and two wooden handles, that supports itself on a point', or 'concave mirrors fitted up in boxes' which might demonstrate some basic optical principles. Simultaneously, they could supply reflecting and achromatic telescopes, along with an array of air pumps of various sorts, electrical machines, or 'An useful and illustrative apparatus, compounded

---

59 Their version of the Atwood apparatus, likely directly from George Adams' shop, does not appear as one of the lots in the Bute auction. On Atwood, see Simon Schaffer, "Machine Philosophy: Demonstration Devices in Georgian Mechanics," in *Instruments*, edited by Albert Van helden and Thomas L. Hankins *Osiris*. Second Series, 9 (1994):157–182, esp. 178–179; Millburn, *Adams of Fleet Street*, 259. The Bute machine seems to have been purchased at auction in 1793 by a Mr. Bond. See Turner, "The auction sale of the earl of Bute's instruments," p. 239, lot 236.

60 The most complete analysis of the surviving sale catalogues is provided by Peter de Clercq. See "Second-hand Instruments in a W. & S. Jones Catalogue of 1795," *Bulletin of the Scientific Instrument Society* 85 (2005), 38–39; "Private Instrument Collections Sold at Auction in London in the Late 18th Century. Part 1: Professional Practitioners and Gentleman Collectors," *Bulletin of the Scientific Instrument Society* 95 (2007): 28–36, esp. 32–33; "Part 2: Instrument Makers and Watchmakers," *Bulletin of the Scientific Instrument Society* 100 (2009): 27–35.

61 DUA, MS 2-726, Correspondence, A 1, Dudley Adams to Dinwiddie, November 27, 1795. Cf. Thomas L. Hankins and Robert J. Silverman, *Instruments and the Imagination* (Princeton: Princeton University Press, 1995), esp. chap. III.

of the luminous conductor, exhausted flask, two jars, exhausting syringe, insulated stand, and wires with balls, &c complete" all of which would allow for demonstrations and further experiments. One of the most notable on offer at the end of the century, clearly designed for the showman, was "A curious collection of working models, to be set in motion by the electrical fluid, consisting of a corn-mill and a three-barrelled water-pump, worked by one crank only; an orrery, shewing the diurnal motion of the earth, age and phases of the moon, &c. and astronomical clock, shewing the aspects of the sun and moon, age, phases, &c. all delicately made of card paper, cork, and wire only, packed in a deal case" all for 2/12/6.[62]

The zones of instrumental exchange were, quite literally, trading zones of the instrumental marketplace frequented by Dinwiddie and his contacts.

Conclusion

Throughout his forty years of lecturing on natural philosophy Dinwiddie had stressed a method that depended on experimental apparatus. While he may now be regarded as unique in so far as he carried instruments as far as the court of the Chinese Emperor, Dinwiddie was then essentially supercargo. In Dinwiddie's long trajectory, the Macartney expedition, however incidentally, marked his subsequent reputation. Yet his career reflected to a great extent the manner in which devices were constructed and circulated in the 18th century. He was, however, hardly the only experimental demonstrator in imperial India. But his approach to performance remained precisely as it had been since his days in Dumfries. He sought out instruments, even built new ones and sold them to the curious. Here he closely mirrors the makers of Fleet Street—from Francis Hauksbee to Martin and Adams. His London connections were maintained by correspondence, whether from Cork or Calcutta. This was critical because instruments produced the requisite theatre that commanded handsome fees. In 1807, when Dinwiddie had returned again to London, he proposed yet another course of lectures. Here he was, after four decades, explicit about the lecturers' method:

---

62 W. and S. Jones, *A Catalogue of Optical, Mathematical, and Philosophical Instruments* (London, [1799]), p. 9. British Library, shelf mark 1609.224 (1).

> By Experiment we put nature out of her course and in her struggle to restore the lost Equilibrium she reveals to us her secrets. A proof of any fact by experiment requires little previous knowledge to understand it, it is an immediate assistant to the senses and therefore *level to the common understanding of all mankind.*[63]

In other words, instruments and experiment revealed what a reliance on mathematics might otherwise obscure.

Instruments made universals. They made evident the same physical principles wherever they were demonstrated. Physics thereby crossed many boundaries, geographic, cultural, social, and intellectual. If Macartney had thought of revealing European—notably British—superiority, he failed. But, the trade in devices also had a secondary effect of reinforcing the global command of a Euro-centric science. As it has recently been pointed out, amid the global revolutions of imperial reach, "go-betweens really exist because transactions exist." Dinwiddie's instrumental trade was part of a global compass where both facts and method were translated from place to place. The experimental method, and the instrumental market upon which it depended, mediated the commercial transactions of philosophers and makers, even amidst the aims of the Board of Trade and "such bloated go-betweens as the English East India Company."[64]

---

[63] DUA, MS 2-726, C 49. 'Natural History and Natural Philosophy', June 1, 1807, sect. 3, p. 3. My italics.

[64] Sanjay Subrahmanyam, "Between a Rock and a Hard Place," in *The Brokered World*, edited by Simon Schaffer, Lissa Roberts, et al., 429–440, esp. 430–431.

"BUNGALLEE HOUSE SET ON FIRE BY GALVANISM": NATURAL AND EXPERIMENTAL PHILOSOPHY AS PUBLIC SCIENCE IN A COLONIAL METROPOLIS (1794–1806)

Savithri Preetha Nair

When in late 1794 the Scottish itinerant lecturer, James Dinwiddie (1746–1815) advertised a course of lectures in natural and experimental philosophy in the *Calcutta Gazette*, a high-ranking Company official remarked: 'An opportunity like the present of acquiring information on subjects of natural and experimental philosophy is so very uncommon in this Country, that I wish to avail myself of it to the utmost, as far as leisure from public business will admit'.[1] A member of the Macartney Embassy to China, Dinwiddie's reputation as a man of science preceded him to Calcutta. The *Calcutta Gazette* of March 27, 1794 updated the public of his movements: 'We hear the *Jackall Cruize* is expected in the course of a few days from China; by that conveyance Dr Dinwiddie and another gentleman [Victor Thibault] of Lord Macartney's Suite are on their way to Bengal'.[2] In 1792, at the height of his reputation as a lecturer in natural philosophy, Dinwiddie had been appointed Astronomer to the Macartney Embassy, which had as its two-fold aim the extension of commerce in the Far East and promotion of British scientific achievement.[3] His chief duties included erecting and regulating a planetarium, constructing and ascending in a balloon, descending in a diving bell and conducting experiments on air, electricity, mechanics and other branches of experimental philosophy.[4] Faced with mounting financial debts (the Embassy having

---

1 Dalhousie University Archives (DUA), James Dinwiddie Collection, MS 2-726, A 52. J. H. Harington to Dinwiddie, undated letter.

2 However, we know that Dinwiddie reached Calcutta only six months later, after an extended stay in China.

3 W. J. Proudfoot, *Biographical Memoir of James Dinwiddie* (Liverpool: Edward, 1868), 26–27.

4 In late 1783, Dinwiddie had released a balloon near Buckingham Gate in London, the ascent of which is reported to have entertained the spectators enormously. M. R. Lynn, *The Sublime Invention: Ballooning in Europe, 1783–1820* (London: Pickering and Chatto Ltd, 2010), 1.

proved a failure), Dinwiddie resolved to break away and head towards Calcutta, in the hope of better prospects in the colonial metropolis.[5]

What kind of consumerist society was Calcutta in the late eighteenth century?[6] Private enterprise 'was the engine of the white town's growth and development'.[7] The city enjoyed a thriving printing industry and many of these were connected with newspapers. At least seventeen were in operation by 1800.[8] The polite society of the white town of Calcutta (which became the administrative capital of Bengal in 1772) spent extravagant amounts of money and effort to maintain living standards appropriate to their social position. Leisure was commoditised; a culture of subscription was prevalent. Rational recreations or leisure activities designed to improve the mind through entertainment, especially of a visual and sensory kind, were an integral part of the consumerist culture of the colonial metropolis. Newspaper advertisements reveal that showmen, ventriloquists, jugglers, magicians, artists, balloonists, model-makers, and mathematical instrument makers vied with each other and traded on the streets and bazaars round Tank square and Government Place in the heart of the city.[9] Private academies to instruct youth in the 'useful branches of literature' operated in the Cossitollah and Bow Bazaar areas.

---

[5] This goes against the claim that Dinwiddie was despatched to India by the Company directors 'at their own expense' to teach natural philosophy at the Fort William College. See K. Raj, *Relocating Modern Science: Circulation and the Construction of Scientific Knowledge in South Asia and Europe* (Delhi: Permanent Black, 2006), 153. In fact, the Company gave it the least priority among the subjects offered at the College, founded for the instruction of junior civil servants.

[6] For a history of the city see J. P. Losty, *Calcutta, City of Palaces* (London: British Library, 1990).

[7] P. J. Marshall, "The White Town of Calcutta under the Rule of the East India Company," *Modern Asian Studies* 34 (May 2000): 314.

[8] Marshall, "The White Town of Calcutta," 324.

[9] Demonstrations by 'Indian jugglers' became part of the popular culture of Regency London and 'a testimony to the imperial power, which had already produced a well-developed and multifarious set of cultural practices by the turn of the nineteenth century. J. Whale, "Indian Jugglers: Hazlitt, Romantic Orientalism, and the difference of view," in *Romanticism and Colonialism: Writing and Empire, 1780–1830*, ed. T. Fulford and P. J. Kitson (Cambridge: Cambridge University Press, 1998), 207. Madras jugglers frequented London in the early decades of the nineteenth century and the act of sword-swallowing was considered one of their most marvellous acts. Dinwiddie records in his journal dated June 12, 1806: 'at 12 o'clock a Madras juggler called by appointment and among many other experiments passed a sword blade down his throat 21 inches- the sword I purchased and will carry it to England- witness to the above T. Rawlins'. He would also be invited to the houses of friends to see their performance, which means that it was as much a popular entertainment in Calcutta's white town as it was in London. DUA, MS-2-726, B71 (April 10–Aug. 5, 1806), entry dated June 12, 1806.

The Old Court House was known for its public entertainments. A French adventurer and proprietor of the 'curious cabinet', Laurency, living on Old Court House Street performed feats with a 'Machine Aerostatique' before paying audiences in the 1790s. Mr Croese living at Cossitollah periodically entertained the public with panoramas[10] and a Mr Maillardet, who resided on Council House Street, exhibited automata brought out from Europe.[11] For vocal and instrumental concerts, the public would visit the Le Gallais rooms and for plays the Calcutta Theatre or Lebedeff's Bengali theatre at Durrumtollah.

Science or natural philosophy in Europe as it was referred to in the eighteenth century was practiced in a variety of spaces, for different purposes and by a range of practitioners. Conscious about the complex and multifaceted nature of enlightenment science, historians have become increasingly attentive to such public sites of scientific practice as pubs, coffee houses, shops, streets and fairs, where learning was combined with entertainment.[12] In colonial non-European settings on the other hand the link between science and the public sphere has rarely been examined. The chief focus of scholarship in colonial science has been the state, that is, on subjects of direct interest to the Company such as natural history, cartography, environment, medicine, and on very rare occasions the exact sciences.[13] In the late eighteenth century, Calcutta was the setting

---

[10] *Calcutta Gazette*, July 7, 1796; Jan. 19, 1797; Jan. 26, 1797; Feb. 2, 1797; and Feb. 23, 1797.

[11] The figures 'are finished in the most superior expensive style having been intended for the late Nabob of Arcot, and consists of a young lady playing a variety of tunes on an organ, with every motion corresponding to nature; man who performs to music, the various and astonishing evolutions of a Sadler's Well Tumbler, with great ease and accuracy. And a singing bird in an elegant snuff box'. *Calcutta Gazette*, July 12, 1797; Jan. 4, 1798; and Feb. 8, 1798. This was in all probability the Swiss automata maker Henri Maillardet best known for his "Draughtsman-writer" automaton and other mechanisms including clocks, which he exhibited in London at the turn of the nineteenth century. In Calcutta, his shop was located at Tank Square.

[12] For example, A. Secord, "Science in the Pub: Artisan Botanists in early Nineteenth-Century Lancashire," *History of Science* 32 (1994): 269–315; L. Stewart, "Other centres of calculation, or, where the Royal Society didn't count: commerce, coffee-houses and natural philosophy in early modern London," *British Journal for the History of Science* 32 (1999): 133–53; and Stewart, "The Laboratory, the Workshop, and the Theatre of Experiment," in *Science and Spectacle in the European Enlightenment*, ed. B. Bensaude-Vincent and C. Blondel (Adershot: Ashgate, 2008), 11–24.

[13] For a sample see K. Raj, *Relocating Modern Science: Circulation and the Construction of Scientific Knowledge in South Asia and Europe* (Delhi: Permanent Black, 2006); D. Arnold, *Science, Technology and Medicine in Colonial India* (Cambridge: Cambridge University Press, 2000); R. Macleod (ed) *Nature and Empire: Science and the Colonial Enterprise*, *Osiris*, Vol. 15, Chicago, 2000; R. Drayton, *Nature's Government: Science, Imperial Britain,*

for sustained inter-cultural encounters between Europeans and Indians.[14] The interactions between Orientalists (mostly Company officials or missionaries) and Indian intellectuals, particularly under the aegis of the Asiatic Society of Bengal are only too well-known. However, what is less-known is that there were others in the city, neither employed by the Company nor attached to missions, but by virtue of having introduced new cultures of consumption to the city, closely interacted with the 'native' society at different levels. At least a few of these men were in the business of science and have fortunately left traces of their activities in the form of publications, images, newspaper advertisements, journals or correspondence. Here, we consider one such individual, the itinerant natural philosopher Dinwiddie, who in the late eighteenth century attempted to fashion a career out of experimentation in Calcutta, the first one to do so in the city. Books alone were insufficient to make sense of the natural world; it was unimaginable without experimentation.[15] This paper argues that Dinwiddie's lecture-rooms, a public space from where the secrets of natural world were diffused to anyone willing to pay for it, signalled the emergence of public science in colonial India.

Experiment is a continuous historical event 'always available for reinterpretation by different groups and different settings'[16] but very importantly, it was not to be limited to the boundaries of the European metropolis. Within months of the invention of the galvanic (or voltaic) pile by Alessandro Volta in March 1800, to establish experimentally the fact of metallic electricity to the non-specialist witness, results of experiments with the pile were reported from a wide range of geographical and social locations in London, Manchester, Sheffield, Cambridge, Pennsylvania, Scotland, France, Italy and Germany.[17] That the colonial city of Calcutta was also witness to similar experiments, thanks to Dinwiddie, has gone entirely unnoticed. Scholarship in the history and philosophy of experimental science has grown substantially in the last couple of decades, forging new

---

and the 'Improvement of the World' (New Haven & London: Yale University Press), 2000; M. Edney, *Mapping an Empire: The Geographical Construction of British India, 1765–1843* (Chicago & London: University of Chicago Press), 1997.

[14] Marshall, "The White Town of Calcutta," 307.

[15] L. Stewart, *The Rise of Public Science: Rhetoric, Technology, and Natural History in Newtonian Britain, 1660–1750* (Cambridge: Cambridge University Press, 1992), xxiv–v.

[16] I. R. Morus, *Frankenstein's Children: Electricity, Exhibition and Experiment in Early Nineteenth Century London* (Princeton: Princeton University Press, 1998), 11.

[17] J. Mertens, "Shocks and Sparks: The Voltaic Pile as a Demonstration Device," *Isis* 89 (June 1998): 300–11; also see M. Rowbottom and C. Susskind, *From Electricity and Medicine: History of their Interaction* (San Francisco: San Francisco Press, Inc., 1984), 50.

perspectives on the socio-politics of experiment and instruments, and of science as public knowledge, entertainment and spectacle.[18] However, almost all scholarship on experimention, a core aspect of modern science, has had Europe as focus and to a marginal extent the 'settler' colonies of the Americas and Australasia, but hardly ever the erstwhile colonial non-European dependencies. Focusing on Dinwiddie's experimental life in Calcutta between 1794 and 1806, this paper aims to unravel the material culture of his practice to capture hitherto unexplored facets of colonial science at the turn of the nineteenth century—its commoditised, experimental and public dimensions. Experiments can hardly be understood without considering 'the surrounding material culture that provides the basic resources' for experimentation, and thus we shall throw light on the activities of Dinwiddie as he exploited local skills and resources and responded to technical, market, social and epistemological demands in the colonial metropolis.[19]

The paper is organised around four sections: we will begin by throwing light on Dinwiddie's 'lecture-rooms' as he called them, the first such "place" for public science in the colonial metropolis, where several social networks converged. We will then explore the links between knowledge, entertainment and commerce by considering Dinwiddie's lectures and experiments. The third section explores the networks, which sustained his career as a natural and experimental philosopher in the city, providing an account of the different categories of philosophically inclined consumers in the metropolis who paid for private tuitions (referred to as 'private company' by Dinwiddie), subscribed to lectures, sought advice on scientific matters and sometimes assisted in setting up experiments. They included women and Indian elites, especially vakeels or 'go-betweens' who mediated between the Company and the native state. And finally we will attend to Dinwiddie's entrepreneurial activities in the city. The income from public lectures and private tuitions alone did not suffice; he was quick to respond to commercial and social demands, exploited local skills and resources, entered into business collaborations and even practiced as a medical electrician for a short duration to sustain his career as an experimentalist in the colonial metropolis.

---

[18] See for example D. Gooding, T. Pinch and S. Schaffer, ed., *The Uses of Experiment: Studies in the Natural Sciences* (Cambridge: Cambridge University Press, 1993 (1989)); J. Golinski, *Science as Public Culture: Chemistry and Enlightenment in Britain 1760–1820* (Cambridge: Cambridge University Press, 1992); Stewart, *The Rise of Public Science*, and the more recent Bensaude-Vincent and Blondel, *Science and Spectacle*.

[19] In so doing, I am indebted to Stewart, *The Rise of Public Science*.

Fig. 1 Council House Street, Calcutta. T. Daniell, Engraving, c. 1798

Fig. 2 Old Court House Street, Calcutta, T. Daniell, Colour Aquatint, c. 1786

A Place for Public Science

A tradition of rational recreation was fairly widespread in the white town of Calcutta in the late eighteenth century, but a culture of public lecturing and experimentation as was common in Europe during this time hardly existed until Dinwiddie's arrival in late 1794. Electricity was fashionable and dominated natural philosophy in the eighteenth century. In colonial India several important publications in natural philosophy were in circulation during the latter part of the century. Amongst the most popular being James Ferguson's *Introduction to Electricity*, Tiberius Cavallo's *A Complete Treatise of Electricity* and his *Essay on Theory and Practice of Medical Electricity*, George Adams's *Essay on Electricity*, Thomas Beddoes on *Factitious Airs*, Jean-Antoine Chaptal's *Chemistry* and journals like the *Philosophical Magazine*. Books and periodicals alone were not sufficient however; a site for viewing replications of crucial experiments, and of seeing novelty was a desideratum in the colonial metropolis. It is this niche that Dinwiddie had exploited. Moreover several members of the polite society of Calcutta, mainly Scottish, were exposed to the culture of public lecturing in Europe, and a few had even attended Dinwiddie's lectures there.

Dinwiddie's lecture-rooms signalled the arrival of a new, rational and visual culture of consumption in the city, one that was centred round spectacular experimentation; it was a modern public space that offered enlightened entertainment to a paying audience. An appropriate site within the cultural geography of the city, from which to diffuse the secrets of the natural world in an enlightening and entertaining fashion, was crucial.[20] The venue for the innumerable experiments performed by Dinwiddie in Calcutta, including the ones with the voltaic pile, was a non-descript rented accommodation in the white town area adapted to the purposes of a lecture-room, laboratory, workshop, and exhibition space. However, he saw to that the lecture rooms were in the heart of the city, close to Government Place and Tank Square. To 'enable Ladies and Gentlemen to judge for themselves, respecting the instruction and entertainment, that may be expected from the Lectures,' on certain notified days the apparatus was exhibited in the rooms.[21] His workshop was where

---

20 'Knowledge was made within the world of goods', Schaffer has remarked in his "The Consuming flame: electrical showmen and Tory mystics in the world of goods," in *Consumption and the World of Goods*, ed. J. Brewer and R. Porter (London & New York: Routledge, 1993), 490.

21 Ibid.

he made economic and technical decisions, and manufactured products for the market on a small scale (to which we shall return in the final section). It was a space (which included the laboratory) where only trusted friends, most of whom were individuals placed high up in the social order, and his artisans and mechanics were allowed access. Resolving practical and technical problems in the workshop went a long way in shaping the polite discourse of science. The outcome of his experiments depended on the process of negotiation between heterogeneous social worlds that converged in that space and not simply on the successful manipulation of the apparatus by the experimenter in the lecture-room.[22] Dinwiddie employed assistants who helped maintain his lecture room and apparatus and hired ingenious craftsmen like George Bertie and Robert Armstrong who could craft apparatus and instruments for use in experiments. While at the Fort William College, where he was appointed professor of mathematics in March 1801, Dinwiddie was assisted by Thomas Wilkins, son of the Orientalist-typographer, Charles Wilkins.[23]

To notify the public of his intention to begin a series of lectures Dinwiddie resorted to the public media. In late October 1794, the *Calcutta Gazette* announced a course of lectures on experimental philosophy to be held by subscription at No. 7, Council House Street, the location of Dinwiddie's first ever 'lecture rooms' in the city.[24] Three years later, he would move to the Old Court House Street (1797–98), and move again within a year to No. 30, Cossitollah Street (1798–1801), which he would share with carpenter Robert Armstrong. On being appointed lecturer in mathematics at the Fort William College in 1801, Dinwiddie would transfer his instruments, chemicals and personal possessions to the College premises. The last three years of his life in India (1803–06) were spent in the Lal Bazaar area, and it was during this period that he would deliver public lectures on galvanism, a most new and fashionable science of the times. The lecture room was often shifted around the rented building, depending on the season. In the early summer, Dinwiddie would choose the front room, it being the coolest part of the house and at the height of the season would simply move the lectures to the early hours of the morning. He would

---

22 Morus, *Frankenstein's Children*, 10.

23 DUA 2-726, B 65, 'Journal', (April 19–July 12, 1801), entry dated May 12, 1801.

24 'Lectures- Doctor Dinwiddie begs leave to inform the Ladies and Gentlemen of Calcutta that he proposes to begin a course of Lectures on Experimental Philosophy at his House No. 7, Council House Street on Monday the 17th of Nov. next.' *Calcutta Gazette*, Oct. 30, 1794.

equip the lecture-room with a 'punka and other contrivances' in the later years to make it more welcoming.[25] On a daily basis the temperature and atmospheric pressure of the room would be recorded; this was especially important because the electricity machine was sensitive to even small changes in the weather. If on the day of the lecture, the monsoon wind blew strong from the south, he would take it as a warning that some experiments would fail. Dinwiddie would in fact soon learn to use the electricity machine as a hygrometer. He also gave great thought to seating arrangements, even dressing the backrests of the seats with cloth coloured with a newly manufactured green dye. Effective lighting was crucial, especially when lectures involved chromatic-experiments. A printed list of the lectures was always available at the lecture-rooms.[26] Dinwiddie's lecture rooms were thus fine-tuned for the most effective demonstrations of experimental philosophy.

Public Demonstrations, 'useful as they are entertaining'

As already stated public demonstrations of natural philosophy were unknown in the colonial metropolis until Dinwiddie's arrival there. In this sense, he had no competitors in the city, but consumers are hardly passive; they exercise discretion over the choice of commodities in the market.[27] Dinwiddie had to persuade audiences through newspapers advertisements that his lectures were entertaining and improving. The dates of his lectures were chosen judiciously so as to avoid clashing 'with the other Entertainments of the Season'. His newspaper articles, an extension of his public lectures, covered a variety of subjects ranging from the Aeolian harp to observations on the transit of Mercury,[28] spreading his fame far and wide, including the French and Danish settlements of Chandernagore and Chinsurah. However, the strength of his audiences hardly expanded with time, in fact it weakened so much so that Dinwiddie remarked in early 1800: 'The spirit for improvement and even taste for

---

25 *Calcutta Gazette*, April 30, 1795.

26 *Calcutta Gazette*, Nov. 20, 1794.

27 For a recent collection that situates science as part of a consumer culture, see *Science in the Market Place: Nineteenth Century Sites and Experiences*, ed. A. Fyfe and B. Lightman (Chicago: University of Chicago Press, 2007).

28 DUA, MS 2-726, B 58, 'Journal' (April 27–June 17, 1800), entries dated May 4, and 28, 1800.

rational amusement seems quite lost in this place—I must therefore leave the country with the trifle I have got'.[29]

Experiment in the age of enlightenment was essentially a public performance; it was designed to be dramatic and entertaining.[30] The *Calcutta Gazette* reported that Dinwiddie's lecture on electricity in late 1794 was attended by a 'numerous and polite audience, who appeared to be highly entertained with the Experiments'.[31] The rooms were open to subscribers every Saturday morning, during the course from eleven till two o'clock, 'in order to explain more fully any particular subjects of the two preceding Lectures'.[32] Dinwiddie constantly added new lectures, apparatus and experiments to his repertoire, and often at considerable trouble and expense.[33] One of his chief epistemological strategies was to steer clear of the scientific controversies raging in Europe at the time and focus instead on presenting a category of results of iconic experiments in an entertaining fashion, in other words to represent natural philosophy 'as a visible and powerful public science'.[34] The experiments by Volta, Nicholson and Carlisle, Cruickshank, Davy, Van Marum, Cavallo, Simon of Berlin, Pepys, Klaproth, and the experiments performed at the Askesian Society and the St. Thomas hospital were all presented to audiences in Calcutta. To make his lectures and exhibitions as entertaining as possible, he took inspiration from local performances of all kinds including those by local ventriloquists, magicians and jugglers, some of who lived in the same locality as him.[35] In late 1798, Dinwiddie was determined to organise a philosophical exhibition, complete with optical experiments and 'a writing and speaking figure'.[36] Two years later he would exhibit for

---

[29] DUA, MS 2-726, B 58, 'Journal' (Feb. 3–April 26, 1800), entry dated Feb. 22, 1800.

[30] G. V. Sutton, *Science for a Polite Society: Gender, Culture and the Demonstration of Enlightenment*, (Colorado: Westview Press, 1995).

[31] *Calcutta Gazette*, Dec. 4, 1794.

[32] *Calcutta Gazette*, Nov. 27, 1794.

[33] Proudfoot, *Biography*, 1868.

[34] P. Fara, " 'A treasure of hidden virtues': the attraction of magnetic marketing," *British Journal for the History of Science* 28 (Mar. 1995): 11.

[35] In exchange for a piece of entertainment by a Madras juggler family, Dinwiddie "shocked" them with an electricity machine, triggering emotions of wonder and curiosity in reverse. DUA, MS 2-726, B 19 Dinwiddie Journal (April 1–May 4, 1795), entry dated April 28, 1795; B-20 (May 9–June 22, 1795), entry dated June 22, 1795.

[36] Incidentally, Dinwiddie referred to a 'Speaking figure in London' and the 'proposition offered to explain the construction of it for five guineas' in his Calcutta journal of 1795. DUA, MS 2-726, Dinwiddie Journal, B-19 (April 1–May 8, 1795), entry dated April 23, 1795.

the first time at a lecture a writing-automaton.[37] In 1799, he added a new dimension to his lectures in an attempt to attract more subscribers; he proposed to deliver six lectures on select 'subjects in experimental philosophy illustrated by a great variety of interesting experiments' and at the conclusion of each, exhibit a few philosophical recreations 'or entertaining experiments, depending on Mechanical, Magnetical, Optical and other principles'.[38]

It was at the Fort William College that Dinwiddie erected a voltaic pile for the first ever time.[39] Soon after he had taken charge as professor of mathematics in March 1801, he gave his inaugural lecture in natural philosophy to the students of the College and six months later, on December 8, 1801 to be precise, Dinwiddie erected his first voltaic pile at the College laboratory.[40] The pile, which consisted of twenty pairs of silver and zinc squares crafted by his workmen, with wet paper used as discs between the metals and erected with the assistance of his friend David Hare, acted powerfully. Dinwiddie was determined to exhibit the pile at his next lecture. The experiment was again a great success and the students had the opportunity of directly experiencing the strong shocks, flashes and sounds produced. Dinwiddie noted in his 'Journal of Galvanism': 'Students much pleased with this new and singular experiment'.[41] He was particularly excited to have discovered a new way of securing 'the attention of young men', who otherwise were uncontrollable. In the syllabus of natural philosophy, which he prepared for the use of the College, Dinwiddie thus included the new science of galvanism and a great number of entertaining experiments. To demonstrate that 'artificial devices could reveal nature itself', Dinwiddie experimented with the pile in a variety of ways.[42] His recreational experiments included the ignition of gun-powder and mercury fulminate with the pile, gold-leaf experiments, and experiments that produced spectacular and dramatic effects such as the 'Bungallee house set on fire by galvanism', a local adaptation of experiments such

37 *Calcutta Gazette*, Jan. 13, 1800. This was most certainly influenced by the 'entertainment called the Androides exhibited lately in London', described in Dinwiddie's journal. DUA, MS 2-726, B 42 Dinwiddie Journal (Sep. 1–Nov. 3, 1797), entry dated Oct. 10, 1797.

38 *Calcutta Gazette*, Jan. 10, 1799.

39 For a history of the College in its early years see T. Roebuck, *The Annals of the College of Fort William from the period of its Foundation on the 4th May 1800 to the present time* (Calcutta: Government Press, 1819).

40 DUA, MS 2-726, D 22 Journal of Galvanism (Calcutta), Dec. 8, 1801.

41 DUA, MS 2-726, D 22 Journal of Galvanism, Dec. 10, 1801.

42 Bensaude-Vincent and Blondel, *Science as Spectacle*, 8.

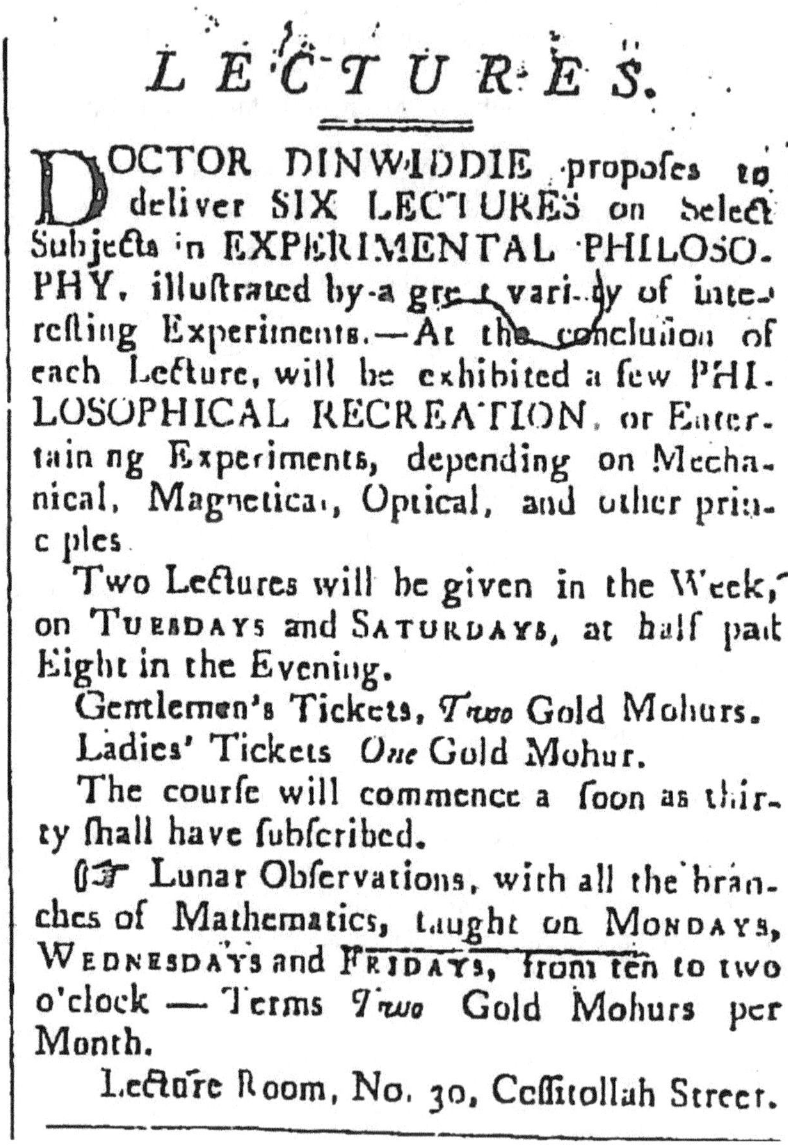

LECTURES.

DOCTOR DINWIDDIE propoſes to deliver SIX LECTURES on Select Subjects in EXPERIMENTAL PHILOSOPHY, illuſtrated by a great variety of intereſting Experiments.—At the concluſion of each Lecture, will be exhibited a few PHILOSOPHICAL RECREATION, or Entertaining Experiments, depending on Mechanical, Magnetical, Optical, and other principles.

Two Lectures will be given in the Week, on TUESDAYS and SATURDAYS, at half paſt Eight in the Evening.

Gentlemen's Tickets, *Two* Gold Mohurs.

Ladies' Tickets *One* Gold Mohur.

The courſe will commence a ſoon as thirty ſhall have ſubſcribed.

☞ Lunar Obſervations, with all the branches of Mathematics, taught on MONDAYS, WEDNESDAYS and FRIDAYS, from ten to two o'clock — Terms *Two* Gold Mohurs per Month.

Lecture Room, No. 30, Coſſitollah Street.

Fig. 3 Advertisements notifying Dinwiddie's lectures in the Calcutta newspapers. *Calcutta Gazette*, 18 Oct 1797 and 10 Jan 1799.

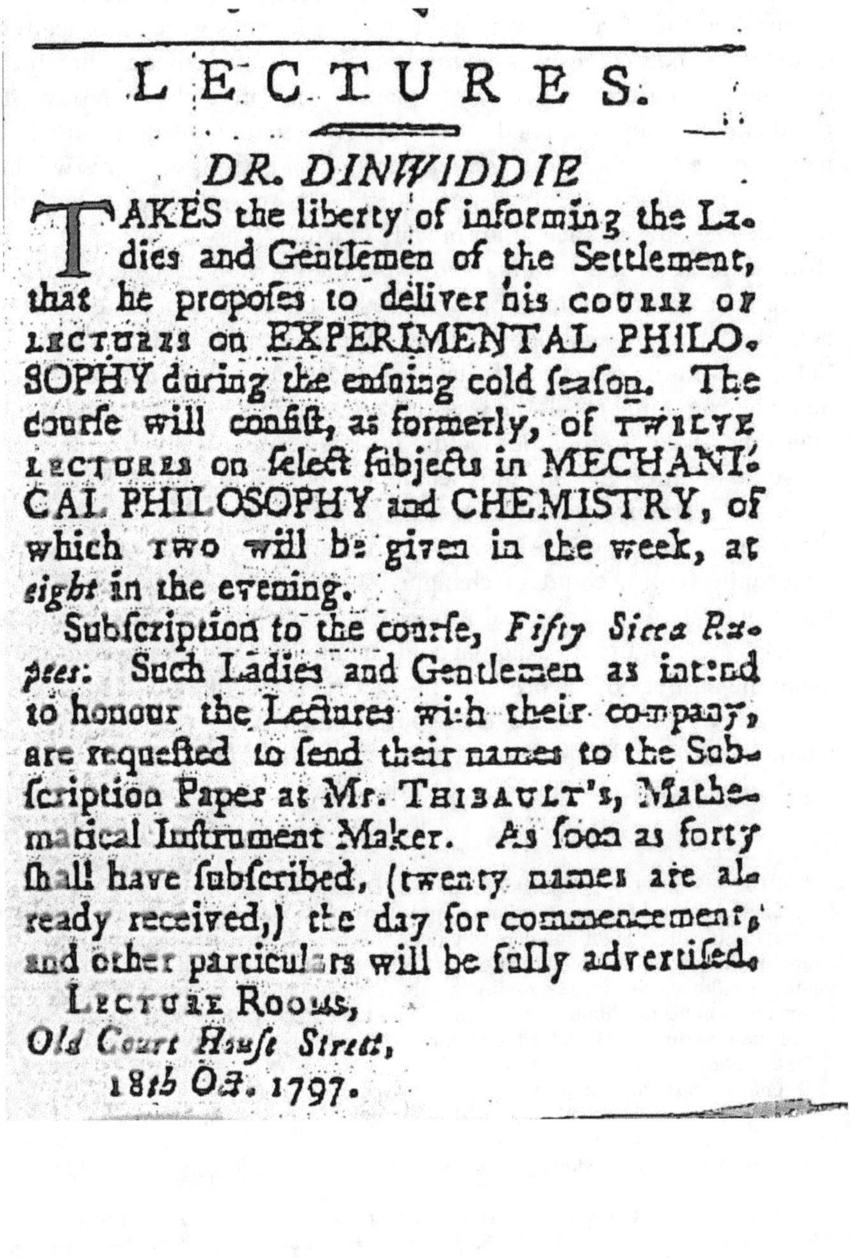

LECTURES.

DR. DINWIDDIE

TAKES the liberty of informing the Ladies and Gentlemen of the Settlement, that he propoſes to deliver his COURSE OF LECTURES on EXPERIMENTAL PHILOSOPHY during the enſuing cold ſeaſon. The courſe will conſiſt, as formerly, of TWELVE LECTURES on ſelect ſubjects in MECHANICAL PHILOSOPHY and CHEMISTRY, of which TWO will be given in the week, at *eight* in the evening.

Subſcription to the courſe, *Fifty Sicca Rupees*: Such Ladies and Gentlemen as intend to honour the Lectures with their company, are requeſted to ſend their names to the Subſcription Paper at Mr. THIBAULT's, Mathematical Inſtrument Maker. As ſoon as forty ſhall have ſubſcribed, (twenty names are already received,) the day for commencement, and other particulars will be fully advertiſed.

LECTURE ROOMS,
Old Court Houſe Street,
18*th* O*ct*. 1797.

as the 'burning schoolhouse', commonly performed by electrical lecturers in Europe.[43]

Two of the chief features of eighteenth-century science were entertainment and utility. As well as making the lectures entertaining, Dinwiddie emphasised their usefulness to industry. The lectures were so conceived as to 'contain an extensive system of philosophical instruction; in which the phenomena are particularly explained, the principles fully investigated and the subject applied to the improvement of agriculture, arts and manufactures'.[44] Dinwiddie assured subscribers that his 'Lectures will be found as useful as they are entertaining'.[45] His second series of lectures were devoted to the ancient art of war, aimed at 'every reader of Ancient History'. Most of the lectures and experiments were designed to cater to the needs of the more important subscribers such as heads of agency houses, merchants and high-ranking Company officials. When his friend Col. James Murray suggested that it would be useful to lecture on the mode of examining fossils, he was quick to respond.[46] On the basis of the questions asked at an earlier lecture on Pneumatics, Dinwiddie decided he would in future give an entire lecture on sound and music.[47] The polite discourse of science was often translated into commercial advantage in the colonial metropolis. Alan Morton has argued that 'ideas in natural philosophy took account of changing social relationships, rather than being only about the physical properties of machines'.[48] This insight is especially relevant to the colonial context, and the Company was quick to make the best use of his presence in the city. Dinwiddie's public lectures also dealt with such subjects as the power of machines and labour-saving technologies including steam-engines, which could profitably be used in sugar manufactories. Dinwiddie's lecture on the strength of timber and

---

[43] DUA, MS 2-726, B 17 Dinwiddie Journal (Dec. 14, 1794–Feb. 14, 1795), entry dated Dec. 14, 1794. The Lal Bazaar, Bow Bazaar and Durrumtollah areas of white town bordered on the 'intermediate town' inhabited by poor whites and Eurasians. Indian thatched or straw huts abounded in the 'interstices of the white town'. Marshall, p. 318. The experiment titled 'Bungallee house set on fire by galvanism' was perhaps inspired by the burning Bengali huts in his neighbourhood of Lal Bazaar. I thank Larry Stewart for bringing to my notice the experiment called 'burning schoolhouse'.

[44] *Calcutta Gazette*, April 30, 1795.

[45] *Calcutta Gazette*, Mar. 5, 1795.

[46] DUA, MS 2-726, Dinwiddie Journal B 19 (1 April–May 8, 1795), entry dated April 17, 1795.

[47] DUA, MS 2-726, Dinwiddie Journal B 21 (June 23–Aug. 27, 1795), entry dated June 26, 1795.

[48] A. Q. Morton, "Concepts of power: natural philosophy and the uses of machines in mid-eighteenth century London," *British Journal for the History of Science* 28 (Mar. 1995): 67.

other local materials was specially prepared for the timber merchants. He had visited timber yards to study the specimens of Sissoo, Saul and timber used in the construction of ships.[49] So who were Dinwiddie's audiences, the 'philosophical consumers' of Calcutta who helped consolidate his authority in the city as a public lecturer and experimentalist?

Consumers of Philosophical Knowledge

There were several sections of the white town of Calcutta in the late eighteenth century who were scientifically inclined and believed in spending time and money for 'improving' themselves. Thanks to the private papers of Dinwiddie, we are in a position to throw light on some of these consumers of philosophical knowledge. One of Dinwiddie's chief sources of income came from the private tuitions he offered in mathematics, astronomy (especially lunar observations) and geometry. He firmly believed that it was 'only in those parts of the science which have been mathematically considered, that natural philosophy can boast of having carried on her investigations with certainty, success and utility'. Without at least 'a moderate share of mathematical knowledge', he warned one could only 'read the flimsy and indeterminate writings of a Nollett or a Priestly and obtain a schoolboy's acquaintance with natural philosophy' but it would be impossible to 'profit by the truths delivered by a De Lambert or a Newton'.[50] After only four years of teaching mathematics, Dinwiddie however concluded that in Calcutta it was hardly worth the trouble. Traders, Company officials, artists, women and at least a few elite Indians, were all part of Dinwiddie's social network in Calcutta. With several of them, he would maintain close personal relationships and spend many an evening engaged in polite conversation on sundry subjects, including galvanism and medical electricity.

a. *Traders*

In the late eighteenth century, Calcutta was regarded as one of the most expensive cities in the world. It was governed largely by wealthy agency houses, merchants, private bankers and high-ranking civil servants, who

---

49 DUA, MS 2-726, Dinwiddie Journal B 22 (Aug. 28–Nov. 5, 1795), entry dated Sep. 8, 1795.

50 Proudfoot, *Biography*, 10.

maintained opulent lifestyles, but were also inclined towards enlightened pursuits and the acquisition of 'useful knowledge' in their leisure hours. The powerful agency houses were centred round individuals involved in remitting profits generated in India to Britain and offered several kinds of services to the servants of the East India Company, including the management of rented apartments on behalf of absentee landlords. And many were Scottish. The Scottish agency house of Fairlie, Fergusson & Co owned ships; Alexander & Co ran the Hindostan Bank and was a major presence in the indigo trade.[51] However, by the 1790s, country trade began to be controlled by robust networks that linked three major centres of commercial activity, London, Canton and Calcutta. Involved in this multinational business activity were David Scott, who founded the agency house of David Scott & Co in London, and William Fairlie, a leading partner in the largest agency house in Calcutta.[52] In 1794, Fairlie, Reid & Co was involved in providing food and bullocks for the army.[53] Fairlie, Reid and Palmer also acted as commissioners, for the Calcutta Exchange Lottery, which indicated their elevated social status in the white town.[54] In 1796, the firm took on the name Fairlie, Gilmore & Co.[55] It was no small matter in Calcutta to know traders like Fairlie, Reid and Gilmore as intimately as Dinwiddie did. Through Fairlie, Dinwiddie would source instruments and laboratory apparatus like glass retorts, receivers, and decanters from Europe and also the most recent publications in experimental philosophy. In return, he would make chests of sulphuric acid on a regular basis for him. In 1802, Dinwiddie borrowed from Fairlie the last volume of the Supplement to the *Encyclopedia Britannica*, 'containing many articles by Prof. Robison', who was professor at the University of Edinburgh.[56] Fairlie & Co. made use of 'Dr Dinwiddie's mechanical and chemical knowledge' to manufacture vitriolic acid, warranted to 'be equal

51 Ibid., 14.

52 B. R. Tomlinson, "The 'Empire of Enterprise': Scottish Business Networks in Asian Trade, 1793–1810", *KIU Journal of Economics and Business Studies*, 8 (2001): 67–83.

53 *Calcutta Gazette*, Aug. 14, 1794.

54 *Calcutta Gazette*, Oct. 22, 1795.

55 *Calcutta Gazette*, April 28, 1796.

56 DUA, MS 2-726, Dinwiddie Journal B 68 (Dec. 22, 1801–Mar. 22, 1802), entry dated Feb. 7, 1802. He was a keen follower of the writings of John Robison, who had become disenchanted with the French Revolution. Ideologically, Dinwiddie was a royalist and a critic of democracy; alluding to the American War of Independence, he warned that all democracies would come to an end sooner or later.

to the very best imported from Europe'.[57] They were hopeful that Dinwiddie's experiments would render its application 'to indigo and other works still more useful and extensive'.[58]

b. *Company Officials*

Letters of introduction from high-ranking Company officials mattered immensely in a colonial setting. Dinwiddie enjoyed a close intellectual relationship with several of the Company's surgeon-naturalists and surveyors, most of who were again Scottish. As soon as he had arrived in Calcutta in late 1794, Dinwiddie was introduced to Governor General Sir John Shore by his friend, the Military Department Sub-Secretary, John Herbert Harrington and by the trader Fairlie.[59] Shore was especially curious to know about the mission to China. On several occasions, Dinwiddie would dine with him conversing on chemistry, light, time and other subjects of mutual interest, even entertaining the latter's young daughter with colourful optical experiments.[60] He often visited the Botanic Gardens at Sibpur, on the banks of the Hooghly, to meet the Company naturalist, William Roxburgh. Their topics of discussion ranged from plants to cochineal, and the use of vitriolic acid in the manufacture of indigo.[61] He was also friends with its founding superintendent Col. Robert Kyd and with the acting superintendent Dr. John Fleming, with whom he discussed dyes. On one occasion, Roxburgh borrowed Dinwiddie's copy of Berthollet on dyeing.[62] Dinwiddie would also communicate with surgeon-naturalists in Bombay and Madras, including Helenus Scott and James Anderson. During a visit to Madras in 1796 with the aim of delivering a course of lectures, he met

57 They had to deal with competitors soon enough. In 1798, Downie and Maitland began selling vitriolic acid 'at one rupee eight annas per pound, in safe packages for carriage'. *Calcutta Gazette*, July 19, 1798.

58 *Calcutta Gazette*, Dec. 7, 1795. Surgeon Helenus Scott of Bombay was interested in rendering 'alkali' an object of commerce as early as 1790. One of the biggest obstacles to this was the hefty 'King's duty', which had to be paid. For his correspondence with Joseph Banks on the subject see British Library (BL): Add Ms 33,979, letter dated Jan. 7, 1790.

59 John Shore would later become a proprietor of the Royal Institution, an institution to which Dinwiddie would return after his Indian residence.

60 DUA, MS 2-726, Dinwiddie Journal B 21 (June 23–Aug. 27 1795), entry dated Aug. 17, 1795.

61 Roxburgh believed that vitriolic acid served no great advantage to the indigo manufactory.

62 DUA, MS 2-726, Dinwiddie Journal B 16 (Oct. 6–Dec. 12, 1794), entry dated Nov. 24, 1794; B 17 (Dec. 14, 1794–Feb. 16, 1795), entry dated Jan. 17, 1795; B 22 (Aug. 28–Nov. 5, 1795), entry dated Sep. 10, 1795; B 30 (July 1–Aug. 1, 1796), entry dated July 9, 1796.

with the educationist Dr. Andrew Bell, famous for his 'Madras Model' of education and purchased from him a Tellurian, a couple of air pumps, compound cylinders, and barometer tubes.

Several of the philosophically inclined Company officials were members of the Asiatic Society of Bengal, founded in 1784 by William Jones for the study of the arts and sciences of the East and chiefly funded by the Company.[63] Dinwiddie himself became a member of the Society as soon as he reached Calcutta and suggested the formation of a 'Physical Class' under its aegis to pursue researches in natural and experimental philosophy, but this was hardly considered with any seriousness at the time. Only months before his arrival in Calcutta, John Shore had replaced William Jones, as President of the Society. Almost on a daily basis, Dinwiddie would meet with fellow Scottish Orientalists, John Bentley (scholar on Hindu astronomy), John Gilchrist (the Hindustani scholar), Major H. T. Colebrooke (Sanskrit scholar), artist Robert Home, Dr James McNabb,[64] Edward Morris, surgeon-naturalist John Fleming,[65] surgeon John Shoolbred,[66] missionary-naturalist William Carey, surgeon William Hunter, J. Howison, Francis Wilford, J. H. Harrington[67] and W. C. Blaquiere, a freemason.[68] Sir J. Anstruther (President, 1799–1805) and Sir Robert Chambers, Chief Justice, who presided over the first meeting of the

[63] The Society was a platform for discussions on the 'history and antiquities, natural productions, arts, sciences and literature of Asia'. For a history of the Society see O. P. Kejariwal, *The Asiatic Society of Bengal and the discovery of India's past 1784–1838* (New Delhi: Oxford University Press, 1988).

[64] McNabb joined as assistant surgeon in Patna in 1799; he was appointed full surgeon in 1806. He maintained accounts with W. Fairlie's agency. James McNabb Papers, BL: MssEur F206/6-7.

[65] Fleming purchased from Dinwiddie several instruments and apparatus including an electrical machine, a pyrometer, a hydrometer, glass retorts, a number of crucibles, a natural magnet and a 'good microscope'. John Kelso to Dinwiddie, undated letter. DUA, MS 2-726. Dinwiddie Correspondence, A 67. John Kelso to Dinwiddie, undated letter.

[66] Shoolbred was the first Superintendent of the Calcutta Native Hospital and remained in charge of it from 1792–1816.

[67] Besides an electricity machine, Harington of the Military department regularly bought from Dinwiddie chemicals like the 'marine acid'. On one occasion, he requested Dinwiddie for a list of short treatises on scientific subjects, which could be read 'for the better understanding of your present course'. DUA, MS 2-726, Dinwiddie correspondence, A 52. J. H. Harington to Dinwiddie, undated.

[68] The Huguenot Orientalist Blaquiere was particularly interested in the *Kalica puran*. See W. C. Blaquiere, "The Rudhiradhyaya or the Sanguinary Chapter, translated from the Calica Puran", *Asiatic Researches*, V (1800): 371–91.

Society, were known to him and often visited his house of experiments.[69] Of all these, Bentley was closest to Dinwiddie, along with David Hare and Fairlie, assisting him to set up experiments; they too were subjected to occasional 'shocks' with the electricity machine and later the voltaic pile. They also shared a common interest in 'Hindoo' astronomy.[70] Bentley had the *Surya Siddhanta* copied out for Dinwiddie, although he suspected that the Sanskrit work was merely a translation of Ptolemy.[71] Colebrooke opposed this view and Dinwiddie was witness to their heated arguments.[72] It was Bentley again who brought to his notice the Hindu text on mathematics, Ganesa's *Ganita Manjari*. Dinwiddie who copied out a collection of magic squares from the book remarked that in the Hindoo algebra there are many 'diaphantine problems'.[73] He sometimes acted as mentor to Colebrooke's young daughter, recommending such books as *Memoirs of the Modern Philosophers* and *La Chaumière Indienne*.[74] Dinwiddie's journals are strewn with 'Indian anecdotes', as he called them, on the Hindu customs and mores, indigenous medicine, and descriptions of nautches and other entertainments to which he was often invited by elite Bengali friends.[75]

---

69 Following the death of William Jones on May 1, 1794, John Shore was elected President of the Asiatic Society (1794–1799), and the members of his committee included Robert Chambers, Justice Hyde, Col. James Murray, Mr. Bristow and Mr. Graham.

70 Several of the members of the Society including Bentley believed that the ancient astronomical texts being discovered in India were being 'cunningly subjected to reverse engineering by the pundits'. S. Schaffer, "The Asiatic Enlightenment of British Astronomy," in *The Brokered World: Go-Betweens and Global Intelligence, 1770–1820*, ed. S. Schaffer et al. (Sagamore Beach: Science History Publications, 2009), 88. Bentley views were published as *A Historical View of the Hindu Astronomy: The Earliest Dawn of that Science in India to the Present Times* (London: Smith, Elder & Co., 1825).

71 DUA, MS 2-726, Dinwiddie Journal B 52 (Dec. 4–Feb. 2, 1799), entry dated Jan. 21, 1799.

72 DUA, MS 2-726, Dinwiddie Journal B 69 (Mar. 29–June 6, 1803), entry dated May 7, 1803.

73 DUA, MS 2-726, Dinwiddie Journal B 32 (Aug. 26–Sep. 24, 1796), entry dated Sep. 22, 1796.

74 *La Chaumière indienne* or "The Indian Cottage" published in 1790 dealt with how a traveller found wisdom in the cottage of an Indian outcast during the French Revolution. Beginning with a critique of schools of thought and scholarly institutions, its author, Bernardin claimed that scholarly societies (by which he meant the Royal Society) often created obstacles to the acquisition and dissemination of knowledge through their jealousy and prejudices. DUA, MS 2-726, Dinwiddie Journal B 68 (Dec. 22, 1801–Mar. 22, 1802), entry dated Feb. 10, 1802.

75 Dinwiddie notes in his journal, 'Sir J. Murray is said to be the almost only English gentleman in the settlement who entertains the natives in his own house, and enters into their affairs, amusements etc'. Proudfoot, *Biography*, 109.

c. *Artists*

In 1794, coinciding with the arrival of Dinwiddie in Calcutta, artist Robert Home declared: 'Now indeed Science hails the claim of a more propitious era'.[76] Like the many traders, Calcutta artists were a philosophically inclined group. At least a few would visit Dinwiddie's house to witness and participate in experiments, and engage in discussions on galvanism.[77] Dinwiddie would even share his accommodation with one or two of them. In the footsteps of the Royal Academy trained artists William Hodges, Johan Zoffany and Ozias Humphry, several artists trained subsequently at the Academy were inspired in the 1780s to travel to India to launch their careers.[78] These included Robert Home and Arthur William Devis who became 'the chroniclers of the new age'.[79] By late 1795, Home had established a successful studio in Calcutta, 'was much employed' and charged handsomely.[80] Home's chief clients were civil servants and their families.[81] Fairlie and Dinwiddie were two of his earliest sitters in Calcutta.[82] Home owned a number of scientific instruments, including an air pump, and sometimes exchanged instruments with Dinwiddie.[83] In 1795, artist Devis forwarded to Dinwiddie through Judge Thomas Brooke, water from the hot springs of Bhirbhum for analysis.[84] The other artists in the Dinwiddie

[76] R. Home, *Select Views of Mysore, the Country of Tippoo Sultan, From Drawings Taken on the Spot by Mr. Home* (London, 1794), v cited in H. De Almeida and G. H. Gilpin, *Indian Renaissance: British Romantic Art and the Prospect of India* (Aldershot: Ashgate, 2005), 58.

[77] DUA, MS 2-726, Dinwiddie Journal B 70 (June 7–Aug. 10, 1803), entry dated July 9, 1803.

[78] For more on these artists see M. Archer, *India and British Portraiture, 1770–1825* (London & New York: Sotheby Park Bernet, 1979); also see P. Pal and V. Dehejia, eds, *From Merchants to Emperors: British Artists and India, 1757–1930* (Ithaca: Cornell University Press, 1986).

[79] De Almeida and Gilpin, *Indian Renaissance*, 139. Devis painted an 'Orientalist' portrait of William Jones, only a month before the latter's death in 1794.

[80] Home's standard charge for a head was 500 sicca rupees (£60).

[81] Archer, *Portraiture*, 311–12.

[82] DUA, MS 2-726, Dinwiddie Journal B 21 (June 23–Aug. 27, 1795), entry dated July 4, 1795; Fairlie's wife, Margaret Ogilvy was the daughter of Calcutta private trader, John Ogilvy, also known to Dinwiddie. I have not been able to trace Home's portraits of Dinwiddie and Fairlie, but an 1802 painting of the Fairlie family by the artist exists in a private collection.

[83] DUA, MS 2-726, Dinwiddie Journal B 70 (June 7–Aug. 10, 1803), entry dated July 9, 1803.

[84] University of Guelph (hereafter JDC, UG), James Dinwiddie Correspondence, XS1 MS A 164, letter no. 1–3. Thomas Brooke to Dinwiddie, Jan. 16, 1795.

network were William and Alexander Taylor,[85] the aquatint artist John Moffat living on Cossitollah Street,[86] Robert Andrews and the engraver Aaron Upjohn, who resided at Sealdah.[87] Of this group, Dinwiddie sat for portraits by William Taylor[88] and Andrews; with the latter for one according to 'a new mode of painting'.[89] The other artists, who might have known Dinwiddie and possibly attended his lectures were the miniature painter Charles Sherreff living on Cossitollah Street,[90] the well-established William Hickey, Balthazar Solvyns at Durrumtollah,[91] John Alefounder behind Solvyns's, Croese who created panoramas, Thomas Taylor and John Brown.[92]

d. *Women*

Among the one hundred and eighty listeners in the audience attending the first lecture, there were several women. Dinwiddie encouraged women to attend by charging them less for a ticket. Women would also sometimes accompany husbands or fathers to see the lecture room and apparatus and be treated to philosophical recreations. It was Dinwiddie's practice of medical electricity, which particularly attracted women to his house.

---

85 For more on painter Alexander Taylor see W. Foster, *British Artists in India, 1760–1820* (London, 1931; reprinted from the Walpole Society's volume 19), 74. In 1802, Taylor advertised that his 'highest finished miniature' was only 200 Sicca Rupees. *Calcutta Gazette*, Oct. 14, 1802. See also DUA, MS 2-726, Dinwiddie Correspondence A 96. Taylor (in Dacca) to Dinwiddie, (Aug. 28, 1798).

86 Moffat living at 29 Cossitollah Street announced in 1798 that he was intending to publish by subscription his 'Views of Hooghly'. *Calcutta Gazette*, Aug. 18, 1798.

87 Upjohn engraved the "Map of Calcutta and its Environs" in 1792. In 1794, he sold his map fixed on rollers at the price of Rs 64. He also sold sun dials to 'infer the Latitude of any particular part of India' from his Sealdah house. *Calcutta Gazette*, June 5, 1794. Two years later, Upjohn advertised his 'neatly engraved *The Hindostannee Horal Diagram* with an explanation, and fixed on paste board for the conveniency of hanging in an office'. *Calcutta Gazette*, May 19, 1796.

88 William Taylor left for North America in 1805; before setting sail, Taylor presented Dinwiddie with Cooper's account of America. JDC, UG, XS1 MS A 164, letter no. 38. W. Taylor to Dinwiddie, dated Oct. 28, 1805.

89 DUA, MS 2-726, Dinwiddie Journal B 70 (June 7–Aug. 10, 1803), entry dated July 6, 1803.

90 In 1800, Sherreff moved to Old Post Office Street, opposite the Company's Printing Office. *Calcutta Gazette*, Aug. 14, 1800.

91 In February 1794, Solvyns announced that he was planning to publish by subscription '250 engravings, descriptive of the manners, customs and dresses of the natives of Bengal'. *Calcutta Gazette*, Feb. 6, 1794.

92 Thomas Taylor, residing at No. 6, Fancy Lane was an engraver, seal-maker and goldsmith. *Calcutta Gazette*, Mar. 26, 1795, Jan. 7, 1796.

e. *Indian Elites*

That the diffusion of modern science from Europe to non-European colonial settings was not a case of passive or a violently coercive process but of a nuanced and productive engagement is demonstrated by the exchanges between Dinwiddie and his Indian students in Calcutta. It shows that it was circulated through reciprocal exchanges albeit under asymmetrical conditions, but exclusively limited to the elite sections of indigenous society.[93] Most of Dinwiddie's pupils were European,[94] but at least a few were Indians. Dinwiddie taught mathematics and astronomy to the local elites Hurry Mohun Thakoor and Gopee Mohun Deb, the former from as early as 1795. Occasionally they would borrow instruments from him; in May 1799, Gopee Mohun borrowed a telescope to observe the transit of Mercury. Despite their active pursuit of self-improvement, it is clear that Dinwiddie did not think of them very highly. He would comment in his journal: 'Nothing can prevail on a native to part with his money for instruction. He is liberal only in Nauches and ridiculous ceremonies—a native who will sometimes lay out half a lac on a festival will grudge to pay a Goldmohur per month for their son at school'.[95] Well-versed in astronomy, Hurry Mohun was once commissioned by a 'person in the country to get a full and perfect account of the spot at present on the sun'.[96] He would spend long hours with Dinwiddie on the terrace of the latter's house observing the skies, performing optical experiments or solving mathematical problems.

Amongst those Dinwiddie tutored at his house were the 'vakeels' or diplomats of the local Rajas, yet another group of philosophical consumers residing in or visiting Calcutta.[97] The vakeel of the Raja of Nepal wished to know on what basis Dinwiddie claimed that it took twenty-four hours for the earth to turn round its axis.[98] Providing an introduction to the subject,

---

[93] See for instance K. Raj, *Relocating Modern Science*, and the more recent L. Roberts, "Situating Science in Global History," *Itinerario* 33, 1 (2009): 9–30.

[94] Amongst his European students were William Rennell, the Cambridge-trained son of the Company's cartographer Major James Rennell, the traders Fairlie and Colvin and the Orientalists H. T. Colebrooke and J. H. Harington.

[95] DUA, MS 2-726, Dinwiddie Journal B 54 (April 29–July 5, 1799), entry dated May 13, 1799.

[96] DUA, MS 2-726, Dinwiddie Journal B 46 (Feb. 11–Mar. 6, 1798), entry dated Feb. 28, 1798.

[97] For 'vakeels' as go-betweens, see K. Raj, "Mapping Knowledge. Go-Betweens in Calcutta, 1770–1820", in *The Brokered World*, ed. S. Schaffer et al., 105–47.

[98] DUA, MS 2-726, Dinwiddie Journal B 42 (Sep. 1– Nov. 3, 1797), entry dated Oct. 1, 1797.

Dinwiddie entertained him with anamorphous images and cylindrical mirrors. One local Raja sent a note to him through his vakeel with just the word 'phosphorus' on it.[99] Another enquired about a manual orrery and yet another wished to purchase Dinwiddie's sextant and be instructed in its use.[100] But of all these encounters with the local elite and go-betweens in Calcutta, the most significant was the one with the Islamic philosopher Tafazzul Husain Khan Kashmiri (d. 1800). Tafazzul Khan was a fine example of the elite Shi'i gentleman scholar of late eighteenth century India.[101] Impressed by his scholarship and personality, Nawab Shuja'u'd-Dowlah of Oudh appointed him mentor to his second son Saadat Ali Khan. In 1788, he was deputed to Calcutta as prime minister to the Nawab of Oudh, Asaf ud Dowla.[102] His contacts in Calcutta included David Anderson, Ralph Broome, Reuben Burrow, Richard Johnson (a former Oudh Resident) and John Shore, members of the Asiatic Society who were present at its inaugural meeting in 1784. They introduced him to all things Europe- its history, sciences and literature and this included classical texts of eighteenth century mathematics and rational mechanics. Within a year of his arrival in Calcutta, Tafazzul had successfully translated, supported by mathematician-astronomer Burrow, Newton's *Principia* into Arabic.[103]

On October 24, 1794, within a month of Dinwiddie's arrival in the city, Tafazzul enrolled himself as a pupil.[104] The chief emphasis of the tuitions was on the "practical". By early February 1795, Dinwiddie had begun to teach Tafazzul lessons in optics and soon progressed to geometry but to his puzzlement, found his student struggling with the construction of chords and tangents prompting a remark in his journal: 'it is somewhat irregular that a man who reads so much theory should be so totally ignorant of practical mathematics'.[105] During the same period, the brother

99 DUA, MS 2-726, Dinwiddie Journal B 50 (June 6–July 29, 1798), entry dated June 13, 1798.

100 DUA, MS 2-726, Dinwiddie Journal B 47 (Mar. 7–April 8, 1798), entry dated Mar. 21, 1798.

101 For more on Tafazzul see F. Robinson, *The Ulama of Farangi Mahall and Islamic culture in South Asia*, (London: Hurst, 2001) and Schaffer, *Asiatic Enlightenment*, 49–101.

102 For a biography of Tafuzzul Husain Khan, see "An account of the life and character of Tofuzzul Hussein Khan," *Asiatic Annual Register* for 1803 (1804), 'Characters', 1–8; Mahomed Ali Khan, ed., *Life of Tuffuzzool Hussain Khan* (Secunderabad: Cheekoty Veerunnah & Sons, 1908); Robinson, *Ulama*, 50–54.

103 Schaffer, "Asiatic Enlightenment," 56–57.

104 That same year, Dring, Cleland & Co sold by public auction 'capital mathematical instruments- highly approved by the late Mr. Burrowes'. *Calcutta Gazette*, July 24, 1794.

105 DUA, MS 2-726, Dinwiddie Journal B 19 (April 1–May 8, 1795), entry dated April 24, 1795.

of the Nawab of Oudh on a visit to Calcutta brought to Dinwiddie, with Tafazzul acting as go-between, a three and a half foot telescope by Gilbert & Gregory for examination.[106] Busy with diplomatic engagements, Tafazzul was forced to discontinue his lessons at one point but resumed classes soon enough, and this time the subject was astronomy. Dinwiddie began by explaining the use of several instruments including the sextant;[107] he was at this time also teaching astronomy to Hurry Mohun, a Hindu. Based on his exchanges with the two students, Dinwiddie inferred that Arabic writers 'began their longitude at the Westernmost land and reckon Eastward' while the Hindus began at the 'Easternmost and reckon Westward'. He also concluded that in the Hindu texts there was no mention of instruments of astronomy, geography or navigation.[108] On a later occasion, Tafuzzul, whose network in Calcutta included several traditional Indian scholars, would convey to Dinwiddie that the Governor General's learned pundit had himself confirmed that 'there was nothing in the Sanskrit books on the subject of Mechanics'.[109] Tafazzul was for some time a collector of information on Hindu astronomy, while at the same time striving 'to integrate Newtonian principles into Indo-Persian cosmology'.[110]

In late 1796, Hunter, the secretary of the Asiatic Society, for whom the Arabic *Principia* represented an acceptance of European intellectual superiority, obtained from Tafazzul, this time with Dinwiddie as go-between, descriptions of astronomical instruments, names of which he wrote out in Arabic.[111] Hunter would speak highly of Tafazzul at an Asiatic Society meeting that year.[112] Dinwiddie was also at this time in correspondence with John Playfair, the Edinburgh mathematics professor, sending him a copy of Jaising's astronomical tables.[113] Over subsequent lessons,

---

106 DUA, MS 2-726, Dinwiddie Journal B 18 (Feb. 15–Mar. 31, 1795), entry dated Mar. 20, 1795.

107 DUA, MS 2-726, Dinwiddie Journal B 22 (Aug. 28–Nov. 5, 1795), entry dated Nov. 2, 1795.

108 DUA, MS 2-726, Dinwiddie Journal B 23 (Nov. 6–Dec. 14, 1795), entry dated Dec. 3, 1795.

109 DUA, MS 2-726, Dinwiddie Journal B 31 (Aug. 2–25, 1796), entry dated Aug. 9, 1796.

110 Schaffer, "Asiatic Enlightenment," 64.

111 DUA, MS 2-726, Dinwiddie Journal B 32 (Aug. 26–Sep. 24, 1796), entry dated Sep. 22, 1796.

112 Schaffer, "Asiatic Enlightenment," 60.

113 DUA, MS 2-726, Dinwiddie Journal B 32 (Aug. 26–Sep. 24, 1796), entry dated Sep. 22, 1796. Since the 1780s, Playfair was involved in the study of ancient Indian astronomy, and engaged in debates with Nevil Maskelyne, the astronomer Royal and his assistant Reuben Burrow, as well as members of the Edinburgh literary world. He was also sending queries to his friends at the Asiatic Society regarding 'the principles of *jyotihśāstra* to be

Dinwiddie would teach Tafuzzul the construction and use of the micrometer, a device used in mechanical trades for precision measurement. He also used his close relationship with Tafuzzul and the latter's influence in the Oudh Court, to explore possibilities of a career in Lucknow as a natural philosopher. In 1797, when it had become certain that Tafuzzul would be forced by the Company to become Asaf ud Dowla's prime minister, Colebrooke suggested to Dinwiddie that he proceed to Lucknow, which would also be a good market for the disposal of his apparatus.[114] The Nawab had under his employ at this time such "enlightened" men as the French general Claude Martin.[115] Unfortunately for Dinwiddie, the relationship between Tafazzul and the Nawab was far from smooth. As a go-between, he was attacked for being 'a servant of the English'.[116] The situation worsened, when a state of political instability descended on Oudh with the sudden death of the Nawab. By mid-1798, Tafuzzul was back in Calcutta as envoy and from this time till his death in 1800, Tafuzzul kept himself busy translating books on European mathematics and natural philosophy into Arabic, with the help of Dinwiddie.

Public lectures and private tuitions were only one aspect of the link between natural philosophy and commerce in the colonial metropolis. These were simply not sufficient for maintaining a decent standard of life in what was a very expensive city in the late eighteenth century. An involvement in entrepreneurial activities in the city, which required the application of philosophical knowledge, especially chemistry, became crucial to Dinwiddie if he wished to earn more; the chief reason for being in the colonial metropolis.

In the Business of Science

Dinwiddie's New Year resolutions for the year 1798 included 'pay the utmost attention to money matters' besides 'Smoke less' and 'prepare

calibrated against Newtonian truths'. Schaffer, "Asiatic Enlightenment," p. 66. Also see J. Playfair, "Remarks on the astronomy of the Brahmins", *Transactions of the Royal Society of Edinburgh*, 2 (1790): 135–92.

114 DUA, MS 2-726, Dinwiddie Journal B 39 (April 4–May 14, 1797), entry dated April 15, 1797.

115 In 1801, Dinwiddie would attend the Calcutta sale of the property of the late General Martin. For a biography of Martin see R. Llewellyn-Jones, *A very ingenious man: Claude Martin in Early Colonial India* (Delhi: Oxford University Press, 1993).

116 DUA, MS 2-726, Dinwiddie Journal B 39 (April 4–May 14, 1797), entry dated May 13, 1797.

chemist chests for sale'.[117] One of the striking features of Dinwiddie's career was his agility to respond to changing market demands, especially in cases where an application of science could bring profitable returns.[118] In several cases the initiative arose from the private entrepreneurs themselves, who would seek advice on technical problems in their indigo, sugar or even chintz manufactories. The Company would also use his expertise 'in several parts of chemistry, mechanics and natural philosophy', which had a certain bearing on their mercantilist interests.[119] Among their demands were the chemical analysis of several raw materials such as indigo, cochineal, saltpetre, fossil alkali, gunpowder, flax and hemp, and the native timbers of the country used in architecture and ship-building. Dinwiddie would also advise them on the placement of lightning-rods and in the regulation of chronometers. Besides the Board of Trade, the Medical, Marine and Military Boards occasionally consulted Dinwiddie on subjects connected with their departments.[120] He also contributed to the Great Trigonometrical Survey of India, which began in 1802 under the leadership of William Lambton.[121] A self-taught mathematician, Lambton's mathematical papers were periodically sent to Dinwiddie between 1799 and 1800 for proof reading. Lambton also sought Dinwiddie's help in constructing surveying instruments like the zenith sector, theodolite, pantograph and artificial horizon.[122] Similarly, the chemical chests carried by the Company naturalist Francis Buchanan on his survey of Mysore in 1800, were prepared by Dinwiddie.[123] Dinwiddie also manufactured nitrous acid for use at the General Hospital in Calcutta.[124]

---

[117] DUA, MS 2-726, Dinwiddie Journal B 44 (Dec. 16, 1797–Jan. 17, 1798), entry dated Jan. 1, 1798.

[118] L. Lunney, "The Celebrated Mr. Dinwiddie: an Eighteenth-Century Scientist in Ireland," *Eighteenth-Century Ireland*, 3 (1988): 69–83.

[119] Proudfoot, *Biography*, 98.

[120] Ibid., 99–100.

[121] In 1799, after the Fourth Anglo-Mysore War, Lambton had proposed that Mysore be surveyed using the new technique of geodesy employed by William Roy in Britain.

[122] DUA, MS 2-726, Dinwiddie Journal B 57 (Nov. 20, 1799–Feb. 2, 1800), entry dated Jan. 13, 1800; Jan. 20–21, 1800.

[123] DUA, MS 2-726, Dinwiddie Journal B 57 (Nov. 20, 1799–Feb. 2, 1800), entry dated Dec. 24, 1799.

[124] The Bombay surgeon Helenus Scott had in a correspondence with the Physician General of Madras James Anderson, published in the *Bombay Gazette*, advocated nitric acid as a substitute for mercury in the treatment of venereal diseases. This was however disputed by several people including surgeon-naturalist John Fleming, who claimed that "Dr. Scott's reasoning is fallacious". It was also believed that Scott's advocacy of nitric acid contradicted some of Thomas Beddoes' facts relating to oxygen. See DUA, MS 2-726, B-28 (May 5–June 4, 1796), entry dated June 2, 1796. Scott was known to Dinwiddie, even prior

Natural philosophy reached the market place in Calcutta in the late eighteenth century in the form of scientific instruments and publications. Dinwiddie's lectures and experiments added to the demand for philosophical instruments; they inspired consumption. The public lecturer and mathematical instrument makers mutually benefited from each other's businesses. Several of them were involved in rebuilding electricity machines with an aim to making them more powerful or capable of producing greater flashes and sparks, or more affordable, or as in some cases making them portable.[125] Victor Thibault, who had begun trading along the Canton-Calcutta route immediately after the dismissal of the Macartney Embassy, set up shop at Tank square in Calcutta on Dinwiddie's recommendation.[126] The subscription paper for Dinwiddie's public lectures was kept at Thibault's. This shop, like several others in the area, functioned as an exhibition room, where the public were free to enter, explore and be entertained, even while being instructed. David Hare, an ingenious Scottish watchmaker, who arrived in Calcutta in 1801, repaired scientific instruments, and helped Dinwiddie with some of his early experiments with the voltaic pile.[127] Yet another mathematical instrument maker, George Bertie, who came out as a steward of an American ship, maintained a shop in Tank square, where he made, cleaned, repaired and 'adjusted' all kinds of mathematical, astronomical and nautical instruments.[128] Bertie not only assisted Dinwiddie in constructing new experimental apparatus but also built the still for his distillery business at Tangrah on the outskirts of the city.[129] Besides these associates, there were those like house-builder and carpenter Robert Armstrong, who

---

to the latter's arrival in India. See JDC, UG, XS1 MS A 164, letter from Helenus Scott to Dinwiddie, undated, 1796.

125 For the European context see L. Pèrez, "Technology, Curiosity and Utility in France and in England in the Eighteenth Century", Bensaude-Vincent and Blondel, *Science as Spectacle*, 11–24.

126 DUA, MS 2-726. See for instance Dinwiddie Correspondence, A 1.33, Thibault to Dinwiddie, letter dated Calcutta, Feb. 27, 1794, and Dinwiddie correspondence, A 99, Thibault to Dinwiddie, letter dated Canton, Feb. 1, 1805.

127 Hare would go on to amass a huge amount of wealth and found the School Book Society, the Hare School and the Hindu College (later renamed the Presidency College). Raja Rammohan Roy, a key figure of the Bengali enlightenment, was a close associate of Hare.

128 *Calcutta Gazette*, Aug. 7, 1800.

129 DUA, MS 2-726, Dinwiddie Journal B 60 (June 18–Sep. 9, 1800), entry dated June 20, 1800. Dinwiddie was at this time experimenting with a new whiskey using the oil of juniper berries.

doubled-up as Dinwiddie's laboratory-assistant and sales-agent.[130] In 1800, he is found sharing Dinwiddie's apartments at 'no. 30, Cossitollah', from where he sold 'Fortification- Billiards ... the most fashionable game' and alcohol and chemicals prepared by Dinwiddie. Dinwiddie entered into several business partnerships during his twelve-year residence in Calcutta, two or three of which were with the entrepreneur John McIntosh, but none were particularly successful. In 1800, Dinwiddie and Intosh began a distilling and brewery business and later a salt-curing establishment selling 'the best quality [salted-beef] the Country produces', from the latter's house in Zigzag Lane, Cossitollah.[131] In late 1801, Dinwiddie signed yet another contract with McIntosh and a Peter Linderman to set up a leather business.[132] Soon after, McIntosh is found advertising kilns constructed 'on a new and improved principle' for drying and preserving flour, which may also have involved Dinwiddie.[133]

Public demonstrations in the late eighteenth century acted as a trigger for the construction of new devices and instruments, and led to the replication of experiments by non-specialists.[134] As new cultural products, the electricity machine and the voltaic pile served to link a wide range of social worlds in India including naturalists, artists, carpenters, blacksmiths, traders, chemists, surgeons and even women and children in Calcutta who were eager to know, experience and contribute to and profit from a new knowledge that blurred the boundaries between electricity and physiology. As soon as Dinwiddie reached the colonial city, he attempted to carve out a part-time career as a medical electrician. One of his earliest patients was William Berrie, clerk and inspector of public granaries, referred by the surgeon Dr Lynd.[135] In 1799, Dinwiddie tried for the first time the Nairne small patent electrical machine for the same pur-

130 DUA, MS 2-726, Dinwiddie Journal B 58 (Feb. 3–April 26, 1800), entry dated Feb. 12, 1800. Armstrong also drew building-plans for a living 'agreeable to the custom of this country', and claimed to have been trained by 'some of the finest surveyors of London'. *Calcutta Gazette*, July 30, 1801.

131 *Calcutta Gazette*, Jan. 21, 1802.

132 DUA, MS 2-726, Dinwiddie Journal B 63 (Dec. 10, 1800–Feb. 5, 1801), entry dated Dec. 23, 1801.

133 *Calcutta Gazette*, Oct. 21, 1802, April 7, 1803, Aug. 11, 1803.

134 Bensaude-Vincent and Blondel, *Science as Spectacle*, 7.

135 DUA, MS 2-726, Dinwiddie Journal B 17 (Dec. 14, 1794–Feb. 14, 1795), entry dated Dec. 17, 1794.

pose.[136] Women would visit his house, sometimes even unaccompanied by a male, to receive shocks for their ailment. The electricity machine was also put to occasional use in Calcutta as a punitive instrument.[137] With his re-invention of the voltaic pile at No. 30, Cossitollah Street, Calcutta in late 1801 Dinwiddie gradually replaced the electrical machine (the pile was more suited to the Indian weather) for the treatment of deafness, blindness, worms, paralysis, rheumatism, night blindness, and even neurological disorders such as the St. Vitus dance. He also attempted to cure himself of deafness and spasms with the pile. By early 1804, he had invented the portable voltaic battery using material that would stand up to the corroding effects of the Indian weather; it was now easy for even the non-specialist to manipulate the pile, with the help of a little instruction. Inspired by Dinwiddie's lectures on medical galvanism, several men began to experiment with it including the Romanian Elias Habesci alias Alexander Gika.[138] This popularity of medical electricity however led Dinwiddie to announce in the local newspaper of his intention to give up the practice. Medical electricity was approached with suspicion in Georgian society, but practitioners like Dinwiddie believed that 'natural philosophy had the potential to improve society by increasing the comfort and happiness of humanity' and recommended its use in ailments, which could not be cured through conventional methods.[139]

---

136 Edward Nairne (1726–1806) was one of the first to design an electrical machine from which negative electricity could be easily obtained. He was a pioneer in the design of powerful electrical machines. Dinwiddie possessed both the large and small Nairne electricity apparatus.

137 The missionary Claudius Buchanan, who would later become the Vice-Provost of the Fort William College, brought to Dinwiddie's lecture room a servant who was guilty of stealing money. Dinwiddie 'gave him a shock from a large Jar, he instantly dropped and lay motionless for sometime on the ground- after recovering he threw himself at his knees and confessed everything'. DUA, MS 2-726, Dinwiddie Journal B 16 (Oct. 6–Dec. 12, 1794), entry dated Dec. 10, 1794.

138 Gika corresponded with Dinwiddie on the use of the galvanic battery in medicine. A liberal Christian, Gika spent the last years of his life in India, where he was known by the title of "Count" and as the author of *The Present State of the Ottoman Empire* (London, 1784). His *Objects Interesting to the English Nation* (1783) was subscribed by several of Dinwiddie's contacts in Calcutta—Dr. Baillie, A. Colvin, H. T. Colebrooke, Robert Chambers, Fairlie, Fergusson, Dr. Fleming, Dr. William Hunter, J. H. Harrington, Charles Ogilvie, and Joseph Queiros.

139 P. Elliott, "'More Subtle than the Electric Aura': Georgian Medical Electricity, the Spirit of Animation and the Development of Erasmus Darwin's Psychophysiology," *Medical History*, 52 (April 2008): 195–220, 195.

Concluding Remarks

Irrespective of their approach, almost all accounts of colonial science in non-European settings at the turn of the nineteenth century have had the Company as their chief focus and on subjects of direct interest to the state such as the field sciences of natural history, cartography, environment, and medicine. What this paper does is to broaden this image of colonial science to include the public in the analysis, and even bring it to the centre-stage, and consider such other spaces of doing science in the colonial metropolis as the public lecture-room. Through an examination of the activities of the lecturer and natural philosopher James Dinwiddie in Calcutta, hitherto unexplored facets of colonial science in the late eighteenth century were captured, namely, its public, experimental and commoditised dimensions. With his arrival in the colonial metropolis, a new culture of rational-recreation had been introduced, one that was centred round public lectures and spectacular experimentation. Dinwiddie's lecture-rooms, which aimed at diffusing the secrets of the natural world to anyone willing to pay for it, signalled the emergence of public science in colonial India.

In the late eighteenth century, a wide spectrum of the white town of Calcutta including merchant-traders, artists, Company officials, women, and very importantly, some indigenous elite believed in spending their leisure hours in 'improving' pursuits. They subscribed to Dinwiddie's private tuitions and public demonstrations and were desirous of acquiring the most fashionable philosophical instruments in the market. With several of them, Dinwiddie maintained close personal relationships, spending many an evening in polite conversations on natural philosophy and with experiments on the electricity machine and the voltaic pile. The paper also demonstrates that the diffusion of modern science from Europe to non-European colonial settings was not a case of passive reception or a coercive imposition but of a nuanced and creative response to the intercultural encounter, albeit under asymmetrical conditions and limited to the indigenous elite.

FROM CALCUTTA TO LONDON: JAMES DINWIDDIE'S GALVANIC CIRCUITS

Jan Golinski

Introduction

"Circulation" has been a pervasive term in historical studies of the sciences in recent years. It captures the ways in which knowledge is made through movements of people, texts, artifacts, and specimens from place to place. It also suggests that these patterns of movement are not simply in one direction—say, outward from a central point of origin to peripheral regions—but are often reversible or cyclical.[1] Historians have long been interested in the role of scientific instruments in the distribution of natural knowledge, since their circulation is evidently a crucial condition for the replication of experiments and the reproduction of phenomena.[2] They have also been considering the roles of those mobile individuals who connect together different cultures or social groups.[3] Such 'go-betweens' often convey instruments as they move from one cultural space to another. They make use of their embodied skills to make apparatus work in new locations, and they can thereby enhance their standing with groups to which the devices are unfamiliar. Instruments, in other words, have a complex relationship to personal itinerancy. They require the aid of human travelers to move from place to place, and they may confer authority on those individuals whose identity is bound up with geographical mobility.

It is hard to think of an individual who fits this role better than James Dinwiddie (1746–1815). Dinwiddie was a Scottish natural philosopher, mathematician, and inventor, whose life was spent in extensive travel

[1] James A. Secord, "Halifax Keynote Address: Knowledge in Transit," *Isis* 95 (2004): 654–72.

[2] Marie-Noëlle Bourguet, Christian Licoppe, and Otto Sibum, eds., *Instruments, Travel and Science: Itineraries of Precision from the Seventeenth to the Twentieth Century* (London: Routledge, 2002); Mario Biagioli, *Galileo's Instruments of Credit: Telescopes, Images, Secrecy* (University of Chicago Press, 2006); Jan Golinski, *Making Natural Knowledge: Constructivism and the History of Science* (University of Chicago Press, 2005), 133–45.

[3] Simon Schaffer, Lissa Roberts, Kapil Raj, and James Delbourgo, eds., *The Brokered World: Go-Betweens and Global Intelligence, 1770–1820* (Sagamore Beach, MA: Science History Publications, 2009).

around the British Isles, China, and India. By way of a remarkable posthumous journey, his personal papers have ended up in Canada, mostly at Dalhousie University in Halifax, Nova Scotia. Dinwiddie mediated between many different worlds, as an itinerant lecturer in Scotland, Ireland and England, a member of Lord Macartney's embassy to China on behalf of the British government in the early 1790s, and then a resident for several years in Calcutta before his eventual return to England.[4] He stands as an embodiment of the great age of the go-between at the turn of the eighteenth and nineteenth centuries, a participant in the establishment of long-distance trade networks, the building of imperial power-structures, and the projection of scientific knowledge on a global scale.

In this paper, I will focus on Dinwiddie's work on galvanism, beginning in 1801, shortly after his appointment as a professor at the new College of Fort William in Calcutta. I will trace how Dinwiddie pursued his galvanic researches in a context of pedagogical display and indigenous practices, and I will also follow him on his return to London in 1807, where he resumed galvanic experiments and attended the demonstrations of Humphry Davy and other lecturers. Dinwiddie's reflections and observations, recorded in his journals and notes, allow us to reconstruct some of the patterns of circulation by which scientific knowledge was made at the intersection of European and Asian cultures in this period. Dinwiddie himself will emerge as a man whose travels mobilized the effects of distance and exoticism to enhance his authority and that of the instruments he deployed. His case helps us to see how such a person could form his distinctive identity in the trading-zones between cultures, and how the equipment he transported derived its meanings from travel and cross-cultural exchange.

Dinwiddie was born in Dumfries in 1746, the son of a farmer. After an education at the Dumfries Academy and Edinburgh University, he began a career as itinerant lecturer in natural philosophy. In the years 1779–1786, he lectured in Ireland, deploying an extensive collection of scientific apparatus to teach mechanics, chemistry, and military sciences. He studied diving bells and hot-air balloons, venturing beneath the sea and making aerial ascents.[5] In 1792, after further lecturing in London and tours in Scotland and Northern England, he was invited to join Lord Macartney's

[4] William Jardine Proudfoot, *Biographical Memoir of James Dinwiddie, LL.D.* (Liverpool: Edward Howell, 1868).

[5] Proudfoot, *Biographical Memoir*, 15–18.

embassy to China. His job was to tend the clocks and astronomical apparatus, including a planetarium designed to impress the Emperor with British mechanical expertise.[6] During the mission's brief residence in Peking and Canton, he also collected botanical specimens and made other scientific observations. After the party left China in September 1794 with its aims largely unaccomplished, Dinwiddie was sent to Calcutta with samples of tea and other Chinese plants for cultivation. He decided to try his fortune there rather than returning to Britain, and found success as a lecturer in mathematics and the sciences to British colonial residents and some Indian students. He interested himself in manufactures and trade in addition to experimental science, contracted his services as a consultant to the East India Company, and was elected a member of the Asiatic Society of Bengal.

Aside from a few months' visit to Madras, Dinwiddie remained in Calcutta until 1807. It was a good place to indulge his fascination with scientific and technical matters. The city was an important port, having been the center of the East India Company's operations for nearly two centuries already. Following the conquest of Bengal in 1757, it was the site of British attempts to build a colonial administration of the territory. As Kapil Raj has documented, Calcutta was a place where many networks of trade, exchange, and migration intersected, where multiple populations jostled against one another, and where many languages were spoken and studied.[7] The Mughal elite was imbued with Persian and Arabic cultures. There were separate educational and legal institutions for Hindus and Muslims, and well-connected communities of Armenian and Portuguese merchants. These were among the complexities of the situation that the small British administration was attempting to subjugate. As the British struggled to enlist the aid of native informants and intermediaries, every aspect of local culture was potentially relevant to their project. Dinwiddie had ample scope for his wide intellectual interests. As his biographer, William Proudfoot, notes: "Astronomical and meteorological observations, governmental surveys, mineralogical discoveries, and discoveries made in

---

6 Proudfoot, *Biographical Memoir*, 26–86; Lord Macartney, *An Embassy to China*, ed. J. L. Cranmer-Byng (London: Folio Society, 2004); Simon Schaffer, "Instruments as Cargo in the China Trade," *History of Science* 44 (2006): 217–46.

7 Kapil Raj, "Mapping Knowledge Go-Betweens in Calcutta, 1770–1820," in *Brokered World*, ed. Schaffer et al., 105–150; idem, "Colonial Encounters and the Forging of New Knowledge and National Identities: Great Britain and India, 1760–1850," *Osiris* (2nd ser.) 15 (2000): 119–34; idem, *Relocating Modern Science: Circulation and the Construction of Knowledge in South Asia and Europe, 1650–1900* (Basingstoke: Palgrave Macmillan, 2007).

translating antient Hindoo manuscripts, formed interesting features in his correspondence." With pardonable exaggeration, Proudfoot declares: "Dr. Dinwiddie . . . became, in a manner, the centre or focus of philosophy in the east."[8]

Calcutta

We can pick up the story of Dinwiddie's life in Calcutta on his fifty-fourth birthday, 8 December 1800. The occasion provoked a series of melancholy reflections in his journal, as he contemplated the circumstances of his life (as he put it) "the wrong side of fifty."[9] Dinwiddie was suffering from the aches and pains of middle age: muscular spasms, an ulcer in his leg, and increasing deafness. The weather was relentlessly hot, his compatriots were succumbing to lockjaw and other tropical ailments, and the hazards of snakes and tigers loomed in the vicinity. Dinwiddie wondered how many more birthdays he would see, and pondered whether he wanted to die in India or should attempt to return to Britain. "Never," he complained, "did a birthday visit me in such pain as this."[10] Reading his journal at this point, one is prompted to reflect that, however fascinating Dinwiddie's life seems to historians, there were times when he experienced it as unadulterated misery.

His personal life was also a cause of unhappiness at this time. He had left a mistress and a daughter back in Scotland, and had not had any contact with them for many years.[11] He had been told that his daughter had married a young gunsmith from Dumfries, who had subsequently enlisted in the army and taken his wife with him to Ireland. "I suppose by this time they have made a grand-papa of you," suggested the correspondent who conveyed this news.[12] The correspondent in question was Joseph Hume, who acted as Dinwiddie's London agent, banker, and supplier of chemicals

[8] Proudfoot, *Biographical Memoir*, 102.

[9] Dalhourise University Archives (DUA), James Dinwiddie Collection, MS 2-726, B 61, September 10–October 21, 1800.

[10] DUA, MS 2-726, B 62, December 8, 1800.

[11] Dinwiddie's daughter is mentioned in Proudfoot, *Biographical Memoir*, 124 and fn. She married James Proudfoot, a gunsmith in Dumfries, whose son wrote his grandfather's biography. Although Dinwiddie never married, he acknowledged his daughter Ann, named after her mother Ann Muir. See: Linde Lunney, "Dinwiddie, James (1746–1815)," *Oxford Dictionary of National Biography* (Oxford University Press, 2004) [http://www.oxforddnb.com/view/article/47299, accessed 23 November 2010].

[12] DUA, MS 2-276, A 59, Joseph Hume to Dinwiddie, May 27, 1796.

and scientific equipment. Hume teased Dinwiddie about his unmarried state: "Perhaps you stay in India to encrease and multiply the species, as I am told this kind of business is well supported in that quarter."[13] Later, Hume pointed out that it was not an unusual thing "for gentlemen in the West and also in the East Indies to do every thing but marry."[14] But, if Hume was suggesting that Dinwiddie should follow the practice of taking an Indian mistress, the hint was not taken up. On the contrary, Dinwiddie's diary records his frequent loneliness in the absence of family and close friends. When a physician whose family had welcomed him into their household departed to return to England, he recorded: "Parting with real friends is painful. I have had my share of this kind of pain."[15]

The thing that gave him hope at this stage, and probably kept him from abandoning India, was the prospect of employment at the new College of Fort William. The project of Lord Wellesley, the colonial governor-general, this was designed to provide a three-year university degree to the teenage recruits of the East India Company, who generally arrived in India without any relevant education or training. It emphasized knowledge of languages on the Oxbridge model—both the classical Western Latin and Greek, and the Eastern equivalents, Sanskrit, Arabic, and Persian—and also provided instruction in modern languages, law, political economy, history and geography.[16] In 1800, Dinwiddie was witnessing the construction of the college and hoping he would be hired to teach mathematics and natural philosophy. He also noted that poor people whose dwellings had been demolished to clear the site for the college were obstructing carriages in protest; he acknowledged they had no other redress for their plight.[17] For him, on the other hand, the new institution offered opportunity. He was appointed to a professorship in January 1801, initially to deliver a course of lectures on mathematics. This never attracted many students, but later offerings on natural philosophy and chemistry were more successful, and a course on natural history was occasionally added. Each course had to be approved in advance by the Provost, David Brown, or the Vice-Provost, Claudius Buchanan—a fellow-Scot and evangelical clergyman, who was close to Wellesley and in day-to-day control of college affairs.

---

13 DUA, MS 2-276, A 59, Joseph Hume to Dinwiddie, August 29, 1800.

14 DUA, MS 2-276, A 59, Joseph Hume to Dinwiddie, July 8, 1802.

15 DUA, MS 2-276, B 64, February 9, 1801.

16 Raj, "Mapping Knowledge Go-Betweens"; idem., "Colonial Encounters"; *Letters of the Marquis Wellesley Respecting the College of Fort William* (London: J. Hatchard, 1812).

17 DUA, MS 2-276, B 63, February 5, 1801.

Though it offered security of employment, the college was also a source of frustration for Dinwiddie. Despite promises by the authorities, he never received adequate compensation for the equipment he purchased for experimental demonstrations.[18] The students, generally younger sons of elite British families, were academically disappointing. Dinwiddie found at least a third of them were bone idle, and many behaved disrespectfully. Lecturing on optics in September 1801, he noted: "Several new attendants[;] some behaved ill laughing and inattentive. The Vice Provost tho' present took no notice of their conduct. Indeed neither the Provost nor vice P[rovost] seems to have any auth[orit]y over them."[19] Though he complained on several occasions to the college administrators, Dinwiddie seems to have had a constant struggle to maintain discipline in his classes. Lecturing on heat two months later, he lamented, "Some of the students behaved very badly.—There is no discipline in the college."[20] Administrators sometimes promised to act, but Dinwiddie was skeptical about their resolve to tackle the problem. Though Buchanan proclaimed his ambition that the college should "enlighten the oriental world, [and] give science, religion, and pure morals to Asia," Dinwiddie noted sardonically that Buchanan himself had recommended students for honors who had never attended the required lectures on Latin and Greek.[21] Dining at the college in February 1802, he recorded that, "At one of the tables the lads were very noisy—and at breaking up were throwing the victuals at one another."[22] The following month he gave a lecture on the chemistry of metals, and paraphrased Adam Smith's description of pin-manufacture—a famous illustration of the division of labor from *The Wealth of Nations*. As he detailed the straightening of the shafts of the pins and the grinding of their heads, he found the students falling about in laughter as they took his comments in an entirely different sense. "The propensity to double entendre among the young men is surprizing—shamefull," Dinwiddie recorded in his diary.[23] It is hard to avoid the conclusion that the stiff and humorless Scot was being ragged mercilessly by his students.

---

18 DUA, MS 2-276, A 21, undated document; Proudfoot, *Biographical Memoir*, 119–120.

19 DUA, MS 2-276, B 66, September 3, 1801.

20 DUA, MS 2-276, B 67, November 19, 1801.

21 Buchanan quoted in Penelope Carson, "Buchanan, Claudius (1766–1815)," *Oxford Dictionary of National Biography* (Oxford University Press, 2004) [http://www.oxforddnb.com/view/article/3831, accessed 23 November 2010]; DUA, MS 2-276, B 69, March 30, 1803.

22 DUA, MS 2-276, B 68, February 6, 1802.

23 DUA, MS 2-276, B 68, March 8, 1802.

In this situation, Dinwiddie received news from Europe of the discovery of the voltaic pile. This was a device assembled from a stack of plates of two different metals (such as silver and zinc) alternating with one another and with pieces of cardboard soaked in acid placed between them. If wires were led from each end of the pile, a constant current of electricity could be drawn off, delivering a significant shock whose magnitude depended on the number of plates in the pile. The pile or battery was publicized by Alessandro Volta in 1800 as an inorganic source of the electrical impulse previously elicited from animal materials by Luigi Galvani. It was quickly replicated throughout Europe as news of the discovery spread, and the electric current was soon used by William Nicholson and Anthony Carlisle to decompose water into its component gases. It seems that Dinwiddie learned of the discovery in one of the scientific publications he received from Britain, perhaps in an issue of *Nicholson's Journal* or in a volume of the *Medical and Physical Journal*, which he is known to have had sent to him.[24] The ease with which he was able to reproduce the device confirms what historians have noted about the readiness with which it was replicated in many locations.[25] Researchers in many places found it relatively easy to assemble the commonly available materials and produce some kind of effect. Dinwiddie had a workman prepare him plates of the required metals. He had earlier organized the manufacture of sulfuric acid at a plant in Bengal, to substitute for the more expensive imported commodity. He assembled his first pile of twenty pairs of metal plates on his fifty-fifth birthday, 8 December 1801, and found it to 'act powerfully.'[26] Within days, he was showing it to his students.

Though the voltaic pile was easily reproduced, arguments about how its actions should be interpreted continued for years among European savants.[27] It was by no means straightforward to decide whether what was called 'galvanism' was identical to the electricity traditionally produced from static generators. Dinwiddie, however, immediately saw the pile as capable of occupying the same social niche as the traditional electrical machine, a device he had deployed throughout his time in India.

---

[24] DUA, MS 2-276, D 22, 'Journal of Galvanism', March 29, 1802; D 34, undated; B 67, October 2–December 21, 1801.

[25] Giuliano Pancaldi, *Volta: Science and Culture in the Age of Enlightenment* (Princeton University Press, 2003), 178–256; Naum Kipnis, "Luigi Galvani and the Debate on Animal Electricity, 1791–1800," *Annals of Science* 44 (1987): 107–142.

[26] DUA, MS 2-276, B 67, December 8, 1801; D 22, 'Journal of Galvanism'.

[27] Pancaldi, *Volta*, 211–256; Jan Golinski, *Science as Public Culture: Chemistry and Enlightenment in Britain, 1760–1820* (Cambridge University Press, 1992), 203–218.

The same day he made his first voltaic battery, he also worked with an electrical machine designed by Edward Nairne, a leading London maker of such apparatus. He noted, however, that static electricity was extremely difficult to manage in the humid Indian climate: "Elect[rica]l Machines in their action are extremely uncertain, in this country."[28] Their effectiveness depended on atmospheric conditions, changing with the season or even the direction of the wind, and sometimes varying with the height of the apparatus above the ground. A couple of weeks later, Dinwiddie was surprised by a sudden and unexplained loss of power of his electrical machine.[29] When they worked, on the other hand, they could be extremely potent. Dinwiddie regularly produced sparks and shocks in the lecture hall. He exhibited an artificial version of the aurora borealis, and used a spark to ignite a gunpowder trail that spelled 'long live the king'—an imperial sentiment that was no doubt welcome in the colonial community and a small-scale imitation of an effect commonly shown in the fireworks displays in which he had long taken an interest.[30] The electrical machine could also be used more directly to consolidate colonial power-relations. Dinwiddie recorded an episode in which a British gentleman suspected his Indian servant of embezzling money,

> ...which the latter denied in spite of a severe flogging. It was arranged to bring the culprit before the [electrical] machine when he was told what would be the consequence if he was guilty; but the fellow still persisted in his innocence. The jar had been previously charged, and when asked to touch it—such was the effect of the shock that he instantly fell down; but soon after got upon his knees, and, in an imploring attitude, confessed everything.[31]

The voltaic battery offered itself as a more reliable and powerful substitute for the electrical machine in these and other roles. In his notes on the pile, Dinwiddie frequently repeated the remark: 'it acts powerfully'.

Though the device required cleaning of the plates as they became corroded—which Dinwiddie called "a troublesom [sic] business"—the voltaic pile was less vulnerable to atmospheric moisture than the Nairne

---

[28] DUA, MS 2-276, B 67, December 8, 1801.

[29] DUA, MS 2-276, B 67, December 21, 1801.

[30] DUA, MS 2-726, B 68, December 23, 1801; 'Notes on Fireworks' (undated). On the general topic, see: Simon Werrett, *Fireworks: Pyrotechnic Arts and Sciences in European History* (Chicago: University of Chicago Press, 2010).

[31] Proudfoot, *Biographical Memoir*, 108.

machine.[32] Measuring the pile's effects by the shocks it delivered to his own body, he also perceived its potency in the classroom. Two days after he made his first pile, he showed it in his lecture, and recorded: "Every student tried it, the shocks and flashes were strong and frequent. Students much pleased with this new and singular experiment."[33] A few weeks later he showed it again: "Every experiment succeeded. Several of the Students after the lecture looked at the apparatus rooms and laboratory."[34] The spectacle of the pile seems to have succeeded in attracting the students' attention, despite their inclination to treat Dinwiddie disrespectfully. His work on galvanism also enhanced his standing in the colony generally. In January 1802, he took a new apparatus of fifty pairs of plates to the house of his friend William Fairlie, a Calcutta merchant, where a "shock was given to a large company."[35] In March of that year, in a lecture at the end of the college term, he demonstrated the analysis of water with a battery of one hundred pairs of plates.[36] He later claimed to have been the first person to have introduced galvanism into India, proudly comparing the size of the apparatus he had assembled with those reported from European centers of research.[37] Though he still complained of being underpaid by the college, Dinwiddie gradually came to feel he was accruing a degree of social status among the colonial elite. At a concert at Government House in June 1803, he recorded with satisfaction an encounter with the governor-general: "I had a 'How do you do Doctor' from His Excellency."[38]

Beyond the walls of the college, Dinwiddie's reputation as a pioneer of galvanism largely rested on his use of the battery for healing. Again, the voltaic pile was a successor to the static electrical generator in this role. Static electricity had been used therapeutically for several decades in Europe, and Dinwiddie had already made use of his electrical machines for this purpose.[39] The voltaic battery had advantages of portability and reliability over the static generator, it proved effective in some cases where the latter method had failed, and it could be used in the rainy season

---

32 DUA, MS 2-726, B 68, January 20, 1802.

33 DUA, MS 2-726, D 22, 'Journal of Galvanism', December 10, 1801.

34 DUA, MS 2-726, B 68, December 31, 1801.

35 DUA, MS 2-726, D 22, 'Journal of Galvanism', January 17, 1802.

36 DUA, MS 2-726, D 22, 'Journal of Galvanism', March 29, 1802.

37 Dinwiddie, letter to *Calcutta Gazette*, reprinted in Proudfoot, *Biographical Memoir*, 113–114.

38 DUA, MS 2-726, B 69, June 4, 1803.

39 Paola Bertucci and Giuliano Pancaldi, eds., *Electric Bodies: Episodes in the History of Medical Electricity* (Bologna: Università di Bologna, 2001).

when static electricity could not be produced. The month after he built his pile, Dinwiddie applied it to a man called Macarthy, who suffered from paralysis on one side of his face.[40] The patient reported some benefit in alleviating his numbness, and the treatment was repeated. Soon, a series of hopeful applicants began to present themselves for galvanic therapy. A Captain Thomas was treated for his deafness, and Dinwiddie also tried the treatment on himself for the same condition. He reported that it produced a curious sound in the ears that was hard to describe.[41] In April 1804, he recorded that he "Galvanized a Servant of Mr. Harrington's for a paralytic affection extending from the elbow to the wrist of the right arm." Even a pile of 120 plates was insufficient at first to restore the patient's feelings, but on repeating the procedure Dinwiddie succeeded in eliciting some sensation.[42] The following month, he "Applied Galvanism to Mrs. Gardiner's right eye, the sight of which she lost all at once in the act of lifting a vessel of water from the floor to the window." The patient's sight was not restored, but she was able to perceive a flash of light in the blind eye.[43]

Dinwiddie's interest in this kind of therapy was as an experimenter; he had no ambitions to become a medical doctor. In June 1804, he declared that he was giving up galvanic healing because he was overwhelmed with too many potential patients.[44] Although he continued to demonstrate the voltaic pile, and even developed a small-scale business in making batteries for use in medical treatment, he abandoned therapeutic practice. He published a note in the *Calcutta Gazette*, describing what had been accomplished in the field by himself and other researchers. Among the diseases "in the cure of which galvanism has been completely successful," he listed: rheumatism, paralysis, headaches, deafness, and "two cases of melancholy madness, and one of confirmed hydrophobia."[45] He was familiar with many of these claims from reading British scientific publications, which summarized the activities of galvanic investigators throughout Europe. He also followed the press reports of the extraordinary career of Giovanni Aldini, nephew to Luigi Galvani, whose experiments in London in 1803 included a spectacular attempt to revive the corpse of an executed

---

40 DUA, MS 2-726, D 22, 'Journal of Galvanism', January 19, 1802.
41 DUA, MS 2-726, D 22, 'Journal of Galvanism', March 5, 1803.
42 DUA, MS 2-726, D 22, 'Journal of Galvanism', April 9, 1804.
43 DUA, MS 2-726, D 22, 'Journal of Galvanism', May 22, 1804.
44 DUA, MS 2-726, D 22, 'Journal of Galvanism', June 12, 1804.
45 Dinwiddie, letter to *Calcutta Gazette*, reprinted in Proudfoot, *Biographical Memoir*, 113–114.

criminal in Newgate Gaol. Dinwiddie does not seem to have been surprised that the attempt failed, but he accepted Aldini's claim to have used galvanism successfully to treat cases of insanity.[46] He also recorded an instance from his own experience of its use to cure a Miss Wittworth of a case of St. Vitus's dance.[47] Otherwise known as Sydenham's chorea, this condition produced uncontrollable jerking movements of the limbs; it was recognized as most common among female adolescents. Galvanism was thought to work more effectively with women than with men, and with the old more than the young. These observations were consistent with what was believed about the characteristics of the nervous system: a heightened sensitivity and a less robust constitution were regarded as typical attributes of the female gender, though by no means restricted to women.

Dinwiddie's deployment of galvanism can be compared with other experimental therapies that were the object of intense scrutiny in the period, such as the respiration of nitrous oxide or 'laughing gas,' explored in a series of celebrated—not to say notorious—experiments by the chemists Humphry Davy and Thomas Beddoes in Bristol in the late 1790s.[48] Galvanism also shared features with mesmerism and the use of the so-called 'metallic tractors'—blades of metal drawn across patients' skin in gestures that were believed to have a healing effect.[49] The ontological status and therapeutic benefits of these remedies were hotly disputed, but like galvanism they were supposed to act directly on the nervous system and thereby to address conditions, such as pain or paralysis, that were experienced as physical but had no obvious organic cause. Experimentation on the investigator's own body was a regular feature of galvanic researches and of inquiries into such other therapies as nitrous oxide.[50] By trying the treatment on himself, the experimenter could more readily assess how other people's responses reflected their personal characteristics. Dinwiddie moved seamlessly between self-experimentation and dispensing the treatment to others, calibrating his patients' reactions by reference to his own experiences of the battery's power. At the same time,

---

46 DUA, MS 2-726, D 22, 'Journal of Galvanism', 23–24, June 10, 1803.

47 DUA, MS 2-726, D 22, 'Journal of Galvanism', [55–56], November 2, 1804.

48 Golinski, *Science as Public Culture*, 166–175.

49 James Delbourgo, "Common Sense, Useful Knowledge, and Matters of Fact in the Late Enlightenment: The Transatlantic Career of Perkins's Tractors," *William and Mary Quarterly*, 3rd ser. 61 (2004): 643–684.

50 Stuart Walker Strickland, "The Ideology of Self-Knowledge and the Practice of Self-Experimentation," *Eighteenth-Century Studies* 31 (1998): 453–471.

he used galvanism as a medium of social interaction. He introduced it into social occasions, offering shocks from the voltaic pile to friends with whom he was conversing or socializing. In this respect, galvanism offered a mode of interpersonal connection, even as it also helped define a new sense of individual identity. In pedagogical displays, in therapeutic applications, and in social gatherings, the voltaic pile thus became a feature of Dinwiddie's persona.

Unlike the pioneers of these other therapies, Dinwiddie was working in Calcutta in a unique situation of interaction between different cultures. He does not seem to have systematically explored the effect of galvanism on Indians' physiology. In cases where the experimental subject was an Indian (as perhaps in the case of Mr. Harrington's servant, mentioned earlier), he did not specifically remark on the fact. But he did investigate the phenomenon in the context of an interest in Indian culture and the connections and comparisons between East and West. He was often scornful about native beliefs and practices that he categorized as "superstitions," such as witchcraft; but he was also fascinated by them. He made notes on Muslim and Hindu dietary practices and legal procedures, collected reports about snake-charmers, and compared ideas about predicting eclipses with Indian astronomers. He was particularly interested in the activities of Indian healers and conjurors, whose performances had similarities with his own. In June 1806, he recorded, "a Madr[as] juggler called by appointment, and among many other experiments passed a sword blade down his throat 21 inches." Dinwiddie purchased the sword as a memento of this 'experiment,' to take with him on his return to England.[51] Indian healing practices were also of interest to him. He mentioned that the natives exposed their bodies to the monsoon rains to alleviate aches and pains, perhaps considering whether he might benefit by doing the same.[52] In May 1801, he witnessed the care of one of his native bearers by a female Indian dentist: "She had two sticks by which she cleaned the teeth making use of several elegant motions and accompanied them with a constant harangue. She concluded with a motion of the instruments across the forehead."[53] The elegant motions and the passes of the healer's instruments over the patient's forehead would have evoked for Dinwiddie the similar gestures of mesmerists or the operators of metallic tractors. He

[51] DUA, MS 2-726, B 71, June 12, 1806.
[52] DUA, MS 2-726, B 65, May 28, 1801.
[53] DUA, MS 2-726, B 65, May 20, 1801.

did not make the comparison explicitly, perhaps because it would have suggested an uncomfortable degree of similarity between Indian and European therapeutics. But he could well have been reminded of what the two had in common. Galvanism had arrived in Calcutta as a Western innovation, the news carried by printed publications arriving on European ships. Dinwiddie had deployed it to enhance his authority among colonials and natives, in the classroom and beyond. But the technique shared features with other experimental therapies and indeed with Indian healing practices.[54] As a performative mode of therapy, it evoked mesmerism, the use of the metallic tractors, and the gestures of the Indian dentist. Dinwiddie seems to have been aware of these connections but preferred not to acknowledge them explicitly. Galvanism, as he used it, borrowed aspects of Indian culture, although it also helped him to stand apart from that culture. It allowed Dinwiddie to appropriate aspects of his situation in Calcutta, even while distancing himself from the supposed ignorance and superstition of the natives.

London

By his teaching, his scientific entrepreneurialism, and his service to the colonial administration, Dinwiddie had by this time become comfortably wealthy. He decided to return to Britain, departing from Calcutta on 15 September 1806 and arriving at Blackwall on the Thames Estuary exactly seven months later.[55] He considered settling in Scotland, making several visits back to his homeland, but ended up staying in London for most of the rest of his life; he died there in 1815. He continued to take an interest in galvanism, studying carefully the way it was deployed by several scientific lecturers in the capital. He was sometimes critical in his comments on them, giving especially close attention to the performances of Humphry Davy at the Royal Institution. He never returned to lecturing himself, but he continued to make use of the voltaic pile in small gatherings as an aid to socialization and an instrument of healing. Doing so, he presented it alongside other exotic goods from his travels. He thereby completed a circuit by which galvanism had arrived in India with the accoutrements

54 For European interest in East Asian medical techniques in an earlier period, see: Harold J. Cook, *Matters of Exchange: Commerce, Medicine, and Science in the Dutch Golden Age* (New Haven: Yale University Press, 2007), 339–377.

55 DUA, MS 2-726, B 72, B 73; Proudfoot, *Biographical Memoir*, 120–121.

of Western exoticism and was then returned to England as one of the curiosities of the East.

Dinwiddie took a voltaic pile with him on the voyage back to England, offering his fellow-passengers galvanic treatment for their ailments. Soon after his arrival in London he ordered a new battery of 120 zinc and copper plates to be made by a Mr. Edmiston, apparently a fellow-Scot working in the capital.[56] He applied the current to himself and claimed to have cured a hydrocele in one of his testicles, which had swollen to nearly three times its normal size. He also galvanized his own legs (presumably the ulcer was still troubling him), and treated a Mr. Murdoch for lumbago and Mr. Curdie for the cramps.[57] Before long, he was attracting a steady stream of visitors for galvanic therapy and had a larger pile of 180 plates made for him. As in India, the voltaic pile was a topic of conversation and a tool for socialization through the medium of healing. In London, as in Calcutta, most of Dinwiddie's circle of friends—at least to judge from their names—were Scots. Galvanic therapy consolidated his links with this group and again substituted to some extent for the absence of a family. He spent Christmas Day 1808 with Mr. and Mrs. Edmiston and a Mr. Morris; he recorded the topics of conversation as: "Galvanism—irritability excitability—Frog galvanized—New metals &c—."[58]

Integral to Dinwiddie's social persona in London was the air of exoticism that attached to him as someone who had returned from the East. This complemented and framed his deployment of galvanism. He displayed the sword he had purchased from the Madras juggler, along with other artifacts and curiosities. In January 1808, he showed Dr. Buchan and another gentleman "some Galvanic Experiments—Costume of India—Chinese foot &c."[59] He collected recipes for curried chicken and other Indian dishes, and perhaps had them served to his guests. He regularly put on demonstrations of voltaic electricity at social gatherings for small private parties at his house. And on these occasions he would also bring out things he had collected in the course of his travels, such as a meteorite and a gold chain. On 10 August 1808, he recorded: "Miss Patoun and three Playfairs and two sons of Mr. P. called at one o'clock to see the Galvanic machine, and other curiosities from the East."[60] Of course, galvanism had

56 DUA, MS 2-726, B 73, May 21, 1807.
57 DUA, MS 2-726, B 74, September 19, October 4–5, 1807; B 75, May 7, 1808.
58 DUA, MS 2-726, B 76, December 25, 1808.
59 DUA, MS 2-726, B 74, January 19, 1808.
60 DUA, MS 2-726, B 75, August 10, 1808.

not originally been a "curiosity from the East" at all; but in Dinwiddie's hands it had become one, its exotic cachet enhanced by involvement with his personal itinerary as a traveler who had returned from distant lands.

Dinwiddie also took a close and often critical interest in how galvanism was used by public lecturers in London. He attended the lectures of John Tatum, George John Singer, Deane Franklin Walker, and others. He also attended meetings of the Askesian Society and took in a lecture at the Spitalfields Mathematical Society.[61] Tatum was Secretary of the City Philosophical Society and lectured in the East End in Dalston Square. Dinwiddie was unimpressed by his technique when he began to attend in the early months of 1808. He noted that Tatum's model powder magazine was 'not in order,' and the spark of his large electrophorus 'scarcely perceptible.'[62] He was appalled by Tatum's pretense of reviving a dead frog by galvanism: "Frog killed by Electr[icity] brought to life by Galvanism!! Convulsed only by Galv[anism]."[63] Dinwiddie was also offended by Tatum's cockney accent; in his notes he ridiculed the lecturer's pronunciation of 'hoxygen,' 'hexperiment,' 'hanimal,' and so on. But he nonetheless returned to Tatum's classes in the winter of 1808–9 and again the following year. Evidently he was learning something from them, though the pleasure of making critical comments on the lecturer's performance was perhaps also part of the appeal. Dinwiddie recorded an occasion when galvanism was applied to convulse the body of a dead hare, and one when Tatum filled the trough of a voltaic battery 'before the company!!'—obviously a serious breach of decorum or showmanship as far as Dinwiddie was concerned.[64] Summing up his observations in January 1809, he remarked: "In public lectures arrangement in the Experiments is an important object—[there is a] great want of this in Mr. Tatum's Lectures."[65]

Attendance at these lectures allowed Dinwiddie to assess his own technique as he contemplated venturing back onto the stage himself. In the event, he never took that step, leaving the crowded and highly competitive

---

[61] DUA, MS 2-726, B 75, May 19, 1808; B77, April 27, 1809. Ian Inkster, "Science and Society in the Metropolis: A Preliminary Examination of the Social and Institutional Context of the Askesian Society of London, 1979–1807," *Annals of Science* 34 (1977): 1–32; Larry Stewart and Paul Weindling, "Philosophical Threads: Natural Philosophy and Public Experiment among the Weavers of Spitalfields," *British Journal for the History of Science* 28 (1995): 37–62.

[62] DUA, MS 2-726, B 75, February 16, 1808.

[63] DUA, MS 2-726, B 75, March 1, 1808.

[64] DUA, MS 2-726, B 79, April 2, 1810.

[65] DUA, MS 2-726, B 76, January 10, 1809.

London lecturing scene to younger men. He was happy to indulge himself as a spectator, however, attending presentations on botany and geology as well as chemistry and electricity, and witnessing a demonstration of a Boulton-Watt steam engine and a display of perpetual motion, along with such non-scientific shows as the phantasmagoria. He saw both Tatum and Davy demonstrate the effects of breathing nitrous oxide, and commented on the bizarre behavior it sometimes—but not always—produced.[66] London in the first decade of the nineteenth century offered an unprecedented range of such scientific spectacles, and Dinwiddie was evidently fascinated by them.[67] He could obviously see the potential for new kinds of pedagogical display and was preoccupied with the issue of how such displays could be managed most effectively.

The lecturer whom Dinwiddie scrutinized most frequently and most closely was Humphry Davy, then at the apex of his reputation among the fashionable elite of the capital for his performances at the Royal Institution.[68] Dinwiddie had learned of the foundation of this new scientific establishment while he was in Calcutta. He began to attend lectures there very soon after arriving in London, and was elected a member in 1810. He recognized the significance of Davy's accomplishment in attracting a "large and genteel company [of] six or seven hundred" to his course on chemistry.[69] He also admired the size and expense of the voltaic battery Davy had had made at the Royal Institution, noting with patriotic pride that the 2,000-plate apparatus—assembled in a series of wooden troughs—was larger than the one constructed under the direction of Napoleon in Paris. Private patronage had laid a sounder foundation for scientific advance than the orders of a tyrant, he concluded: "The mind was roused to energy by the cheering voice of public freedom and popular applause."[70] He was awed by the mobilization of such resources, which extended beyond even his highest ambitions.

---

66 DUA, MS 2-726, B 77, December 13, 1808; E 6, January 28, 1809.

67 John Brewer, "Sensibility and the Urban Panorama," *Huntington Library Quarterly* 70 (2007): 229–249; James Chandler and Kevin Gilmartin, eds., *Romantic Metropolis: The Urban Scene of British Culture, 1780–1840* (Cambridge University Press, 2005); Celina Fox, ed., *London: World City, 1800–1840* (New Haven: Yale University Press, 1992); Richard D. Altick, *The Shows of London* (Cambridge, MA: Harvard University Press, 1978).

68 David Knight, *Humphry Davy: Science and Power* (Oxford: Blackwell Publishers, 1992); Golinski, *Science as Public Culture*, 188–235.

69 DUA, MS 2-726, B 76, December 17, 1808.

70 DUA, MS 2-726, E 9, February 23, 1811.

Dinwiddie became a regular at Davy's lectures, turning up for his courses on chemical philosophy and geology as well as those on chemistry. He faithfully wrote down Davy's remarks about his discoveries and his sometimes controversial theories. Dinwiddie's is a good record of what Davy was saying in the lecture theatre, though he admitted to missing things sometimes because of his deafness. He noted Davy's conjectures that the chemical theory of phlogiston still had some plausibility, that there was ultimately only one kind of matter, and that volcanoes and earthquakes were caused by the newly-discovered alkali metals—all of which doctrines Dinwiddie himself doubted but dutifully recorded.[71] When it came to the lecturer's technique, however, Dinwiddie allowed himself to be more explicitly critical. He appreciated that Davy had to make a pitch to his audience to sponsor scientific research, but he clearly thought the lecturer had gone too far when he suggested there only needed to be one publicly supported institution, and that the Royal Institution could deliver instruction more cheaply than Edinburgh University professors.[72] When Davy launched into an oration on the necessity for public support of chemistry in January 1811, Dinwiddie commented critically: "This resembled a Methodist sermon rather than a Lecture on Chemistry."[73]

Dinwiddie had a particularly good eye for the theatrical dimension of Davy's performances, as one would expect from a man who had been an accomplished scientific performer himself. He noted when Davy's rhetoric verged on the declamatory, when the speaker became 'a little elevated' or long-winded.[74] He complained when he got diverted off the topic or spoke with his back to the audience. He mocked Davy's courtesies to women, as when the lecturer promised that if he succeeded in making diamonds from charcoal he would present them to the ladies in attendance. Davy, Dinwiddie noted, "said he always had great regard to the obs[ervatio]ns of the Ladies!!"[75] Indeed, the glamorous young man's appeal to women had been much remarked upon by his contemporaries, who often ridiculed his dandyism in preening himself before a largely female audience. Dinwiddie seems to have been well aware of that aspect of Davy's reputation.[76]

---

71 DUA, MS 2-726, B 77, January 15, 1809; E 6, March 11, 1809; E11, March 13, 1811.
72 DUA, MS 2-726, E 6, April 29, 1809.
73 DUA, MS 2-726, E 7, January 19, 1811.
74 DUA, MS 2-726, E 7, January 1, 1810.
75 DUA, MS 2-726, E 7, February 19, 1810; E 11, February 9, 1811.
76 On this, see: Jan Golinski, "Humphry Davy's Sexual Chemistry," *Configurations* 7 (1999): 15–41.

Dinwiddie was of course particularly interested in Davy's deployment of the voltaic battery in his lectures. He noted occasions when it worked well, producing powerful or spectacular effects. But he also mercilessly recorded occasions when it did not work properly or was not well managed, when it was apparently out of order or its effects could not be seen. An episode in early June 1810 was a case in point. Davy announced that he would be using the newly constructed battery of 2,000 plates, but he began by demonstrating the old one of half that size. Dinwiddie noted: it "did not act well, some defect in the connection." Evidently there was a fault in the wiring that linked the lecture theatre with the neighboring room in which the battery was assembled. When the whole assembly of 2,000 plates was connected, "It did not act. Mr. D. went into the trough room several times, the noxious gas became too powerful for the audience, several left the room when at length Mr. Davy was advised to adjourn the lecture till Saturday next."[77] This was a chaotic and embarrassing scene, obviously, with members of the public driven from the lecture theatre by the obnoxious fumes released by the new battery. Fortunately, Davy recovered himself and reasserted control of his instrument a few days later. The audience reassembled on Saturday 9 June, when the 2,000-plate battery had been placed in the basement laboratory downstairs, with wires leading up through the floor into the lecture-theatre. This arrangement worked much better: the wires communicated the electric current, while noxious fumes were kept away from the spectators. Dinwiddie remarked: "With these brilliant experiments Mr. Davy concluded his Lectures for the season."[78]

Dinwiddie's sardonic observations on Davy and other London lecturers no doubt reflected a degree of melancholy, even a tinge of jealousy. As an old man, he had to accept that he could no longer expect to be successful with a metropolitan audience; certainly he could never hope to attract the adulation accorded to Davy. This gives his criticisms a particular edge and acuity that is valuable for the historian. It also illuminates the contrast between the scenes in London and in Calcutta. In India, when Dinwiddie first introduced it, galvanism was an exotic novelty. The voltaic pile functioned more reliably than static electrical generators, and it could yield sparks and shocks to hold the attention of students at a lecture or even to frighten a recalcitrant native servant. It also provided a potent,

[77] DUA, MS 2-726, E 7, [undated, June 1810].
[78] DUA, MS 2-726, E 7, June 9, 1810.

robust, and mobile source of electric current for medical therapy. Dinwiddie grasped its potential as a therapeutic device similar in its effects to other experimental methods of the era. He seems also to have recognized its parallels with certain indigenous forms of healing, though he doubtless would have insisted on the superiority of Western 'science' over Eastern 'superstition.' Galvanic therapy took advantage of the proximity of Indian healing practices, even while its advocates proclaimed its distance from—and superiority over—native ignorance.

In London, the field of practitioners was much more crowded and the equipment much more commonplace. At this point in his life, Dinwiddie had little chance of muscling onto the stage as a lecturer in the capital. London presented a variety of scientific shows and other displays that had risen to a new level of spectacular popularity in the years he had been away. Dinwiddie could only look on with admiration at the fashionable crowds that attended the most successful lecturers and envy the laboratory resources they commanded. He did, however, find a way to deploy galvanism in smaller gatherings and he continued to use it therapeutically. He presented the voltaic pile in the context of artifacts and curiosities of the East, evoking his own exotic past to enhance its credentials. This was to enact a curious pattern of circulating knowledge: an instrument that had arrived in Calcutta as a product of the most recent Western science was returned to the West as an exotic import from the East. The actual materials of the battery had not come far, since Dinwiddie had a new one made when he arrived in London, but his persona conferred upon it the exoticism of his own travels. To this extent, his success with the voltaic pile was one of the fruits of Dinwiddie's extraordinary philosophical itinerary.

There are also broader implications of Dinwiddie's story. Fully understood, it can contribute toward overcoming what have been called the 'distortions of a bipolar vision' affecting the history of science.[79] In this vision, 'Western' and 'non-Western' forms of knowledge are seen as polar opposites, defined as mutually exclusive, with the former expressing a unique degree of rationality not shared by the latter. Dinwiddie's uses of the voltaic pile suggest a more complex and ambivalent picture. Information about the pile was circulated from West to East, but the artifact was made and deployed by Dinwiddie in a context in which he was simultaneously studying Indian culture and struggling to establish his authority with

[79] "Introduction," in *Brokered World*, ed. Schaffer et al., ix–xxxviii (quote on xv).

colonists and natives. He appropriated aspects of the situation around him, even as he used the pile to enhance his standing in the colony. When he took it back to England, he presented the pile as an Eastern curiosity, using its exoticism to increase its cachet. He thus continued to make use of the apparatus for healing in small gatherings, even though he could not attain a place on the lecturer's stage in London. In short, the device's uses and meanings were determined by its back-and-forth motions between Europe and South Asia. Rather than a bipolar picture of Western science dominating the East, the story indicates how scientific knowledge was implicated in the multiple encounters and entanglements of European and other cultures in this period. Dinwiddie's travels and those of the apparatus he took with him give evidence of the remarkable productivity of the patterns of circulation that connected East and West.

It would be hazardous to try to generalize from Dinwiddie's case. His journey was a unique one, the range and duration of his travels forming a distinctive feature of his personal identity. Clearly, the voltaic pile could be deployed successfully without its user sharing Dinwiddie's well-traveled persona. On the other hand, the rapid and widespread replication of the pile in Europe has tended to obscure the specific itineraries of people and artifacts by which this was accomplished. Dinwiddie's story shows how complex these paths could be, and how they could extend well beyond European territory. In his hands, the voltaic pile was made meaningful by association with its user's mobility, the aura he possessed by virtue of travel from distant places. And, in its turn, the galvanic apparatus contributed in no small part to enhancing this personal authority. It became one of the accoutrements of the exotic individual that was James Dinwiddie.

PART TWO

CIRCULATION BEYOND DINWIDDIE

BRINGING EASTERN SCIENCE TO THE WEST: PORTUGUESE VOYAGES OF INTELLECTUAL DISCOVERY

Arun Bala*

Introduction

It is often assumed that the Portuguese voyages into the East were largely a matter of geographical discovery. The notion that the Portuguese could have transmitted key scientific ideas from the East, which could have crucially influenced early modern natural science, is often not entertained. Such perceptions do double injustice—first, they fail to adequately recognize the role that the Iberian West played in the birth of modern science; second, they ignore the crucial contributions the East made to scientific ideas that developed in the early modern West. This paper looks at one key influence on modern science from the East that is likely to have been mediated by the Portuguese voyages—the impact of the Indian Kerala School of mathematical astronomy on modern astronomy in the period of the 17th century scientific revolution. In particular it looks at the development of infinite series expansions of circular and trigonometric functions and the proto-calculus used by Indian astronomers. A little more than a century after the arrival of the Portuguese in the East similar ideas came to transform European conceptions of both mathematics and astronomy.

Many historians assume that transmissions of such ideas must have occurred largely through the Jesuits, who were sent as missionaries to the East but also trained in the mathematical and astronomical sciences. In a recent study George Gheverghese Joseph writes:

> It is our conjecture that between 1560 and 1650 knowledge of Indian mathematical, astronomical, and calendrical techniques accumulated in Rome and diffused to neighbouring Italian universities like Padua and Pisa, and to wider regions through Cavalieri and Galileo, and through visitors to Padua like James Gregory. Mersenne may have also had access to knowledge from

* I wish to express my deepest gratitude to Bernard Lightman and Gordon McQuat for their insightful comments and suggestions to both expand and tighten the arguments developed here.

> India acquired by the Jesuits in Rome and, via his well-known correspondence, helped to diffuse this knowledge throughout Europe.[1]

But the paucity of available textual evidence for transmission in the archives of Jesuit communications back to Europe has also led others to argue that ideas paralleling the Indian discoveries of the Kerala School could have been independently rediscovered in early modern Europe. This point has been emphasized by David Bressoud, who writes:

> There is no evidence that the Indian work on series was known beyond India, or even outside Kerala, until the nineteenth century.... by the time these series were rediscovered in Europe, they had, for all practical purposes, been lost to India. The expansions of the sine, cosine, and arc tangent had been passed down through several generations of disciples, but they remained sterile observations for which no one could find much use. No. Calculus was not discovered in India.[2]

However, the debates hitherto have presupposed that the transmission of these mathematical ideas from India to Europe, had it occurred, must have been mediated by scholars, especially the Jesuits since they were the only Europeans trained in mathematical astronomy who went to the East. This paper argues that it is also possible to suppose that the transmission of ideas to Europe could have been mediated not by theoretical ideas carried by European scholars, who inherited them from their Indian counterparts, but by computational practices and techniques, carried by European craftsmen who had learnt them from Indian craftsmen they worked with. The Indian craftsmen themselves would have come to learn these practices and techniques, directly or indirectly, from Indian scholars. These could have been adopted in an opportunistic and *ad hoc* fashion by European artisans, navigators and cartographers in the East to be subsequently transmitted back to Europe. They could then have inspired

---

[1] George Gheverghese Joseph, *The Crest of the Peacock: Non-European Roots of Mathematics* (Princeton: Princeton University Press, 2011), 438. He had already referred to this possibility in his second edition of the book more than a decade earlier when he conjectured that mathematics from medieval India, especially Kerala in southern India, may have influenced developments in European mathematics in the sixteenth and seventeenth centuries. See Chapter 9 (p. 249) in Joseph, *The Crest of the Peacock* (Princeton: Princeton University Press, 2000). Much earlier the same possibility had been suggested by A. K. Bag *Mathematics in Ancient and Medieval India* (Varanasi/Delhi: Chaukhambha Orientalia, 1979), 285.

[2] David Bressoud, "Was Calculus Invented in India?" *The College Mathematics Journal* 33(1) (2002): 12.

new theoretical thinking there which ultimately led European scholars to develop notions similar to those associated with the Indian school.

The important mathematical developments in India with which we are concerned took place between the 14th and 16th centuries, and were the achievements of a line of mathematical astronomers beginning with Madhava (1350–1425), and continuing with Paramesvara (c. 1370–1460), Nilakantha Somayaji (1444–1544) and Jyesthadeva (c. 1500–1575), who have now come to be labeled as belonging to the Indian Kerala School of mathematics.[3] Although the Kerala achievements were localized in the south of India their antecedents lay in nearly a millennium of research in North India. In particular the Kerala School built on the tradition started by Aryabhata in his treatise *Aryabhatiya*, which he composed in 499 CE in Kusumapura, now identified as Patna in the North Indian state of Bihar. The *Aryabhatiya* not only presented a new cosmology incorporating the use of epicycles derived from Greek astronomical traditions, but also introduced new mathematical innovations including the decimal place value system with zero as a place holder for powers of ten, the trigonometric notion of sine, and the value of π to five decimal places. His work was developed over the next few centuries by a chain of mathematical astronomers, including Varahamihira (505–587), Bramagupta (598–668), Mahavira (800–870), Shridhara (870–930) and Shripati (1019–1066) culminating in the work of Bhaskaracharya (1114–1185). In his most important work entitled *Lilavati* (The Beautiful) Bhaskaracharya introduces the concept of arithmetical infinity which he defines as any positive quantity divided by zero, e.g. 5/0.

Bhaskara's naturalisation of the notion of infinity, and his adoption of notions of proto-calculus, paved the way for the development of infinite series expansions of circular and trigonometric functions, such as π, sin$x$, cos$x$ and inverse tan$x$. This 'passage to infinity' was taken by Madhava who, as we saw, pioneered the Kerala School. The seminal findings of the school go beyond ancient Greek achievements in mathematics where we do not find any systematic study of infinite series, or infinite series representations for π and the trigonometric functions. For example in ancient Greek mathematics, Archimedes had computed the circumference of a circle by approximating the circle to a polygon with a large number of

3 For a detailed study of the Kerala School of mathematics see George Gheverghese Joseph, *A Passage To Infinity: Medieval Indian Mathematics from Kerala and Its Impact*.

sides. Adding the length of all these sides would give an approximation to the circumference, and the larger the number of sides the closer would be the approximation to the actual circumference. He used a technique now described as the method of exhaustion.

By contrast, the Indians used the method of rectification for computing π. The Indian method of rectification grew out of the geometrical approach for representing numbers when solving problems in arithmetical progressions developed by Nilakantha.[4] Jyesthadeva uses the method to derive the Madhava-Gregory inverse tangent series in the *Yuktibhasa*, a seminal work he wrote c. 1550 which summed up the previous achievements of the Kerala School over two centuries. According to Ramakrishnan, this derivation not only shows how trigonometric functions can be expressed as infinite series, but also "involves several techniques including the ideas of integration and differentiation."[5]

Joseph distinguishes the approaches of the method of exhaustion and the method of rectification as follows:

> An unusual aspect of the Kerala approach to the derivation of a number of infinite series is their use of the method of direct rectification. The method of direct rectification of an arc of a circle involves summation of very small arc segments and reducing the resulting sum to an integral. This is an interesting geometric technique different from the "method of exhaustion" used in Iṣlamic and European mathematics. In the Kerala case, you are subdividing an arc into *unequal* parts, whereas in the Islamic and European case there is a subdivision of the arc into *equal* parts.[6]

These discoveries were made possible in India because the discovery of the decimal place-value number system with zero made Indian mathematics capable of representing trigonometric functions as an infinite series. In 1814 Laplace described the significance of this number system as follows:

> The ingenious method of expressing every possible number using a set of ten symbols (each symbol having a place value and an absolute value) emerged in India. The idea seems to be simple nowadays that its significance and profound importance is no longer appreciated. Its simplicity lies in the way

---

[4] For a more extensive discussion see Mallayya, V. Madhukar, "Geometrical Approach to Arithmetical Progressions from Nilakantha's *Aryabhatiyabhasya* and Sankara's *Kriyakramakari*,." in *Proceedings of The International Seminar and Colloquium on 1500 Years of Aryabhateeyam*, (Kochi, India: Kerala Sahtra Sahitya Parishad, 2002).

[5] P. Ramakrishnan, "Some Aspects of Development of Kerala Mathematics," in *Proceedings of The International Seminar and Colloquium on 1500 Years of Aryabhateeyam*, (Kochi, India: Kerala Sahtra Sahitya Parishad, 2002), 138.

[6] Joseph, *The Crest of the Peacock*, 427.

> it facilitated calculation and placed arithmetic foremost amongst useful inventions. The importance of this invention is more readily appreciated when one considers that it was beyond the two greatest men of Antiquity, Archimedes and Apollonius.[7]

However, it must not be assumed that these developments in Indian mathematics were motivated by purely theoretical interests. Actually they were also driven by the need for more precise computations in making astronomical predictions. These were seen as imperative for addressing the agricultural and astrological concerns of Kerala society. In the case of agriculture there was a need to have accurate knowledge of the solar calendar in order to determine the seasons so crucial for the agrarian management of the economy. The calendar was also important to determine dates for religious festivals, which came to be closely associated with agricultural activities such as sowing and harvesting. Astrology also generated a pressing demand for precise astronomical knowledge since it was called upon to determine the auspicious times for many religious rituals such as the naming ceremony, the first feeding, initiation, marriage and other rites of passage. Casting horoscopes to determine the destiny of a person in a society that took the law of karma for granted also assigned an important role for astrological knowledge closely linked to the positions of the stars and planets at the time of birth.[8]

There was also a wider socio-cultural context driving the need for more accurate astronomical calculations. From the middle of the first millennium there emerged a vast expansion in maritime trade that linked South India to the Middle East and Africa in the west, and Southeast Asia and China in the east. Impetus for such trade followed the invasion of the central Asian Huns in the fifth century that not only traumatized China, India and the Mediterranean world, but also rendered unstable and dangerous overland trading routes between these regions. This pan-Asian oceanic trade continued to expand in the centuries following making the Indian Ocean a second corridor for trade across civilizations, and one in which South India came to play a prominent role. It led to greater demand for accurate astronomical computing techniques for navigation which may have been another factor that could have encouraged the mathematical developments of the Kerala School. Indeed interest in astronomy was

[7] Georges Ifrah, *The Universal History of Numbers: From Prehistory to the World Order* (New York: John Wiley & Sons, 2000), 361.

[8] Joseph, *A Passage To Infinity*, 10–12.

given a great boost during the reign of Ravi Varma [844–885 CE] when he established an observatory with a giant armillary sphere in his capital city of Mahodayapuram (present-day Kodungallur). The astronomers who worked in his observatory, and hence had close links with the court, are known to have worked within the Aryabhatan tradition.[9] Hence it is reasonable to suppose that the Kerala mathematicians built on the achievement of their North Indian counterparts and were motivated to extend the work in the north not only for the traditional reasons of agriculture and establishing religiously propitious times, but also to support more accurate navigation computations.

However, this still leaves open the question: Is it is reasonable to assume that the discoveries of the Kerala School were transmitted to Europe simply because they were discovered earlier in India, and the Portuguese arrival there created a route for such transmission more than 150 years before similar discoveries were made in Europe? Needham has claimed that this is a natural presumption since, as he put it, "the details of any transmission are difficult to observe." Indeed he considered it incumbent on those who wished to claim independent discovery to demonstrate their case.[10] However, evidence for the *possibility* of transmission is not sufficient to establish transmission, and it is still reasonable to require that those who make such claims must provide more positive support for it.

Almeida and Joseph[11] undertook a project to make a stronger case going beyond Needham's appeal to priority of discovery and possibility of communication by looking for documentary evidence of transmission. They conducted a three year study of communications of Jesuit missionaries in Kerala back to Europe. Their study included manuscript letters and reports from Jesuit missionaries to their headquarters in Rome at the *Archivum Romanicum Societate Iesu* (ARSI); manuscript correspondences in the Gregorian University Archives in Rome of Jesuit scientists in India such as Rubino, Ricci and Schreck to Clavius and Grienberger in Europe; and manuscript correspondences of earlier Jesuit missionaries to the Malabar up to 1568 in the Ajuda Library in Lisbon. They also investigated correspondences from the Jesuit mathematician Borri found in the University of Coimbra archives, since he was the only missionary who went to

[9] Joseph, *A Passage to Infinity*, 13.

[10] Joseph Needham, *The Grand Titrations: Science and Society East and West* (London: Allen & Unwin, 1969), 83.

[11] D. F. Almeida and Joseph G. G., "Eurocentrism in the History of Mathematics: The Case of the Kerala School." *Race and Class* 45 (2004): 45–59.

the Malabar and returned to Europe. In addition, they looked at the works of Jesuit mathematicians in Rome who could have come across Kerala mathematics, including the works of Clavius and Grienberger. Finally, they examined Indian scientific manuscripts in libraries in Europe sent back by the Jesuits, especially Malayalam and Tamil palm leaf manuscripts in the Vatican library and in Leiden University archives. Despite their detailed and painstaking survey, they found no direct evidence for any documentary transmission of the discoveries of the Kerala School mathematicians to Europe.[12]

Nevertheless, they find the absence of documentary evidence both puzzling and surprising. It leads Joseph to argue that "[Kerala mathematics] will be the first major case of scientific development in the post-Ancient era that has remained localized in its place of origin, and that despite the existence of a direct corridor of communication to Renaissance Europe."[13]

Kerala Mathematics: Transmission as Computing Techniques

The absence of textual evidence after intensive search could also imply that the search for documentary evidence—at least in this case—may be methodologically flawed. The Indian discoveries may not have been transmitted to Europe as a system of theoretical ideas (i.e. as propositional knowledge involving 'know-that' claims), but rather as a system of computing techniques (i.e. as skills-based knowledge involving 'know-how'). If this were the case then we could suppose that it was not the Jesuits who carried the Indian knowledge to Europe but European craftsmen who had acquired this knowledge in the form of computing techniques and practices from Indian craftsmen they came in contact with or worked with. Such a supposition is quite reasonable since there is ample evidence that in activities such as navigation, ship-building, and the building of forts and missionary schools in the East, the Portuguese worked closely with local artisans, navigators, cartographers and architects.[14]

Such a conjecture becomes even more plausible when we consider that the Kerala School of mathematicians were also astronomers concerned with finding better rules of computation for astronomical predictions.

[12] See Joseph, *A Passage To Infinity*, 178–188.

[13] Joseph, *A Passage To Infinity*, 187.

[14] Lotika Varadarajan, *Indo-Portuguese Encounters: Journeys in Science, Technology and Culture*, (New Delhi: Indian National Science Academy, 2006).

These Indian scholars saw their mathematical rules more like the rules of grammar discovered by the ancient Indian grammarian, Panini, often considered the father of modern linguistics, than the axioms or theorems of Euclidean geometry. They defined correct practices rather than true beliefs. Hence they saw these rules as abstracted from experience in the same fashion Panini abstracted grammatical rules from linguistic experience, and constantly refined his rules by strategic fine-tuning to give better computational accuracy.

Such a flexible and pragmatic search for better computational techniques led Indian mathematicians to many discoveries, including the Indian number system, trigonometry, negative numbers, infinite series representations of irrational numbers, logarithms, and mathematical series representations of circular and trigonometric functions. The intimate connections between Indian linguistics and Indian mathematics is emphasized by Joseph:

> Panini offered what must be the first attempt at a structural analysis of a language. On the basis of just under four thousand sutras (i.e. rules expressed as aphorisms), he built virtually the whole structure of classical Sanskrit language, whose general shape hardly changed for the next two thousand years.... an indirect consequence of Panini's efforts to increase the linguistic facility of Sanskrit soon became apparent in the character of scientific and mathematical literature. This may be brought out by comparing the grammar of Sanskrit with the geometry of Euclid—a particularly apposite comparison since, whereas mathematics grew out of philosophy in ancient Greece, it was, partly an outcome of linguistic developments in India.[15]

Joseph elaborates this point by noting that while the geometry of Euclid begins with definitions, axioms, and postulates before going on to derive an extensive structure of connected theorems, Panini started with some general concepts, vowels and consonants, nouns, pronouns, and verbs, and so on—about one thousand seven hundred basic building blocks. With these he constructed compound words in a process similar to the way mathematical functions are specified today. As a result, the linguistic theoretical structure for Sanskrit he developed came to shape the structure of mathematics and its modes of reasoning in India. According to Joseph the algebraic character of ancient Indian mathematics can be seen as simply a by-product of its advanced linguistic tradition.

---

[15] Joseph, *Crest of the Peacock*, 314–316.

However, the approach also led Indian mathematicians to ignore offering deductive demonstrations of their mathematical discoveries since they assumed that these were more like inductive generalizations of grammatical rules discovered from linguistic experience. Bronkhorst emphasizes this point:

> The question of the absence of proofs in Indian geometry acquires a different dimension once it is clear that Euclidean and classical Indian geometry may not really concern the same objects. The objects of classical Indian geometry—its triangles, circles, spheres etc.—are no mere abstractions but things present in the outside world.... As such they resemble the objects of grammatical analysis far more than do Euclidean diagrams. The objects of grammar—sounds, words, phrases—are there in the outside world, and grammar is therefore a science which deals with objects whose existence is quite independent of the volition of the grammarian.[16]

This epistemological divide in the orientation of Indian and Greek mathematics also suggests that we may be looking in the wrong direction when we seek evidence for transmission of Kerala mathematics in the writings and works of Jesuit missionary scientists to scholars in Europe. In the first place the knowledge offered by the Kerala mathematicians would not be deemed worthy of communication as mathematical knowledge. It would be seen as simply offering utilitarian rules-of-thumb—the kind of practical knowledge that would have only value for cartographers, mariners and calendar makers. But such a view would not have stopped the Jesuits from using their mathematical training, and their knowledge of Indian languages, to translate the mathematical techniques developed by the Indians for use by European craftsmen who accompanied them. In short, although there would have been hindrances to the transmission of the Kerala School discoveries to European scholars as acceptable mathematical knowledge, there would have been no obstacle for them to be adopted as good technical knowledge by European craftsmen.

Hence, the Indian mathematical discoveries may have reached Europe as computing rules-of-thumb via craftsmen rather than as a corpus of theoretical principles. Even in India these discoveries of the Kerala School were seen as valuable only because they yielded knowledge of techniques needed to make astronomical calculations—calculations used to devise calendars for predicting the arrival of the monsoons, determining religious

---

[16] Johannes Bronkhorst, "Panini and Euclid: Reflections on Indian Geometry," *Journal of Indian Philosophy* 29 (2001): 55–56.

holy days, and making charts for navigators and mariners. The scientist and historian of science Roddam Narasimha has described the Indian approach to mathematics as 'computational positivism'—mathematics is designed essentially to make calculations that would fit observation with increasing precision.[17] He sees the approach as largely inspired by the algorithmic orientation of Indian grammar—one less concerned with theories and models, and more focused upon fine-tuning algorithms to yield better predictions.[18]

However, it is important to note that such a process of transmission of the Kerala mathematical discoveries to Europe by craftsmen, had it occurred, is not likely to be the direct communication of the various infinite series discovered by Indian mathematicians. It is more likely to have been only a truncated version of these infinite series that involved a few initial terms. This follows from the way these discoveries were passed down to craftsmen within India in the form of a finite series involving a few initial terms (how many depending on the degree of precision required and the rapidity of convergence of the series). Moreover, it would have been likely that the mediating role between the Indian scholars and the Indian craftsmen would have been played by the royal house in Kerala. It was the custom of the Indian mathematical astronomers to present their results to royal houses which supported them, where it would have been adapted into finite series approximations before transmission to Indian craftsmen. It is these truncated series that might have been received by European counterparts of Indian craftsmen as they worked alongside them.[19]

Let us look at one example of how such a process could have occurred. Take the infinite series expansion for sine. It is generally assumed to have been discovered by Isaac Newton in 1670 and is written as follows:

$$\sin \theta = \theta - \frac{\theta^3}{3!} + \frac{\theta^5}{5!} - \frac{\theta^7}{7!} + \ldots \text{ (Newton Series)}$$

However, Madhava gave the equivalent of the same series centuries earlier as follows:

> Multiply the arc by the square of the arc, and repeat [any number of times]. Divide by the product of the square of the radius times the square

[17] Roddam Narasimha, "The Indian Half of Needham's Question: Some Thoughts on Axioms, Models, Algorithms and Computational Positivism," *Interdisciplinary Science Reviews* 28(1) (2003): 60.

[18] Narasimha, "The Indian Half of Needham's Question," 60–63.

[19] For more details see Joseph, *A Passage To Infinity*, 26–31.

of successive even numbers increased by that number [multiplication being repeated the same number of times]. Place the arc and the results so obtained one below the other and subtract each from the one above. These together give the *jiva* . . .[20]

Madhava's instructions for generating the infinite series for sine are given in terms of the *jiva* which is the Indian equivalent for sine, such that

$$jiva \equiv r\sin\theta$$

where r is the radius of a circle and θ, the angle subtended by an arc s, say.

Then Madhava's algorithm requires us to calculate the *jiva* as the sum of an infinite series. By the first sentence of the Madhava algorithm the numerator of the nth term t_n of the series is given by multiplying the arc s by its square s^2, this multiplication being repeated, say n times, to obtain $s^{2n} \cdot s$.

The second sentence of the Madhava algorithm asks us to divide the above numerator of the nth term by a denominator derived by multiplying the square of the radius r^2 by $(2^2 + 2)$ and then $(4^2 + 4)$, $(6^2 + 6)$ and so on ("square of successive even numbers increased by that number") repeating the process n times, and then multiplying the terms together. This gives the denominator as $(2^2 + 2) \cdot (4^2 + 4) \cdot \ldots [(2n)^2 + 2^n] \cdot r^{2n}$.

Then the nth term t_n becomes:

$$\frac{s^{2n} \cdot s}{(2^2 + 2) \cdot (4^2 + 4) \cdot \ldots [(2n)^2 + 2^n] \cdot r^{2n}}$$

From the third sentence of Madhava's algorithm we get the following terms:

$$(s - t_1),\ (t_2 - t_3),\ (t_4 - t_5), \ldots$$

The last sentence gives the *jiva* as the sum of the above terms:

$$jiva = (s - t_1) + (t_2 - t_3) + (t_4 - t_5) + \ldots$$

Substituting:

(1) $jiva \equiv r\sin\theta$
(2) $s = r\theta$, so that $s^{2n+1} / r^{2n} = r\theta^{2n+1}$, and since
(3) $[(2k)^2 + 2k] = 2k \cdot (2k + 1)$, it follows that
(4) $(2^2 + 2)(4^2 + 4) \ldots [(2n)^2 + 2n] = (2n + 1)!$.

[20] Quoted in C. K. Raju, "Computers, Mathematics Education, and the Alternative Epistemology of the Calculus in the Yuktibhasa," *Philosophy East & West* 51(3) (2001): p. 350.

By cancelling *r* from both sides, we are led to the Newton series for sine:

$$\sin \theta = \theta - \frac{\theta^3}{3!} + \frac{\theta^5}{5!} - \frac{\theta^7}{7!} + \ldots..$$ [21]

It is significant that Madhava's instructions for generating the infinite series are given as iterative rules. They allow the series to be prolonged indefinitely to any number of terms we desire by repeated application of the rules. Moreover, succeeding terms are smaller than previous ones, and the value of the sum of the series converges. Hence, for practical purposes, craftsmen concerned with only approximate values to a desired level of accuracy may take only a few initial terms of the series, how many depending on the desired approximation, to calculate sine values. Moreover, the terms of the Madhava and Newton series would have been identical but for the fact that the Indian *jiva* and Western sine are different, since *jiva* = $r\sin\theta$. Consequently, knowledge of the Madhava series approximations used by Indian craftsmen could easily have been adapted to Western contexts. The new truncated series would then have arrived in Europe as a rule-of-thumb used by European craftsmen involving only a few initial terms. It could then have served as an inspiration for European mathematicians to construct an infinite series expansion for sine very similar to what had been discovered for *jiva* by their Indian counterparts.

If so, the most important reason for European interest in these Indian computation techniques would have been their need for more precise astronomical predictions needed to solve problems associated with long-distance navigation and the calendrical sciences. According to Joseph, the Jesuits were very concerned with augmenting their knowledge of the Indian sciences by including *jyotisa* (astronomy/astrology) in the curriculum of the Jesuit colleges they set up on the Malabar Coast. He argues that even Matteo Ricci, who lived for four years in India before proceeding to China, was very interested in the Indian calendar. Ricci is known to have written a letter to Maffei requesting the assistance of an "intelligent Brahmin or an honest Moor" to enhance his understanding of Indian ways of recording and measuring time. Also Jesuits working on the Malabar Coast had close relations with the royal court of Cochin, which played an important role in supporting and linking the scholars of the Kerala School with Indian craftsmen. Therefore, it is quite possible that the

---

[21] See Raju, "Computers, Mathematics Education, and the Alternative Epistemology of the Calculus in the Yuktibhasa," *Philosophy East & West* 51(3) (2001) for detailed discussion of this derivation and its implications.

Jesuits had access to the discoveries of the Kerala School through the royal household.[22]

This shows that the transmission of the computing techniques discovered by Indian scholars to their European counterparts could have been mediated by Indian and European craftsmen. The European craftsmen would then have adopted them from Indian cartographers, navigators and calendar makers as better computing techniques or rules-of-thumb for computing the values of trigonometric functions. We know that the most significant contribution of the Kerala School was moving from the finite procedures of ancient mathematics and treating their limit passage to infinity. However, for practical computation it is necessary to also move away from the infinite series representations to finite series approximations. These finite series transmitted to Europe could have inspired European mathematicians to subsequently reconstruct the original infinite series discovered by Indian scholars.

This may have been facilitated by the breakdown around 1550 in Europe of the sharp separation maintained in the preceding medieval period between the liberal and the mechanical arts. Following 1550 some learned European scholars developed an increasing interest in the mechanical arts because they had acquired economic significance as a result of their utility in navigation, cartography, mining, metallurgy, surveying, mechanics and gunnery. According to Zilsel it resulted in a loosening of the divide between scholars and superior craftsmen, so that the methods of the craftsmen came to increasingly inform the knowledge of scholars. With Galileo and Bacon this process had proceeded to the point where their theoretical perspectives came to be fully shaped by the mechanical orientations and interests of craftsmen. Indeed Zilsel conjectures that the physical laws of modern science could themselves be seen as rooted in rules-of-thumb originally adopted by craftsmen.[23]

Hence, it is quite possible that some of the rules-of-thumb European scholars learnt from European craftsmen could have been transmitted to these craftsmen by their Indian counterparts who themselves may have inherited their practices from Indian scholars. However, the European scholars would not have recognized their indebtedness to either Indian craftsmen or Indian scholars. This cannot be attributed simply

---

[22] Joseph, *Crest of the Peacock*, 440–443.

[23] Edgar Zilsel. "The Social Origins of Modern Science." Edited by D. Raven, W. Krohn and R. S. Cohen in *Boston Studies in the Philosophy of Science* v. 200. (Dordrecht, Netherlands: Kluwer Academic Publishers, 2003), 14–15.

to Eurocentric prejudice since the process of rediscovery is like to have precluded the possibility of such recognition. Firstly, European mathematicians would not have received the rules for generating the infinite series directly from their Indian counterparts. Instead they would have constructed it from a finite series used as a practical rule-of-thumb by European craftsmen. Secondly, the rule-of-thumb they began with would also have been modified to fit Western trigonometric functions that were different from their Indian equivalents. Hence, the influence from India would have been rendered invisible to European mathematicians, both because of the adaptations made by European craftsmen and the extensions made on these adaptations by European scholars.

Consequently, it is not surprising that European thinkers, even if they had been inspired by Indian ideas, would trace these new mathematical ideas to the ancient Greeks. The nearest equivalent to the "passage to infinity" known to them was the method of exhaustion developed by Leucippus, Democritus and Antiphon. This method had been put on a more rigorous basis by Eudoxus, culminating in the brilliant discoveries of Archimedes. Archimedes had used the method of exhaustion to calculate the areas and volumes of various geometric figures by summing them up as part of an infinite series. He deployed the method to find an approximation for the area of a circle that led to a value for π more accurate than any known before. He also used it to compute areas of ellipses, volumes and surface areas of spheres and cones, and volumes of segments of revolution of paraboloids and hyperboloids. Given that these Greek achievements were the closest approaches to the infinite series they discovered, early modern European mathematicians would naturally have come to treat them as the *only* pioneering achievements upon which their new discoveries were made.

Some Circumstantial Evidence for Transmission

However, this historical consciousness may be false because there exists strong circumstantial evidence for a dialogical exchange between India and Europe that transformed mathematical traditions in both these areas. These include significant epistemological and methodological breaks in the mathematical traditions of India and Europe in the 16th and 17th centuries. Let us first look at the changes in the Indian tradition. What we find is that, following the arrival of the Portuguese in India, the discoveries of the Kerala School came to be written in a form that broke both with the traditional language of communication and epistemological style of presentation in India. This radical change can be seen in the *Yuktibhasa* of Jyesthadeva. He is one of the major figures in the lineage of the Kerala School and was born in 1500, not only immediately after the arrival of the Portuguese, but also in the same area where they set up Jesuit schools. He wrote the *Yuktibhasa* about 1550, in which he summarized the accomplishments of the Kerala School. However, he did not write the work in Sanskrit, as his teachers in the past had done, but in Malayalam, the regional language of Kerala. Moreover, unlike earlier Indian writers, who only gave the mathematical discoveries and techniques without proofs, Jyesthadeva offers detailed demonstrations of the theorems and algorithmic rules he presents in the *Yuktibhasa*.[24]

These changes suggest that Jyesthadeva must have been addressing an audience quite different from that of his predecessors. His predecessors addressed their work to pupils who could read Sanskrit, and with whom they worked closely so that the proofs of theorems could be given informally, since results were what were important. It is likely that his new audience was not conversant with Sanskrit but only Malayalam—otherwise Jyesthadeva would not have expended the effort of translating mathematical ideas from one linguistic medium into another. Secondly, it is also likely that his new audience expected not only to be given the mathematical results of the Kerala School, but also demanded proofs for them. Hence, his work cannot have been written to interest calendar makers, navigators or craftsmen in the region, since such people would not be interested in proofs. If he was not writing for his fellow scholars or local craftsmen who else could Jyesthadeva have been addressing? It is possible to suspect that his audience must have been those studying astronomy in the missionary schools of the Portuguese. Having been educated in the traditions of Western mathematics these students would not have accepted the results of the Kerala School at face value without proof. Moreover, these students would have no knowledge of Sanskrit. Previous Indian mathematicians may have shown the proofs to their disciples without including it in their texts, but such a method could not be adopted with the students in the Jesuit schools.

---

[24] Sarma, K. V., K. Ramasubramaniam, M. D. Srinivas and M. S. Sriram, (tanslators), *Ganita-Yukti-Bhasa* (*Rationales in Mathematical Astronomy*) *of Jyesthadeva: Volumes I and II*, (New York. Springer, 2009).

Why would Jyesthadeva want to show the achievements of the Kerala School to the Indian students in the Jesuit schools? It is likely that the competition initiated by the alternative astronomical tradition introduced by the Portuguese was sufficient motivation for him. Hence, the dramatic reorientation of language and epistemology in the presentation of the Indian discoveries of the Kerala School by Jyesthadeva can be seen as the direct outcome of a dialogical encounter between European and Indian traditions of mathematical astronomy.

This encounter would also have been a serious concern for the newly converted Catholic students in the Jesuit schools. This is because the ritual calendars the Jesuits brought with them to India had celebrations of feasts and holy days closely linked to the agricultural cycles and seasons adapted to the European context. By contrast the dates for such rituals in India, say concerning harvest times and the timing of the monsoon seasons, were more fittingly defined by the Indian calendars, normally given in the traditional *panchanga*. In adapting to the Indian context, the Jesuits permitted the Church calendar to be contoured to "the indigenous social world of the Catholics with its patterns of material production, indigenous ideas and ritual practices."[25]

A parallel methodological reorientation within the European tradition also suggests that Indian influences may have entered Europe. As we have seen in their approach to problems associated with infinite sums Europeans began to deploy the method of direct rectification used earlier by the Indians rather than the ancient Greek method of exhaustion so effectively developed by Archimedes.

Such methodological and epistemological reorientations in mathematics in both India and Europe in the early modern era furnish circumstantial evidence of mutual influence. It would also explain why many of the mathematicians who made seminal contributions to the emergence of modern calculus, such as Fermat, Roberval and Cavalieri, emerged within five years of each other at the turn of the 17th century, and five decades after the publication of Jyesthadeva's *Yuktibhasa* in India.

Thus, it is quite possible that the discoveries of the Kerala School may have reached Europe as a body of approximate series for trigonometric and circular functions. These could have inspired and enabled European mathematicians to reconstruct the infinite series from which they had

[25] See Rowena Robinson. "Two Ritual Calendars in Southern Goa," In Lotika Varadarajan (ed.), *Indo-Portuguese Encounters: Journeys in Science, Technology and Culture*, (New Delhi: Indian National Science Academy, 2006), 555–556.

been abstracted. However, their rational basis in Indian mathematics would have been known to neither the craftsmen who carried the approximate series into Europe, nor the scholars who reconstructed the infinite series in Europe.[26] The European mathematicians would see their work as extending the method of exhaustion of the ancient Greeks—the only method that they knew involving a "passage to infinity" that antedated their work.

Establishing Transmission Mediated by Craftsmen

However, the above discussion has only shown strong circumstantial evidence for the transmission of the mathematical discoveries of the Kerala School to Europe, but has not conclusively established it. Of course, the transmission of these discoveries as know-how and computation techniques through craftsmen and technicians would also explain the failure to find documentary evidence of transmission in Jesuit communications. Indeed we know that many other transmissions also took place without scholarly mediation. Consider the case of the origins of gunpowder, printing and compass which puzzled Bacon near the dawn of the modern era:

> It is well to observe the force and virtue and consequences of discoveries. They are to be seen nowhere more conspicuously that in those three which were unknown to the ancients, and of which the origin, though recent, is obscure and inglorious; namely printing, gunpowder, and the magnet. For these three have changed the whole face and state of things throughout the world, the first in literature, the second in warfare, the third in navigation; whence have followed innumerable changes; insomuch that no empire, no sect, no star, seems to have exerted greater power and influence in human affairs than these mechanical discoveries.[27]

We now know how these transmissions were originated in China and came to Europe largely through the corridor of communication created by the Mongolian empire.

---

26 Moreover, the Portuguese state saw all knowledge of maps and map-making, as well as navigation and calendar techniques, as secrets of state that were prohibited from being widely disseminated. This would also have precluded Jesuits from documenting this knowledge in the letters sent back to Europe which were intended for wide circulation. Hence, even if the Jesuits participated in communicating the know-how of Indian techniques to Portuguese craftsmen they travelled with, they would not transmit this knowledge in documented form back to Europe.

27 Quoted in Joseph Needham, *Science and Civilization in China.* Vol. 1 (Cambridge: Cambridge University Press, 1954), 19.

But even the discovery of documentary evidence in European 16th century manuals used for navigation, map-making and calendar construction, involving the use of approximate series analogous to the Kerala School discoveries, would not conclusively establish transmission. The Europeans could have made these discoveries independently. Hence, it may be concluded that unless we have direct translations of Indian texts, or direct acknowledgment of Indian sources for these discoveries, we cannot conclusively establish transmission.

However, such a conclusion becomes questionable if we look at a parallel case of another transmission of Indian mathematical discoveries into Europe—the Indian number system. What is the evidence that the number system was transmitted to Europe rather than independently rediscovered there? It is possible that the Europeans arrived at the decimal-place number system with zero on their own, since many of its component elements were known to them. The use of decimal numbers was already well-established with the Greeks, and the place-value system was known to the Babylonians. Only the invention of zero was needed, and that too had been discovered by the Mayans without contact with the Indians. Hence the independent discovery of zero was all that was needed for the Europeans to rediscover the number system after the Indians. Therefore, it might appear that nothing precludes the possibility of parallel discoveries by Europeans and Indians of the currently prevalent number system.

But this is not the case. Convincing evidence for transmission can be found in the striking similarities in the signs for numbers used by both Europeans and Indians. Since the choice of number signs is based on purely arbitrary and conventional decisions, and identical numbers can be represented by quite different number signs, we have to conclude that the parallels in the signs for numbers in Europe and India can only be the outcome of transmission. These "accidental" parallels in number science cannot easily be explained away as independent achievements since they could have been different without affecting the efficiency of the number system. Hence, it is the accidental similarities between the two number systems in India and Europe which furnished the strongest circumstantial evidence for transmission, and not the crucial similarities that made the number system function so effectively.[28]

[28] However, we now know that Indian numerals came to be used in Alexandria sometime in the 5th century AD before entering Europe and that they did not get transmitted through scholarly roots, but through traders in harbors and ports. See Karl Menninger,

Can one use the above insight to establish the transmission of the discoveries of the Kerala School to Europe? To do so we raise the following questions: Were the earliest infinite series for π, sin$x$, cos$x$, and tan^{-1}x discovered in Europe similar to those known in India? Are there other series possible which were, in fact, not discovered first? If so, would it be reasonable to suppose that it was the approximate series expansions of π and various trigonometric functions discovered in India which inspired the infinite series discoveries in Europe? I leave these as open questions for discussion awaiting answers.

Broader Implications

The approach I am proposing is closely linked to the Zilsel thesis that sets out to explain why modern science emerged in Western Europe in the 17th century but not elsewhere. Edgar Zilsel attempts to answer what has since been labeled the Needham Question as follows: the emergence of early capitalism weakened the belief in authority that had previously kept apart academically trained scholars who saw themselves as rational thinkers from what Zilsel labels "superior craftsmen." The academics saw the craftsmen as possessing experimental skills but lacking the more theoretical knowledge that came with methodological intellectual training. By contrast, academics despite their conceptual advantages, lacked practical skills. Thus there developed a confluence of scholarly theoretical knowledge and craft-based empirical practices. Using the exemplar of William Gilbert as a case study, Zilsel details how Gilbert's study of magnetism was nurtured by the work of artisans such as Robert Norman, a navigator and compass maker.[29]

However, Zilsel does not explain why the contributions of artisanal knowledge has often been overlooked. In her recent study *The Body of the Artisan: Art and Experience in the Scientific Revolution*, the historian of science Pamela Smith argues that, although what she calls "artisanal epistemology" shaped science in the 16th century, natural philosophers in

Number Words and Number Symbols: A Cultural History of Numbers (Cambridge, Mass.: MIT Press, 1970), 406.

29 Edgar Zilsel, *The Social Origins of Modern Science*, eds. D. Raven, W. Krohn and R. S. Cohen in *Boston Studies in the Philosophy of Science* v. 200. (Dordrecht, Netherlands: Kluwer Academic Publishers, 2003), 71–95.

the late 17th century came to distance themselves from acknowledging its influence. She writes:

> In the late sixteenth and early seventeenth centuries, a type of person emerged who began to call himself a "new philosopher" and to declare himself a knower of nature. This represents a process of transition by which the artisanal epistemology was incorporated into a new discourse about nature. But in the late seventeenth century, the new natural philosophers expressed an ambivalence toward the role of the body and the senses that came to a head during the institutionalization of the new philosophy into the university curriculum. These late-seventeenth-century new philosophers were unsettled by the involvement of the body in cognition, and they sought to control the bodily dimension of empiricism at the same time that they began to distance themselves from artisans and practitioners. This is the trajectory I trace over the course of *The Body of the Artisan*.[30]

The trajectory Smith traces can explain why histories of science have often ignored the influence of craft traditions on early modern science. It may also have masked the transmission of scholarly discoveries from India to Europe mediated by craftsmen. If so the scholar-craftsmen dialogue may be more complicated than Zilsel had assumed. The meeting of craftsmen with scholars may have been particularly fruitful because European craftsmen had new knowledge they had acquired from different parts of the world as a result of their voyages of geographical discovery. Consider other discoveries in Europe that simply happened to occur after the opening of the corridor to China which paralleled discoveries already known to the Chinese. The equatorial mounting for telescopes used by Tycho Brahe, and now the standard mounting for such instruments, was the same mounting used by Chinese astronomers. The equal angle projection introduced in Europe by Mercator in 1548, as well as the use of equatorial coordinates in astronomy adopted by sixteenth century European astronomers, was known to the Chinese since the time of Zhang Heng in the second century.[31]

One could also point to other discoveries made in Europe after contact with China, such as the circulation of blood discovered by Harvey, and

[30] Pamela H. Smith, *The Body of the Artisan: Art and Experience in the Scientific Revolution* (Chicago: The University of Chicago Press, 2004), 20.

[31] For a more detailed discussion of these Chinese discoveries see Colin Ronan, *The Cambridge Illustrated History of the World's Science* (Cambridge: Cambridge University Press, 1983), especially Chapter 2 on Chinese Science. See also Arun Bala, *The Dialogue of Civilizations in the Birth of Modern Science* (New York: Palgrave Macmillan, 2006), 143–144.

properties of magnets discovered by Gilbert. What is important to note is that all these discoveries could have been easily transmitted from China, and they constitute the sort of knowledge that could be acquired by a European captain of a ship working on maps and stellar coordinates; a doctor on a ship listening to his Chinese counterpart explaining an illness as due to poor circulation of blood; a ship's guide speaking of the properties of a magnet and the lines of geomancy it followed, and so on. All these furnish circumstantial evidence that the Portuguese voyages of geographical discovery could also have been voyages of intellectual discovery.

ANTHOLOGIZING THE BOOK OF NATURE: THE ORIGINS OF THE SCIENTIFIC JOURNAL AND CIRCULATION OF KNOWLEDGE IN LATE GEORGIAN BRITAIN

Jonathan R. Topham*

Writing in the preface to a new monthly journal of science in 1813, the Scottish chemist Thomas Thomson observed that the "superiority of the moderns over the ancients" consisted "not so much in the extent of their knowledge [...] as in the degree of its diffusion".[1] This advance in the circulation of knowledge, he averred, was to a significant extent a consequence of the inception of moveable-type printing. More especially, it had been promoted by the periodical publications which existed in such profusion in Britain, France, and Germany, and most particularly by the new kinds of commercially produced "philosophical" journals that had emerged during the last quarter of the eighteenth century and began to be called "scientific" journals from the turn of the century. As modern historians have shown, these pioneering journals, such as the botanist and agriculturalist François Rozier's *Observations sur la physique* (f. 1771), were explicitly motivated by a desire to overcome the inadequacies of learned transactions in conveying the current state of scientific knowledge. Learned society publications were too slow to appear, Rozier argued, often too limited in their readership because written in the vernacular, and too extensive and expensive for many to acquire. New journals such as Rozier's aimed to bring together all findings within their respective domains, from across the Western world, the better to promote the progress of science.[2] It was not until the 1790s that such commercial scientific journals began to be produced in Britain (although, as we shall see, they were preceded by

* I am grateful to Geoffrey Cantor and to the editors of this volume for their helpful comments on earlier drafts of this chapter.

1 [Thomas Thomson], "Preface," *Annals of Philosophy* 1 (1813): [1].

2 James E. McClellan III, "The Scientific Press in Transition: Rozier's Journal and the Scientific Societies in the 1770s," *Annals of Science* 36 (1979): 425–49, esp. 433. See also Thomas Broman, "Periodical Literature," in *Books and the Sciences in History*, ed. Marina Frasca-Spada and Nicholas Jardine (Cambridge: Cambridge University Press, 2000), 225–38, and Maurice Crosland, *In the Shadow of Lavoisier: The "Annales de Chimie" and the Establishment of a New Science*, British Society for the History of Science Monographs, 9 ([n.l.]: British Society for the History of Science, 1994).

several commercial periodicals of medicine and of agriculture and the arts). The new British journals, Thomson claimed, likewise served to advance science not only by facilitating "the publication of useful discoveries" but also by circulating "the valuable dissertations of foreigners through Britain, which might otherwise remain in great measure unknown to us".[3] Taking this claim as my starting point, my object in this essay is to examine the motivation behind the foundation of the eight scientific journals established in Britain between 1793 and 1806, examining them in relation to the theme of the international circulation and appropriation of knowledge. In particular, I will argue that these pioneering journals of science were initially conceived of as anthologies, intended to draw together in an affordable and manageable form the scientific findings of the entire Western world.

The inception of commercial scientific journals at the end of the eighteenth century is one of the most iconic developments of this transformational period in the 'invention' of modern science. Indeed, scientific journals are so distinctive of modern science that they seem to be a necessary part of it. This perception has, however, stunted the historical study of their development. Since scientific journals are such a necessary part of modern science, why should their origin require explanation? Or why, indeed, should we expect scientific journals in the past to be different from those at the present day? Of course, merely to make these questions explicit is to show them to be naïve. As James Secord has recently argued, the later dominance of the scientific paper should not blind us to the extent to which both the scientific paper and the scientific journal were undergoing complex processes of development in the first half of the nineteenth century. Although the Royal Society's *Catalogue of Scientific Papers*, first published in the 1860s, retrospectively canonized contributions to early scientific journals such as William Nicholson's *Journal of Natural Philosophy* (f. 1797) and Alexander Tilloch's *Philosophical Magazine* (f. 1798) as being of a piece with the scientific papers of the newly professionalizing scientific specialties of that later period, this significantly distorts the historical record.[4] It was far from clear in the 1790s either that the primary purpose of a scientific journal should be to provide opportunities for authors to stake claims to scientific discovery or that having an

[3] [Thomson], "Preface," 3 and 4.

[4] Royal Society of London, *Catalogue of Scientific Papers (1800–1863)*, 6 vols (London: Royal Society of London, 1867–72).

account printed in a scientific journal was the appropriate way to stake such claims. Conversations in society meetings and other settings, correspondence, lengthy treatises—all were means of staking credible claims to knowledge. Only as a result of subsequent developments did the scientific periodical come, later in the nineteenth century, to take on this key role in science.[5]

My object in this chapter, then, is to re-expose the uncertain beginnings of commercial scientific journals in Britain, focusing on the distinctive way in which they were motivated or justified, not as the means of staking claims to scientific priority, but as international anthologies of scientific discovery, drawing together materials from a wide range of printed sources, as well as from original contributors. As I have argued elsewhere, these new journals constituted a significant element in a complex communication network by which Continental science came to circulate in Britain.[6] In this chapter, I restrict my attention to considering how, within the particular circumstances of the late eighteenth-century British book trade, such circulatory practices came to seem attractive to a number of individuals. The chapter is divided into two main sections. I begin by looking at the communicative context of the sciences in late-eighteenth-century Britain, asking what kinds of print media were available and how they compared to those on the Continent. This section draws on Thomas Thomson's retrospective review of scientific journalism, both to expose the continuities between the new journals and what went before, and to point out the ways in which events in Britain differed from those in France and Germany. The next section looks at the eight new scientific journals founded in Britain between 1793 and 1809. I show that these were introduced by somewhat marginal, often practically oriented men, and were to a significant extent motivated by financial considerations. Moreover, I argue that they were explicitly justified by a desire to present an anthology of discovery, drawn from a wide range of expensive publications, which would constitute a commercially attractive product to cash-strapped readers. By concentrating and making available the findings of

---

5 James A. Secord, "Science, Technology and Mathematics," in *The Cambridge History of the Book in Britain*, Vol. 4, *1830–1914*, ed. David McKitterick (Cambridge: Cambridge University Press, 2009), 451–59. See also Alex Csiszar, "Broken Pieces of Fact: The Scientific Periodical and the Politics of Search in Nineteenth-Century France and Britain," PhD diss., Harvard University, 2010.

6 Jonathan R. Topham, "Science, Print, and Crossing Borders: Importing French Science Books into Britain, 1789–1815," in *Geographies of Nineteenth Century Science*, ed. David N. Livingstone and Charles Withers (Chicago: University of Chicago Press, 2011), 311–44.

the learned of all Western nations, editors hoped to produce an ongoing encyclopaedia, so much more attractive for purchase because it was constantly updated. Finally, in my concluding remarks, I briefly consider the importance of such practices for the circulation of science in Britain and beyond, urging the need for more systematic study of these sources as providing valuable evidence of the processes of international interchange and encounter.

Periodicals and Science in Late Eighteenth-Century Britain

We began with Thomas Thomson's history of the scientific journal, and it is helpful to continue with it, since it provides a revealing contemporary perspective. Before the seventeenth century, Thomson reported, the learned had been obliged to rely on "epistolatory correspondence", which had many inadequacies. He continued:

> The first periodical work of science which made its appearance in Britain was the *Philosophical Transactions*, begun in 1665, and continued for many years, in numbers, published monthly, quarterly, or annually, as materials were more or less copiously supplied. The *Journal des Sçavans* in Paris, and the *Leipsic Acts* in Germany, were somewhat similar in their plan. About the middle of the 18th century the Philosophical Transactions altered their form, and came to be published only in volumes. From that period they have consisted entirely of original papers, and have taken no notice of the discoveries made by foreigners, nor of the scientific books which have made their appearance in different countries. Thus Britain no longer possessed a periodical philosophical journal. The *Monthly and Critical Reviews* indeed had commenced, and were conducted with considerable spirit; but being entirely confined to criticisms on books, they could scarcely be considered as registers of the discoveries of science. Perhaps the *Gentleman's Magazine*, which [...] contained a great deal of philosophical as well as miscellaneous information, would have been entitled to rank as a philosophical journal, had not the greater number of its columns been filled with articles of belles lettres and antiquarian research.[7]

Thomson considered that the new specialist "philosophical journals"—including his own in addition to Nicholson's *Journal* and the *Philosophical Magazine*—served to advance science by facilitating "the publication of useful discoveries" and by circulating "the valuable dissertations of

[7] [Thomson], "Preface," 1–2.

foreigners through Britain, which might otherwise remain in great measure unknown to us".[8]

The teleological rhetoric of Thomson's history was clearly intended to accentuate the novelty of the new specialist journals. However, the new journals had more in common with existing periodicals than this outline suggests. Certainly, the *Philosophical Transactions*, which in the second half of the eighteenth century were issued twice a year, did not typically contain extracts, abstracts, and reviews of foreign scientific publications. Nevertheless, Britain did possess a vigorous and expanding periodical press, which provided readers with ready access to a wide range of scientific findings. As Thomson identifies, the monthly review journals, of which the *Monthly* (f. 1749) and the *Critical* (f. 1756) were two of the most prominent and long-lived, provided extensive summaries of and critical commentaries on new books, both British and foreign. Moreover, contrary to Thomson's claim, such publications often provided reviews of other periodicals. The *Philosophical Transactions*, for instance, were routinely reviewed by the *Monthly* and *Critical*, usually with a list of all contributions, and often with lengthy synopses of those considered particularly interesting. This was later extended to include the transactions of the new learned societies formed outside of London or dealing with specialist subjects. The transactions of Continental societies were not so regularly reviewed, but they were nevertheless often given the same synoptic treatment, as were the increasing numbers of specialized commercial periodicals which, as we shall see, began to be produced both in Britain and on the Continent.

While the reviews provided regular synopses of scientific publications, both British and foreign, the magazines—of which the *Gentleman's Magazine* (f. 1731) was the archetype—also functioned as repositories of scientific knowledge, combining occasional extracts from, or accounts of, scientific publications with readers' own scientific observations. The *Gentleman's* innovative application of the term 'magazine' (that is, 'storehouse') to a periodical publication was intended to convey a sense of its encyclopaedic ambitions. Such magazines reflected an Enlightenment conceit that members of the educated public could contribute to the amassing of observations and experiments and assist in the progress of knowledge. Contributions ranged from natural history to the practical arts and from meteorology to agriculture, and, as Roy Porter has shown,

8 [Thomson], "Preface," 3–4.

the many medical contributions often came from highly respected practitioners (fig. 1).[9] On its semi-centenary in 1782, the *Gentleman's Magazine* claimed: "There has scarce a new Subject been started, a new Invention introduced, or a Discovery of any Kind, either by Land or Sea, of which a satisfactory Account is not to be found in the Gentleman's Magazine".[10] Similarly, in 1785, the *European Magazine* (f. 1782) began a regular section providing "an account of the discoveries and improvements which are daily made in philosophy, chemistry, astronomy, mechanics, &c. [...] principally extracted from the Transactions of the various learned Societies, from other foreign periodical publications, and from the communication of our correspondents."[11] Yet, as Thomson correctly identified, these magazines' purview was general rather than special.[12]

While the reviews and magazines of the later eighteenth century provided substantial scope for the communication of scientific findings, the last quarter of the century witnessed the gradual emergence of new kinds of more specialized periodicals. In part, this can be seen as a consequence of the newly competitive conditions of the book-trade in these years. Prior to 1774, the London booksellers who dominated British publishing had successfully prosecuted their claims that perpetual copyright was enshrined in common law, and they had combined together to secure control over books that were often decades or even centuries old. The House of Lords ruling against perpetual copyright in 1774 overthrew the certainties of this monopoly, so that booksellers had now to find new products to appeal to new markets, including novels, anthologies, pocket-books, and magazines. In such a market, magazines had a particular attraction. The amount of investment required to produce a 48-page octavo magazine issue was much smaller than that required for the average book, and the bookseller

[9] Roy Porter, "Lay Medical Knowledge in the Eighteenth Century: The Evidence of the *Gentleman's Magazine*," *Medical History* 29 (1985): 138–68, and "Laymen, Doctors and Medical Knowledge in the Eighteenth Century: The Evidence of the *Gentleman's Magazine*," in *Patients and Practitioners: Lay Perceptions of Medicine in Pre-industrial Society*, ed. Roy Porter (Cambridge: Cambridge University Press, 1986), 283–314.

[10] Anon., "[Advertisement]," *Gentleman's Magazine* 52 (1782): ii.

[11] *European Magazine* 13 (1785): 409. The *Annual Register* (f. 1758) similarly included a regular section of 'Useful Projects', as well as a section of 'Natural History'. The role of such publications in the development and communication of science in the eighteenth century clearly warrants concerted attention.

[12] On the subsequent transformation of such Enlightenment miscellanies into recognizably and self-consciously 'literary' magazines in the years following the Napoleonic wars, see Geoffrey Cantor et al., *Science in the Nineteenth-Century Periodical: Reading the Magazine of Nature* (Cambridge: Cambridge University Press, 2004), 10–13.

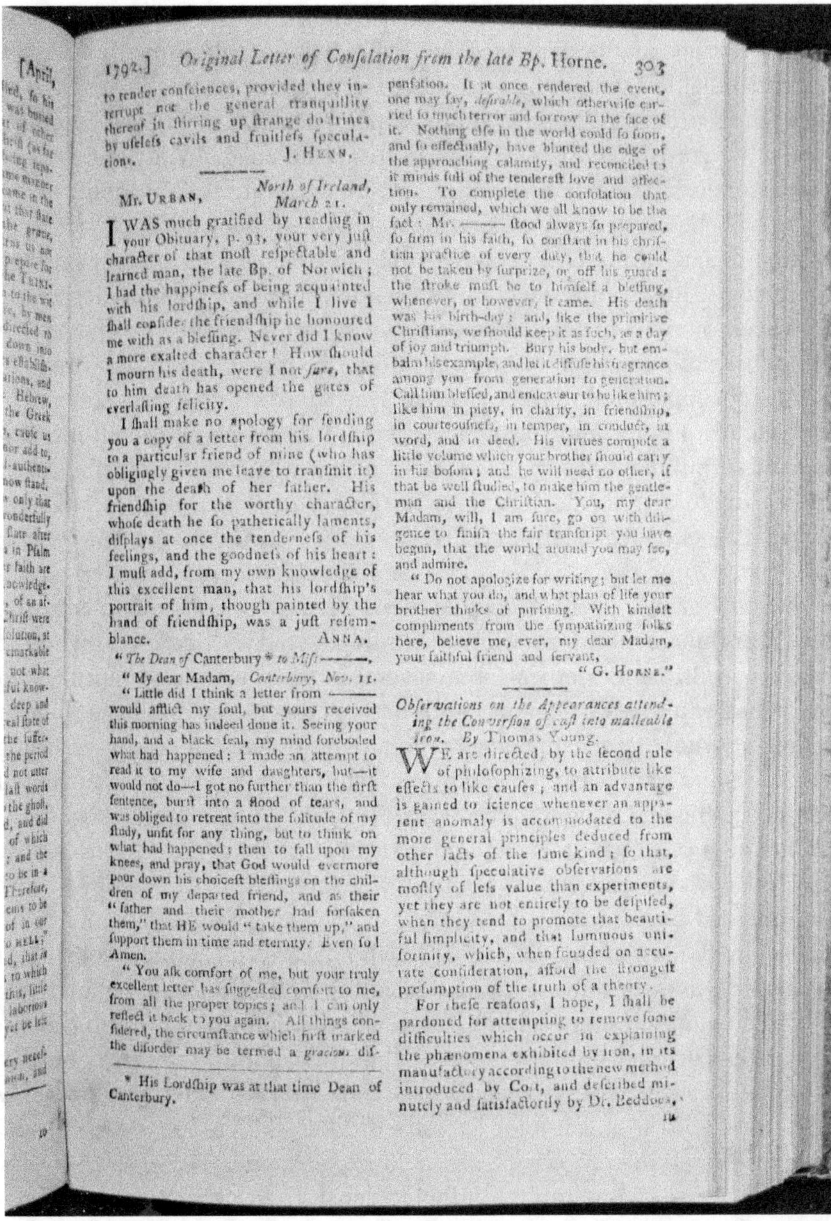

1792.] *Original Letter of Conſolation from the late Bp.* Horne. 303

to tender conſciences, provided they interrupt not the general tranquillity thereof in ſtirring up ſtrange doctrines by uſeleſs cavils and fruitleſs ſpeculations. J. Henn.

Mr. Urban, *North of Ireland, March* 21.

I WAS much gratified by reading in your Obituary, p. 93, your very juſt character of that moſt reſpectable and learned man, the late Bp. of Norwich; I had the happineſs of being acquainted with his lordſhip, and while I live I ſhall conſider the friendſhip he honoured me with as a bleſſing. Never did I know a more exalted character! How ſhould I mourn his death, were I not *ſure*, that to him death has opened the gates of everlaſting felicity.

I ſhall make no apology for ſending you a copy of a letter from his lordſhip to a particular friend of mine (who has obligingly given me leave to tranſmit it) upon the death of her father. His friendſhip for the worthy character, whoſe death he ſo pathetically laments, diſplays at once the tenderneſs of his feelings, and the goodneſs of his heart: I muſt add, from my own knowledge of this excellent man, that his lordſhip's portrait of him, though painted by the hand of friendſhip, was a juſt reſemblance. Anna.

"*The Dean of* Canterbury* *to Miſs* ———.

"My dear Madam, *Canterbury, Nov.* 11.

"Little did I think a letter from ——— would afflict my ſoul, but yours received this morning has indeed done it. Seeing your hand, and a black ſeal, my mind foreboded what had happened: I made an attempt to read it to my wife and daughters, but—it would not do—I got no further than the firſt ſentence, burſt into a flood of tears, and was obliged to retreat into the ſolitude of my ſtudy, unfit for any thing, but to think on what had happened; then to fall upon my knees, and pray, that God would evermore pour down his choiceſt bleſſings on the children of my departed friend, and as their "father and their mother had forſaken them," that HE would "take them up," and ſupport them in time and eternity. Even ſo! Amen.

"You aſk comfort of me, but your truly excellent letter has ſuggeſted comfort to me, from all the proper topics; and I can only reflect it back to you again. All things conſidered, the circumſtance which firſt marked the diſorder may be termed a *gracious* diſpenſation. It at once rendered the event, one may ſay, *deſirable*, which otherwiſe carried ſo much terror and ſorrow in the face of it. Nothing elſe in the world could ſo ſoon, and ſo effectually, have blunted the edge of the approaching calamity, and reconciled to it minds full of the tendereſt love and affection. To complete the conſolation that only remained, which we all know to be the fact: Mr. ——— ſtood always ſo prepared, ſo firm in his faith, ſo conſtant in his chriſtian practice of every duty, that he could not be taken by ſurprize, or off his guard: the ſtroke muſt be to himſelf a bleſſing, whenever, or however, it came. His death was his birth-day: and, like the primitive Chriſtians, we ſhould keep it as ſuch, as a day of joy and triumph. Bury his body, but embalm his example, and let it diffuſe his fragrance among you from generation to generation. Call him bleſſed, and endeavour to be like him; like him in piety, in charity, in friendſhip, in courteouſneſs, in temper, in conduct, in word, and in deed. His virtues compoſe a little volume which your brother ſhould carry in his boſom; and he will need no other, if that be well ſtudied, to make him the gentleman and the Chriſtian. You, my dear Madam, will, I am ſure, go on with diligence to finiſh the fair tranſcript you have begun, that the world around you may ſee, and admire.

"Do not apologize for writing; but let me hear what you do, and what plan of life your brother thinks of purſuing. With kindeſt compliments from the ſympathizing folks here, believe me, ever, my dear Madam, your faithful friend and ſervant,

"G. Horne."

* His Lordſhip was at that time Dean of Canterbury.

Obſervations on the Appearances attending the Converſion of caſt into malleable Iron. By Thomas Young.

WE are directed by the ſecond rule of philoſophizing, to attribute like effects to like cauſes; and an advantage is gained to ſcience whenever an apparent anomaly is accommodated to the more general principles deduced from other facts of the ſame kind; ſo that, although ſpeculative obſervations are moſtly of leſs value than experiments, yet they are not entirely to be deſpiſed, when they tend to promote that beautiful ſimplicity, and that luminous uniformity, which, when founded on accurate conſideration, afford the ſtrongeſt preſumption of the truth of a theory.

For theſe reaſons, I hope, I ſhall be pardoned for attempting to remove ſome difficulties which occur in explaining the phænomena exhibited by iron, in its manufactory according to the new method introduced by Cort, and deſcribed minutely and ſatisfactorily by Dr. Beddoes,

Fig. 1 A page from the *Gentleman's Magazine* for 1792, containing "Observations on the Appearances attending on the Conversion of cast into malleable Iron" by the eighteen-year old medical student and later renowned natural philosopher, Thomas Young (1773–1829). Reproduced from the *Gentleman's Magazine* 62 (1792): 303, by kind permission of Leeds University Library.

could moderate the number of copies produced in line with sales. Moreover, if a magazine proved successful, the ongoing sales provided a regular and reliable income. In fact, however, the earliest of the specialized magazines predated the 1774 ruling, a point which reflects the extent to which the growing market for print in this period had already encouraged entrepreneurial speculation, both in Britain and on the Continent.

Two of the earliest specialized periodical markets were for magazines related to agriculture and the arts (from the 1760s) and to medicine (from the 1770s)—literary products aimed at identifiable occupational groups. However, there were no analogous attempts to find or create a market for periodicals of natural philosophy or natural history, subjects for which no obvious occupational market existed. This situation contrasts somewhat with that in France and Germany, where, at the same period an increasingly large number of specialized scientific magazines appeared, including the agricultural reformer François Rozier's *Observations sur la physique* (1771), the chemical lecturer Lorenz Crell's *Chemisches Journal* (1778), Lavoisier's *Annales de chimie* (1789), and Crell's student, Friedrich Gren's *Journal der Physik* (1790). Yet while there were no British equivalents, the new Continental journals were not only treated to extensive synoptic reviews in British periodicals, but several of them were also republished in English editions. Prominent in this activity was radical publisher Joseph Johnson, whose circle included several prominent chemists, including Priestley, and who was involved in publishing a wide range of translated works.[13] Johnson issued a translation of the first volume of the *Annales de chimie* in 1791, which at 3*s.* 6*d.* undercut the imported French original by 2*s.* 6*d.* However, while the translation was intended to be ongoing, it did not proceed beyond the first volume. In the same year Robert Baldwin, the scion of a large book-trade family who became a leading publisher of scientific works, began to issue a translation of Crell's journal, which ran to three volumes before folding. Then, in 1795, Johnson also published a translation in English and French of the one volume of Swiss naturalist Johann Caspar Fuessli's *Archiv der Insectengeschichte*, a decision probably actuated by his friendship for Fuessli's brother, the painter Henry Fuseli, who translated it.

These translations imply a growing sense that a market might exist in Britain for specialist scientific magazines. At the same time, however,

[13] Gerald P. Tyson, *Joseph Johnson: A Liberal Publisher* (Iowa City: University of Iowa Press, 1979).

their rapid failure was not encouraging. Moreover, as Josiah Wedgwood's secretary and chemical assistant Alexander Chisholm observed, writing anonymously in the *Monthly Review* in 1792, while the translator of the *Annales de chimie* had promised "the speedy and *complete* communication of discoveries" in chemistry, the translation only told half the story; English readers also needed access to Rozier's journal, which emanated from the opposing party.[14] Merely translating Continental journals was unlikely to meet the needs of British science, and over the following decade and a half at least eight attempts were made to commence a British journal which, like its Continental counterparts, would draw together the most important new work within a manageable compass. The next section examines these eight journals in turn, drawing out from their rhetoric and practice the manner in which they were intended to anthologize scientific discovery.

Anthologizing the Book of Nature: The Commercial Origins of the Scientific Journal

Of the commercial "philosophical" journals commenced in Britain during the 1790s and 1800s, the only two to have attracted attention are those that survived the period and came to play a significant role in the history of science, namely, the *Journal of Natural Philosophy, Chemistry, and the Arts* and the *Philosophical Magazine*.[15] This section places these more famous journals in the context of others that failed or which came to be associated exclusively with the practical arts, highlighting the extent to which the several publications were begun with a similar sense that a scientific anthology would prove extremely useful to the book-buying public, and that this utility would ensure the new journals' commercial attractiveness. As we shall see, the journals were commenced by men somewhat at the margins of learned science, most of whom shared a background in the practical arts, and several of whom were dissenters. The new journals

14 [Alexander Chisholm], "The Annals of Chemistry, Vol. I," *Monthly Review* 2nd ser. 8 (1792): 169. On Chisholm's involvement in private translation activities in this period, see Larry Stewart, "Assistants to Enlightenment: William Lewis, Alexander Chisholm and Invisible Technicians in the Industrial Revolution," *Notes and Records of the Royal Society* 62 (2008): 17–29.

15 S. Lilley, "'Nicholson's Journal' (1797–1813)," *Annals of Science* 6 (1948–50): 78–101; W. H. Brock and A. J. Meadows, *The Lamp of Learning: Two Centuries of Publishing at Taylor & Francis* (London and Philadelphia: Taylor & Francis, 1998), ch. 4.

were also all motivated in part by the desire of their editors to secure a steady income from publications that were their own property. As the previous section indicates, these new "philosophical" journals of the 1790s drew on pre-existing journalistic practices in the general monthly magazines and reviews, in the specialist medical and agricultural journals of recent years, and in Continental scientific journalism. Yet the rapid success of some of them signalled that they had found a considerable market which not only yielded a significant income for several of the editors over the following two decades, but soon had some of the country's leading publishers interested.

John Aikin's Memoirs of Science and the Arts *(1793–94)*

The first of these new periodicals was also the most ambitious. Beginning in January 1793, the *Memoirs of Science and the Arts* promised to provide notices of every article in all the principal learned transactions throughout the world. Since by the editor's own estimate these amounted to more than fifty titles, it is not surprising that the project appears to have collapsed under the weight of its own ambition before a second volume was complete. The anthologizing ambition was spelt out in detail in the anonymous preface to the first volume. Relating the progress of the arts and sciences in modern times to the impetus given to individual research and publication by learned societies and their transactions, the preface considered that these had now become so extensive and so varied in their places and languages of publication, that the "private scholar" found it "an extremely difficult matter to obtain and peruse them", especially in the "detached part of Europe" that was Britain. By confining itself solely to learned transactions, the new journal was intended "to afford more complete information concerning them, than can be done in Journals which embrace the whole of literature". The only merit the *Memoirs* claimed for themselves was that of "utility".[16]

The *Memoirs* were in large part the production of John Aikin (1747–1822), a physician and rational dissenter, whose father had been the classics tutor at Warrington Academy.[17] Aikin had for over twenty years

16 Anon., "Preface," *Memoirs of Science and the Arts* 1 (1793): ii and iii.

17 The attribution to Aikin was made in a contemporary advertisement (*Oracle and Daily Advertiser*, 1 November 1800, p. 1c). Also, when reissued in 1797 they were described as being "by a member of the Royal College of Physicians", a description that applies to Aikin. See [John Aikin], *The Philosophical Transactions Abridged; or, Memoirs of Science and the Arts. From the Works of the Learned and Œconomical Societies of Europe, Asia, and*

combined medical practice with a wide range of literary productions—including not only works on medical biography and *materia medica*, but also a variety of books for children—almost all published by Joseph Johnson, of whose circle he was a significant member. His medical practice in Great Yarmouth having been irreparably damaged by his support for reformist causes following the French Revolution, he removed to London in 1792.[18] This brought him into closer contact with dissenting friends, including Johnson, his own sister Anna-Laetitia Barbauld in Hampstead, and old acquaintances from Warrington, including Joseph Priestley and Gilbert Wakefield, who were working at the new dissenting academy in Hackney. Indeed, Aikin became secretary of a Literary and Philosophical Society initiated by Priestley, which may have been the London Philosophical Society founded in 1794, or a separate body in Clapton.[19] It seems highly likely that it was in these circles of rational dissent that the plan for the new journal arose; Aikin told his friend John Haygarth that he was engaged in the plan "with a few literary persons", and that he was chiefly responsible for "the medical and natural history departments".[20]

That the new publication was not issued by Johnson may reflect the publisher's interest in the *Analytical Review*, a journal which contained extensive accounts of learned transactions, both British and foreign. Instead the *Memoirs* were published by Robert Faulder, John Egerton, and John Sewell (the publisher of the *European Magazine*)—all well-established booksellers—although Sewell was replaced by the reform-minded nonconformist Charles Dilly in the second volume. Like the transactions it

---

America (London: printed for the proprietors, and sold by James Wallis; Edinburgh: James Dickson; Dublin: J. Archer; Manchester: Clarke & Co.; Newcastle: Charnley & Son, and J. Bell; and Liverpool: Jones, [1797]).

18 Tyson, *Joseph Johnson*, p. 66; A[rthur?] A[ikin?], "Dr. John Aikin," *Gentleman's Magazine* 93 pt. 1 (1823): 85–89; and Lucy Aikin, *Memoir of John Aikin, M. D., with a Selection of His Miscellaneous Pieces, Biographical, Moral and Critical*, 2 vols (London: Baldwin, Cradock, and Joy, 1823).

19 Aikin, *Memoir*, 1: 152 and 163. Robert Schofield considers Aikin's epistolatory reference to such a society being instituted "here" as being likely to refer to the "very friendly and innocent club" Priestley reported he had "found" in Clapton. See Joseph Priestley, *The Present State of Europe Compared with Antient Prophecies; A Sermon, Preached at the Gravel Pit Meeting in Hackney, February 28, 1794, Being the Day Appointed for a General Fast* (London: J. Johnson, 1794), xiv; and Robert E. Schofield, *The Enlightened Joseph Priestley: A Study in His Life and Work from 1773 to 1804* (University Park, PA: Pennsylvanian State University Press, 2004), 305–06. However, Aikin's "here" very probably relates to London, suggesting that this might have been the Philosophical Society of London. See Ian Inkster, "Science and Society in the Metropolis: A Preliminary Examination of the Social and Institutional Context of the Askesian Society of London, 1796–1807," *Annals of Science* 34 (1977): 9.

20 Aikin, *Memoir*, 1: 155.

abridged, the *Memoirs* had a quarto format, possibly in part to accommodate its plates. The first number was apparently issued in January 1793, and the intention was to continue it in 2*s.* monthly numbers.[21] However, the preface issued with the first number warned of the current difficulty of obtaining Continental publications, and it is clear that the monthly schedule soon slipped. In December 1793, an 'Address to the Public' blamed the lack of regularity and the restricted content on the embattled state of the Continent, which had led to the suspension of scientific activity and difficulties in obtaining foreign publications. Indeed, of the twenty volumes of transactions abridged in the first year, only three were Continental publications. However, claiming that these difficulties had now been largely surmounted, the address promised a continuation in twice-yearly parts. The first of these, priced 10*s.* 6*d.*, appeared the following year, but no more were issued.[22] Aikin's daughter was unsure about the "causes of failure", but it seems clear that the intended comprehensiveness of the *Memoirs* was impractical on a number of grounds.[23] Moreover, sales had clearly not lived up to expectations, since the great remaindering bookseller James Lackington was soon offering half-price copies from his "Temple of the Muses" in Finsbury-Square, and in 1797 the remaindered copies were offered for sale under a different title.[24]

John Wyatt's Repertory of Arts and Manufactures *(1794–1862)*

Even before Aikin's journal ceased publication, a new monthly had begun with a similar purpose, but a rather more contracted sense of what would be "useful". The *Repertory of Arts and Manufactures* was commenced by another London-based émigré from the industrial heartlands of the Midlands. Its founder, John Wyatt (1752–1818), was the scion of a famous dynasty of Staffordshire inventors and architects, whose father developed some of the earliest textile machinery and later worked for Matthew Boulton. In the 1760s, both John and his brother were apprenticed to Boulton, who

---

21 Aikin, *Memoir*, 1: 92; *Analytical Review* 16 (1793): 529. In the copies I have examined, the first volume is separated into two parts, each with a separate titlepage; however, it is unclear at what stage the shift from monthly to biannual publication was made. The first volume sold for £1 1s. See [John Aikin], "[Review of *Memoirs of Science and the Arts*]," *Monthly Review* 2nd ser. 13 (1794): 238.

22 Anon., "Address to the Public," *Memoirs of Science and the Arts* 1 (1793): [i]; and *Analytical Review* 19 (1794): 499.

23 Aikin, *Memoir*, 156.

24 *Oracle and Daily Advertiser*, 3 October 1798, p. 1b; [Aikin], *Philosophical Transactions*; and *Universal Magazine* 101 (1797): 382.

was like an adoptive father to them. John continued to work for Boulton for a number of years, initially as a clerk, then as his London agent between 1776 and 1778, later executing commissions for him on the Continent. He was also involved with a number of mechanical projects, chairing the committee of mechanics of the Society of Arts for a spell, and later successfully projecting the iron bridge over the Thames at Southwark.[25] In April 1794, however, he published a prospectus for his new monthly magazine, to be published in octavo and priced 1*s.* 6*d.* Directed at "Artists and Manufacturers of every Class" as well as "Country Gentlemen and Farmers", the magazine was intended "to promote and extend the Knowledge of the useful Arts".[26] The first issue of the new magazine appeared in June 1794, and an advertisement published in the first volume declared that one of its "principal objects" was

> to establish a vehicle, by means of which new discoveries and improvements, in any of the useful Arts and Manufactures, may be transmitted to the public; particularly to Artists, Manufacturers, and others, who, from various circumstances frequently attending those discoveries and improvements, (such as their being announced in a bulky or expensive publication, or in a foreign language,) might otherwise have but little chance of ever becoming acquainted with them.

Thus, while the magazine was to contain official patent specifications and contributions from readers, it was also to contain materials derived from the transactions of learned societies. These would be selected on the grounds of their practical utility, excluding the "speculative or other matter" foreign to the purpose of the *Repertory of Arts*.[27]

According to his friend and long-time printer, John Nichols, it was Wyatt who "projected" the magazine. The name of the well-established booksellers George and Thomas Wilkie appeared on the prospectus, and early advertisements announced that the journal was "printed for" the Wilkies and for the leading wholesaler George Robinson.[28] Nevertheless,

---

25 See *DNB*; John Martin Robinson, *The Wyatts: An Architectural History* (Oxford: Oxford University Press, 1979), 13–17; D. G. C. Allen and John L. Abbott, *The Virtuoso Tribe of Arts and Sciences: Studies in the Eighteenth-Century Work and Membership of the London Society of Arts* (Athens, GA, and London: University of Georgia Press, 1992), 361; and Christopher Davy, "Southwark Iron Bridge," *Mechanics' Magazine* 16 (1831–32): 258.

26 [John Wyatt], [*Prospectus for "The Repertory of Arts and Manufactures"*] ([n.l.: n.p.], [1794]). A copy of the prospectus is in the John Johnson Collection, Bodleian Library, Oxford.

27 Anon., "Advertisement," *Repertory of Arts and Manufactures* 1 (1794): i and iii.

28 John Nichols, *Literary Anecdotes of the Eighteenth Century*, 9 vols (London: for the author, 1812–16), 9:191n. Wilkie's name alone was given in the prospectus, but Robinson's

Wyatt rapidly made the *Repertory* the basis of a personal publishing business, and in 1799 he took sole responsibility, selling the *Repertory* from an address in Fleet Street before moving his office in 1803 to a more gentrified address in Hatton Garden.[29] That Wyatt was able to establish a publishing business around the new journal is a testament to the commercial success which it achieved. The editor of the *Monthly Review*, Ralph Griffiths, greeted the first volume enthusiastically, emphasizing the "the novelty and importance" of the "compilement".[30] The *British Critic* was also animated in its praise, and after just one volume was able to report that the *Repertory* had achieved a "very extensive circulation".[31] By 1797, it was reported to be selling a thousand copies per number and by May 1800 the print run had risen to 1500 copies.[32] Moreover, despite the war, some, at least, of these copies were going overseas. In 1801, for instance, Wyatt was advertising agents in Dublin (Dugdale) and Hamburg (Remnant); and in 1817 in Antwerp (Vandervaeren) and Hamburg (Puther and Bessu).[33]

Such commercial success attracted emulation. Looking back in 1806, Wyatt noted that his had been the first of the new periodical publications "devoted exclusively either to Philosophical objects, or the improvement

was added in the advertisement in *The Times*, 19 December 1794, p. 2a. There is no reference to the *Repertory* in Robinson's surviving copyright records; see G. E. Bentley, "Copyright Documents in the George Robinson Archive: William Godwin and Others, 1713–1820," *Studies in Bibliography* 35 (1982): 67–110. A few other prominent booksellers were also listed on the title pages; however, these are an uncertain guide to the journal's publishing history, since some numbers were reissued on several occasions with variant imprints.

29 *The Times*, 8 April 1796, p. 2a and 3 January 1799, p. 2b. Between 1808 and 1811, the *Repertory* was sold by Wyatt's son Walter Henry (1781–1849)—later an inventor and manufacturer—from the "Repertory and Patent Office, 9 Pickett-Street, Temple Bar", perhaps because the father was busy forwarding his plans for Southwark Bridge. However, on reverting to the Hatton Garden office of the now sexagenarian Wyatt, publication of the journal began to be handled by the prominent periodical wholesalers Sherwood and Co., and later also by the medical specialist Thomas Underwood. See *The Times*, 10 January 1812, p. 2a and 8 February 1814, p. 1e. After Wyatt's death in 1818, the journal continued under the management of the family, though renamed *Repertory of Patent Inventions* in 1825. In the 1830s, it was sold to James Shirley Hodson (a neighbouring publisher in Hatton Garden) and continued as a going concern until 1862.

30 [Griffiths, Ralph], "[Review of *Repertory of Arts and Manufactures*]," *Monthly Review* 2nd ser. 16 (1795): 147.

31 *British Critic*, 5 (1795): 535.

32 C. H. Timperley, *Encyclopaedia of Literary and Typographical Anecdote*, 2 vols (New York and London: Garland Publishing, 1977), 2:795; and John Bowyer Nichols to C. Taylor, 26 May 1800, Royal Society of Arts, AD/MA/104/10/302.

33 *The Times*, 8 August 1801, p. 1d and 17 July 1817, p. 2b.

of Arts and Manufactures", and that the success of the *Repertory* and its chiefly practical focus had prompted others, "both at home and abroad", to begin periodicals "of a more speculative or theoretic tendency".[34] Indeed, the continuities between the *Repertory* and the other periodical compilations from learned transactions are unmistakeable. Admittedly, Wyatt's first volume contained twenty-two patent specifications, but these accounted for less than a quarter of the pages. Of the remaining thirty-two articles, six were original, two came from the *Annales de Chimie*, four from agricultural reports, and the remaining twenty were extracted from the transactions and memoirs of the Society of Arts, the American Philosophical Society, the Manchester Literary and Philosophical Society, the Royal Irish Academy, the Royal Society, and the academies of sciences of Paris and Turin. Moreover, while these were certainly practical in orientation, they nevertheless included papers by Benjamin Franklin, Claude Berthollet, Joseph Banks, Thomas Henry, and Jean-Antoine Chaptal. In later volumes, the proportion of original articles increased. Nevertheless, by combining these with extracted articles, Wyatt intended his journal "to form a perpetual Register of useful Discoveries".[35] Indeed, by 1815, the magazine's printer, John Nichols, could suggest that to manufacturers the work would "be found of much greater utility than an Encyclopedia, because it gives the improvements in each branch as they arise or become known, while an Encyclopedia details only what was universally known at the time of writing the article."[36]

William Nicholson's Journal of Natural Philosophy, Chemistry, and the Arts *(1797–1813)*

One of the titles seeking to emulate Wyatt's success was the *Journal of Natural Philosophy*, begun in March 1797 by William Nicholson (1753–1815; fig. 2). Yet while the new publication bore a significant resemblance to Wyatt's *Repertory*, its editor was from the start anxious to point out the differences. Like his journal, however, Nicholson's life bore notable parallels with Wyatt's. The son of a London solicitor, he had been an East India Company midshipman before coming to work in the late 1770s as a commercial agent on the Continent for Josiah Wedgwood. On his return

34 [John Wyatt], "Preface," *Retrospect of Philosophical, Mechanical, Chemical, and Agricultural Discoveries* 1 (1806): [iii].

35 *Morning Chronicle*, 8 August 1801, p. 1d.

36 Nichols, *Literary Anecdotes*, 9: 192n.

Fig. 2 William Nicholson (1753–1815), editor of the *Journal of Natural Philosophy*. Reproduced from the *European Magazine* 62 (1812): 83f, by kind permission of Leeds University Library.

to Britain he began to find work as an author, and the *Introduction to Natural Philosophy* which he published with Joseph Johnson in 1782 established his reputation both with the publishers and more widely. Through Wedgwood, he became secretary of the General Chamber of Manufacturers of Great Britain in 1784, and in the same year he also became one of the secretaries of a philosophical club meeting at some of London's coffee houses (the Coffee House Philosophical Society), through which he came into contact with leading natural philosophers.[37]

Nicholson's new journal was designed to build on his established reputation in both the sciences and the practical arts. Its subtitle, as printed in Nicholson's prospectus of 1 March 1797, stated that it would give "an Account of the Present State, New Publications, and Discoveries" in natural philosophy, chemistry, and the arts, and in "the various Departments of Manufacturing Industry". The prospectus laid great emphasis on the character of Nicholson as editor. Noting that, unlike most British periodicals, Continental journals of "Literature, of Science, and of the Arts" had not infrequently been published under the names of "men of character and ability", Nicholson argued that the latter practice made a work worthy of receiving signed correspondence from reputable men and of being quoted "by other Authors of credit". Moreover, he considered that his own credentials as a cultivator of science and as someone with practical experience in the arts and manufactures fitted him to conduct a journal on such a basis. Nevertheless, the prospectus acknowledged that his accounts of new discoveries would be extracted or abridged from "the Acts of Academies" and other publications, as well as from "observation, enquiry, and correspondence".[38]

The preface to Nicholson's first number gave a more extended justification of the value of such anthologizing practices in a journal of science, asserting the importance of accurately conveying the most important scientific discoveries over and above providing content that was entirely original.[39] Nicholson asserted that, given "the very limited circulation of academical Transactions, from their price, their number, their extent, distance of publication, difference of language, labour of perusal, and the efforts of mental abridgment ... even the best memoirs they contain must continue unknown to a very large class of men of science". In such circumstances, "public utility" demanded that discoveries "buried from the knowledge of the world" should be placed above original observations, even when they were several years old. What was most important, however, was that the selection of materials be discriminating. By the end of

---

[37] See *DNB*; the manuscript biography by Nicholson's son in the Bodleian Library, Oxford (MS Don.e.125) sheds little light on the *Journal of Natural Philosophy*.

[38] [William Nicholson], [*Prospectus for "Journal of Natural Philosophy"*], ([London: G. G. and J. Robinson], 1797), 1 and 2. A copy of the prospectus is in the John Johnson Collection, Bodleian Library, Oxford.

[39] [William Nicholson], "Preface," *Journal of Natural Philosophy* 1 (1797–98): [iii].

the first volume, Nicholson was able to boast that his work was not "an indiscriminate compilation of things nearest at hand, nor a loose temporary record of transactions". Rather, he considered that the journal would become "the Repository of Discoveries in Science and the Arts". Of course, this was also Wyatt's expressed ambition, and while Nicholson was careful not to refer to the rival journal by name, he clearly intended to establish the superiority of his own claims as an editor over those of Wyatt.[40]

The proportion of original communications in Nicholson's journal was from the start higher than in Wyatt's, due in part to his scientific connections and his greater involvement in metropolitan research activity, for example through the Coffee House Philosophical Society. He invited contributions not only from "Philosophers" but also from "manufacturers, and others", and it was no doubt the latter whom he particularly sought to reassure when he promised to be scrupulous in respecting the "rights" of those who communicated with him verbally or in writing, including in "that first and most sacred property which men hold in the products of their own understanding". At the end of the first year, Nicholson reported that nearly half of the papers published were original, a third were new translations from Continental publications, and only around a sixth were reports or abridgements of papers already published in British "academical collections". A similar proportion was reported in 1802, but by now Nicholson claimed that he was struggling to find space for translated and abridged materials, as a result of the large number of valuable original communications he was receiving.[41]

Like Aikin and Wyatt, Nicholson was financially reliant on his editorial labours and was able to make a reasonable profit from the considerable sales of his journal. Some of Nicholson's earliest publications had been handled by Johnson, but by the time he started the *Journal* all of his new publications were published by Johnson's "invariable and confidential friend", George Robinson.[42] However, while the *Journal* was for a spell published on a half-profits basis, Nicholson soon became sole proprietor, employing Robinson as a commission publisher.[43] Nicholson found initial

[40] For instance, he commented: "it would not be difficult to point out imperfections in the works of others, and promise to avoid them". [Nicholson], "Preface," iv.

[41] [William Nicholson], "Preface," *Journal of Natural Philosophy* 2nd ser. 1 (1802): [iii]; and [William Nicholson], "Preface," *Journal of Natural Philosophy* 2nd ser. 2 (1802): [iv].

[42] [William West], "Annals of Authors, Artists, Books, and Booksellers. Letter XI. Mr. Johnson, of St. Paul's Church-Yard, and His Literary Connexions," *Aldine Magazine* 1 (1839): 205.

[43] Bentley, "George Robinson Archive", pp. 93–94 and 106–107; *Journal of Natural Philosophy* 2nd ser. 33 (1813): 152.

sales of around seven hundred and fifty copies disappointing, but circulation seems to have increased to give a healthy income.[44] Perhaps in consequence, Nicholson was keen to assert that his name and character gave the *Journal* a value far beyond a mere commercial compilation. Some review journals disagreed, resenting his encroachment on their territory. The reviewer in the Tory *British Critic*, for instance, asserted in August 1798 that the new journal consisted largely of abridged articles extracted from publications which the *Critic* had already reviewed, and that many of the original articles were "of a trifling nature". While allowing that some who were interested in following the progress of arts and sciences did not have access to new scientific publications, the *Critic* was unwilling to give its "entire approbation" to a work which professed "to extract, and, as it were, to monopolize the beauties of all the modern scientific publications".[45] At the other end of the political spectrum the *Critical Review* was also sensitive to the potential encroachment on its own territory, claiming that Nicholson's accounts of books were "too general and indiscriminate".[46]

The *Critical* was nevertheless generally enthusiastic about the new journal, seeing it as supplying a long-standing need for something like the Continental scientific journals and claiming that Rozier's *Journal de physique* had often hitherto given the first information to British readers of discoveries made in Britain. Moreover, the reviewer considered that the particular value of Nicholson's name was not so much the promise of accuracy, as Nicholson claimed, as the promise of discrimination in the selection of materials. Johnson's *Analytical Review* also warmly welcomed the new journal as the first "well directed attempt" to follow the model of Continental journalism in Britain, complimenting Nicholson on the *Journal*'s management. It was "not a hasty compilation of such materials as the editor could most readily lay hold of, but an able, judicious, and well meant endeavour, to propagate and improve scientific knowledge".[47] In the *Monthly Review* Cambridge mathematician Robert Woodhouse reflected at length on the reconfiguration of scientific print. A general "scientific journal" like the *Journal des sçavans*, he averred, was no longer

44 Timperley, *Encyclopaedia*, p. 795; [William Nicholson], "Advertisement," *Journal of Natural Philosophy* 1 (1797–98): [v]. An agreement was drawn up in December 1799 allowing Robinson to reprint a thousand copies each of the early numbers to make up sets. See Bentley, "George Robinson Archive," 106.

45 Anon, "[Review of *Journal of Natural Philosophy*]," *British Critic* 12 (1798): 116–17.

46 Anon, "[Review of *Journal of Natural Philosophy*]," *Critical Review* 2nd ser. 26 (1799): 283.

47 Anon, "[Review of *Journal of Natural Philosophy*]," *Analytical Review* 28 (1798): 364 and 377.

practicable given the expansion of the objects of scientific enquiry. Nicholson's *Journal* was to be welcomed as "the first of its kind" to provide a specialized focus.[48]

Alexander Tilloch's Philosophical Magazine *(1798–)*

The founder of the *Philosophical Magazine*, Alexander Tilloch (1759–1825), was also heavily involved in the mechanical arts, although he was not, like Wyatt and Nicholson, intimate with Boulton, Wedgwood, or other members of Birmingham's Lunar Society. The son of a Glasgow tobacco merchant and magistrate, he matriculated at the University of Glasgow without graduating, and subsequently spent some time in his father's line of business. He also independently invented a process for stereotyping, entering into partnership with the printer, Andrew Foulis the younger, and patenting and briefly employing the technique. In 1787 he moved to London, becoming in 1789 one of the co-proprietors and co-editors of the first daily evening paper, the *Star*, which he edited until ill health prevented his doing so around 1821. He also continued his involvement in the practical arts, devising a new method of printing banknotes, patenting other mechanical inventions, and serving for a spell as chairman of the Committee on Correspondence and Papers at the Society of Arts.[49] Furthermore, Tilloch's philosophical interests led to his becoming a member of several societies frequented by practically-oriented men, including the London Philosophical Society (f. 1794), the Askesian Society (f. 1796), and the Mineralogical Society (f. 1799).[50] However, while he later became a corresponding member of learned societies as far afield as Edinburgh, Dublin, and Munich, Tilloch was given to understand that his election to the Royal Society would be black-balled on the grounds that he was a newspaper and journal editor, indicating that his status was in some ways marginal.

Heavily involved in the practical arts and in philosophical pursuits, and already experienced as an editor, Tilloch saw in the success of Wyatt's and

---

[48] [Robert Woodhouse], "[Review of *Journal of Natural Philosophy*]," *Monthly Review* 2nd ser. 29 (1799): 301–03.

[49] Anon, "Society of Arts," *Universal Magazine* 2nd ser. 11 (1809): 344. In addition to the *DNB* see Anon., "Memoir of Alexander Tilloch," *Imperial Magazine* 7 (1825): 208–22; [Thomas Byerley?], "Alexander Tilloch, LL.D.," *Literary Chronicle* 7 (1825): 141–42; and Anon., "Alexander Tilloch, LL.D.," *Annual Biography and Obituary* 10 (1826): 320–34.

[50] Brock and Meadows, *Lamp of Learning*, 90–97; Inkster, "Science and Society"; and Paul Weindling, "The British Mineralogical Society: A Case Study in Science and Social Improvement," in *Metropolis and Province: Science in British Culture, 1780–1850*, ed. Ian Inkster and Jack Morrell (Philadelphia: University of Pennsylvania Press, 1983), 120–50.

Nicholson's journals an invitation to capitalize on his interests to produce a publication that would make available to "the Public in general ... the improvements which are daily made in the different branches of human knowledge". His prospectus acknowledged the existence of similar publications but promised that the *Philosophical Magazine* would operate on an altogether larger scale. Tilloch proceeded to outline his sources. First, there were the "transactions of the [Continental] Academies, works too dear to be purchased by readers in general, and written, besides, in languages with which few are acquainted." Through a Continental correspondence network, and despite the war with France, Tilloch promised to secure such publications promptly, and to "select and translate" from them. Secondly, he would mine the "great variety" of Continental "Journals". Thirdly, Tilloch promised extracts from "the transactions of the Public Societies in Great Britain" and, fourthly, information about newly patented machines. Only "in last place" did Tilloch list his ambition to include "original communications on Scientific Subjects," noting that he had received "assurances of assistance from some of the most distinguished characters in the kingdom as Philosophers and Mechanics". In addition, he offered "short notices of New Discoveries and Improvements", obituaries of men eminent in science or the arts, and notices and extracts from new books of importance.[51]

Tilloch's forthright anthologizing ambition was encapsulated in a punning Latin epigraph to the prospectus which subsequently, for many years, adorned the title-page: "The spider's web [*textus*] is no whit the better because it spins it from its own entrails; and my text no whit the worse because, as does the bee, I gather its components from other author's flowers".[52] When the first monthly issue of the *Philosophical Magazine* appeared in June 1798, price 2*s*., it fulfilled Tilloch's promises by providing translations and abstracts of work extracted from Continental transactions, journals, and books. Of seventy-seven articles in the first volume, fifty-five were from taken such sources, and a further seven were taken from British publications (fig. 3). Only sixteen were original, and some of those were descriptions of published work. In addition, a regular

---

[51] [Alexander Tilloch], "Prospectus of a New Periodical Work, Intitled, The Philosophical Magazine," *Star*, 29 June 1798, p. 1a–b.

[52] "Nec aranearum sane textus ideo melior, quia ex se sila gignunt. Nec noster vilior, quia ex alienis libamus ut apes." I am indebted for the translation to Ann Moss, "The *Politica* of Justus Lipsius and the Commonplace-Book," *Journal of the History of Ideas* 59 (1998): 424.

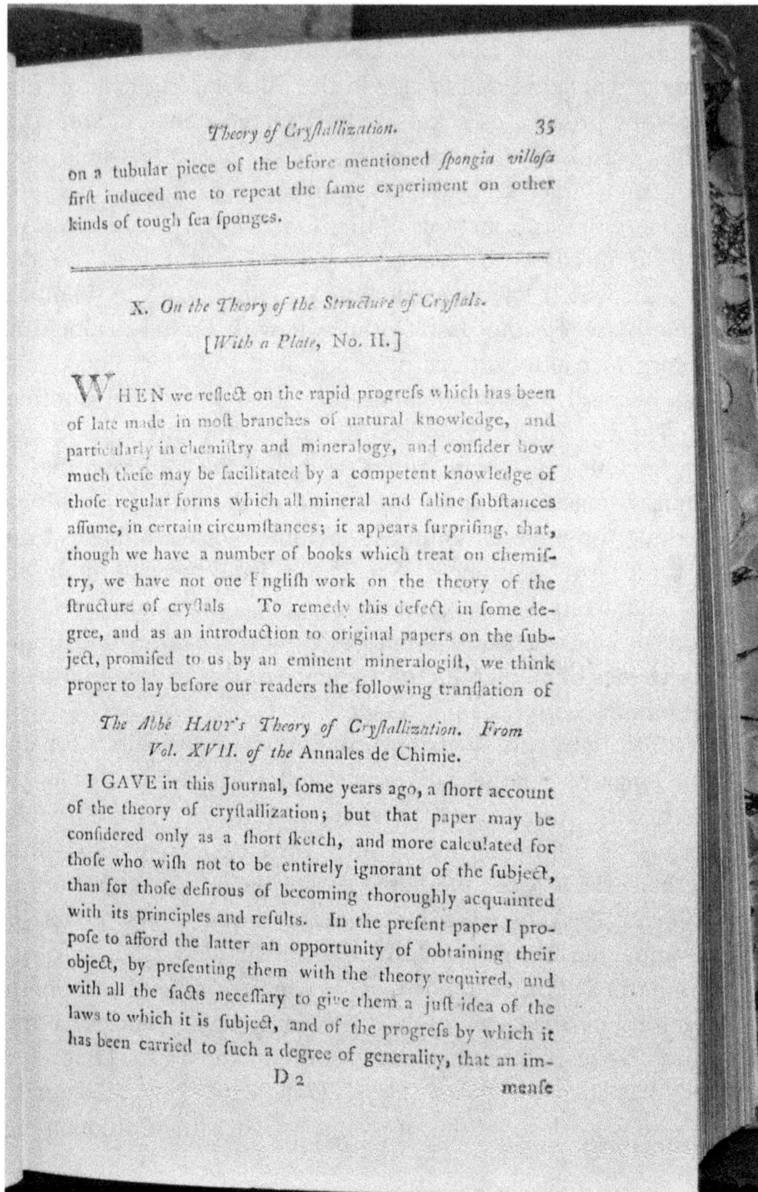

Theory of Cryſtallization. 35

on a tubular piece of the before mentioned *ſpongia villoſa* firſt induced me to repeat the ſame experiment on other kinds of tough ſea ſponges.

X. *On the Theory of the Structure of Cryſtals.*

[*With a Plate*, No. II.]

WHEN we reflect on the rapid progreſs which has been of late made in moſt branches of natural knowledge, and particularly in chemiſtry and mineralogy, and conſider how much theſe may be facilitated by a competent knowledge of thoſe regular forms which all mineral and ſaline ſubſtances aſſume, in certain circumſtances; it appears ſurpriſing, that, though we have a number of books which treat on chemiſtry, we have not one Engliſh work on the theory of the ſtructure of cryſtals. To remedy this defect in ſome degree, and as an introduction to original papers on the ſubject, promiſed to us by an eminent mineralogiſt, we think proper to lay before our readers the following tranſlation of

The Abbé HAUY'*s Theory of Cryſtallization. From Vol. XVII. of the* Annales de Chimie.

I GAVE in this Journal, ſome years ago, a ſhort account of the theory of cryſtallization; but that paper may be conſidered only as a ſhort ſketch, and more calculated for thoſe who wiſh not to be entirely ignorant of the ſubject, than for thoſe deſirous of becoming thoroughly acquainted with its principles and reſults. In the preſent paper I propoſe to afford the latter an opportunity of obtaining their object, by preſenting them with the theory required, and with all the facts neceſſary to give them a juſt idea of the laws to which it is ſubject, and of the progreſs by which it has been carried to ſuch a degree of generality, that an im-

D 2 menſe

Fig. 3 In an early issue of the *Philosophical Magazine*, Alexander Tilloch sought to compensate for the inadequacy of English literature on the theory of crystals by translating a paper from the *Annales de chimie* by French mineralogist René Just Haüy. Reproduced from the *Philosophical Magazine* 1 (1798): 35, by kind permission of Leeds University Library.

"Intelligence" section allowed Tilloch to summarize further Continental work. Tilloch's first, self-congratulatory preface reiterated that his "grand Object" had been "to diffuse Philosophical Knowledge among every Class of Society, and to give the Public as early an Account as possible of every thing new or curious in the scientific World, both at Home and on the Continent".[53]

This emphasis on the inclusive nature of his readership was also evident in Tilloch's prospectus, which described the work as being "for the use of the common ranks as well as of the learned", and suggested that it would be valuable to "young persons" wishing to acquire knowledge of "the philosophical sciences". As with Wyatt and Nicholson, he laid considerable emphasis on the practical importance of the magazine to those engaged in the "arts and manufactures", but he also reported his intention to include articles of a more theoretical cast which might ultimately prove to be of practical value. The journal was to cover "Natural and Experimental Philosophy, Chemistry, Economics, Natural History, Mechanics, Geography, Statistics, Astronomy, Meteorology, Antiquities; and, in a word, every thing that relates to Science, the Liberal and Fine Arts, Trade, Manufactures, and Agriculture". Tilloch again echoed Nicholson in emphasizing that "New Discoveries, or Inventions" were not to be the journal's overriding preoccupation, and that, where necessary, articles would return to important but relatively unknown materials of an older vintage.[54]

Like the *Repertory of Arts* and the *Journal of Natural Philosophy*, the *Philosophical Magazine* was the property of its editor, for whom it became a valuable source of income. More than the other editors, however, Tilloch assembled a list of prominent booksellers to act as commission publishers, with Richardsons and Cadell and Davies initially heading a list of nine. Significantly, the list included William Remnant in Hamburg, and booksellers were soon added in Dublin, Edinburgh, and Glasgow, indicating the extent to which the journal was intended to function as a communication hub. The first volume was greeted by the *Monthly Magazine* as "very promising" and by 1804 it was clear to contemporaries that the *Philosophical Magazine*, like the Nicholson's *Journal*, was there to stay, "fixed on the firm rock of public approbation and support".[55] Production figures support this assessment. By 1803, the print-run stood at 1250 copies per month,

53 [Alexander Tilloch], "Preface," *Philosophical Magazine* 1 (1798): [iii].

54 [Tilloch], "Prospectus".

55 *Monthly Magazine* 6 (1798): 52; D., "Cursory Character of the Magazines," *Universal Magazine* 2nd ser. 1 (1804): 361.

and while this had fallen to 1000 copies by 1813, it continued at that level until 1827, and it seems clear that Tilloch secured a significant profit at this period.[56] So successful was the journal by 1805 that leading publishers John Murray, Archibald Constable, Longmans, and Richard Phillips were all in the market to buy it. Murray thought it "one of the most respectable and best-selling of our periodical publications", and it was only the extent of the stock in hand that scuppered the deal. In the event, Tilloch retained ownership of the magazine until his death in 1825, when it was purchased as a going concern by his co-editor Richard Taylor.[57]

Thomas Garnett's Annals of Philosophy *(1801–02)*

Notwithstanding the increasingly full market, a further attempt to found a philosophical journal was made in June 1801, when Thomas Garnett (1766–1802) "and other gentlemen" published the first volume of their *Annals of Philosophy*.[58] Garnett's journal was distinctive in being an annual, more in keeping with the *Annual Register* (1758–), which had long included sections for "Useful Projects" and "Natural History", than with the monthly magazines. It nevertheless had much in common with its competitors, as Garnett did with the other early editors. The son of a Westmoreland yeoman, Garnett had early studied mathematics and natural philosophy while apprenticed to the celebrated Sedbergh surgeon and mathematician John Dawson. His scientific education was broadened while studying for the Edinburgh M.D., and, practicing as a physician in Bradford and Harrogate, he began to lecture and publish on natural philosophy and chemistry. Following his marriage in 1795 he established a growing reputation as an itinerant lecturer, and was appointed the first professor of natural philosophy at Anderson's Institution in Glasgow in 1796. He moved to London in 1799, having been invited to become the first professor of natural philosophy and chemistry at the Royal Institution. However,

---

56 Printing Ledgers of R. Taylor & Co., 1803–10 and 1804–15; Printing Journals of R. & A. Taylor, 1815–26 and 1827–54 ('Journal 1829'), Taylor & Francis archive, St Bride Printing Library, London. Even making some rather conservatively questionable assumptions about printing costs, Brock and Meadows estimated an annual profit of £264; Brock and Meadows, *Lamp of Learning*, 101.

57 Thomas Constable, *Archibald Constable and His Literary Correspondents: A Memorial*, 3 vols (Edinburgh: Edmonston and Douglas, 1893), 1: 353; John Murray papers, National Library of Scotland, MS 41908, p. 152; Brock and Meadows, *Lamp of Learning*, ch. 4.

58 The earliest advertisement I have located (announcing "this day is published") is *Morning Chronicle*, 29 June 1801, p. 2a; the earliest review dates from November 1801 (Anon, "Annals of Philosophy", Critical Review 2nd ser. 33 (1801): 252–58).

a combination of ill health and growing estrangement from Count Rumford and the institution's managers led to Garnett's resignation in mid-June 1801. It was at precisely this juncture that he advertised his *Annals of Philosophy*, and the scheme was clearly part of a wider occupational strategy which involved resuming medical practice and lecturing on his own account from his premises in nearby Great Marlborough Street.[59]

As the Royal Institution professor, Garnett was officially expected to edit the *Journals of the Royal Institution* (1800–03), which were intended to publish details of its scientific investigations on a weekly or at least fortnightly basis, at a price of 3*d.* or 6*d.*, depending on the number of pages. In the event, only two numbers appeared during Garnett's professorship, and these were edited by Rumford.[60] Garnett's own journal, however, was rather different in intent. As with the other new journals, the stated objective was to produce, for reasons of utility, "a general and concise view of the scientific discoveries of the year, with references to the original works in which such discoveries were published". Garnett continued:

> As it is in the power of very few, on account of indispensable occupations, to read every publication on the sciences and arts that comes from the press, it is certainly desirable to have a work, which, in the compass of a volume, will give a detail of all that has been done, and which, placed in a library, will serve as an index, pointing out the sources from whence satisfactory information on each subject may be obtained.

The intended work would thus be useful even to those based in the metropolis, but yet more so to those living in the country, and especially to those "in America, and the East and West-Indies", who would, "[b]y means of a small annual volume [...] become acquainted with the state of science and of the arts in every part of the world".[61] The opening section provided accounts of new scientific discoveries in natural philosophy, natural history, and chemistry, abstracted from learned transactions, other

---

59 On Garnett see *DNB*; Henry Bence Jones, *The Royal Institution: Its Founder and Its First Professors* (London: Longmans, Green, 1871), 162–79; Thomas Garnett, *Popular Lectures on Zoonomia or the Laws of Life, in Health and Disease* (London: for the benefit of the author's children, 1804), v–xxii; and S. G. E. Lythe, *Thomas Garnett (1766–1802)* (Glasgow: Polpress, 1984). The published outlines of Garnett's new lecture courses on chemistry and natural philosophy were advertised alongside the *Annals*. See *Morning Chronicle*, 10 November 1801, p. 2a.

60 Jones, *Royal Institution*, 141, 153–56, 181–83, and 188; *Archives of the Royal Institution of Great Britain in Facsimile: Minutes of Managers' Meetings, 1799–1900*, ed. Frank Sulloway, 15 vols in 7 (Menston: Scolar Press in association with the Royal Institution of Great Britain, 1971–76), 1: 112, and 2: 42, 183, and 185.

61 [Thomas Garnett], "Preface," *Annals of Philosophy* 1 (1801): [i]–ii.

scientific journals, and books. In addition, however, Garnett's *Annals* purported to provide a listing of the entire output of the British press for the year, as well as of some Continental literature, together with notes. A third section, headed "Miscellaneous", contained original contributions relating to improvements in practical and fine arts and in agriculture, together with obituaries.

The *British Critic* greeted the new journal with the observation that Garnett's justification of it in terms of "the necessity of gathering the scattered branches of scientific knowledge" into a "small compass, which may be useful to such inquisitive persons as have neither time nor opportunity for acquiring more extensive information", had become the typical pattern for "similar periodical publications". Nevertheless, it welcomed the plan as "excellent" and the execution as accomplished, concluding that the *Annals* contained information "incomparably more extensive than that of any other similar publication".[62] The potential encroachment on the territory of existing reviews was again a concern, and the *Critical Review* considered the review of new books "totally unnecessary". Nevertheless, the reviewer agreed that there was "ample room" in the market for a "collection of this kind":

> In an *extensive* literary journal much must be passed over hastily that might require the minuteness of detail, the abstruser and drier parts of science must be slightly touched in a work that "must please to live;" and the reviewer will sometimes "omit what he despairs of adorning." The scientific analyst, while he labours on the same materials with the journalist, is not impeded by the same difficulties: he has models in the *miscellanea curiosa*, and in the various abridgements of memoirs, to guide him, and he may fill up the picture, of which the reviewer can only sketch the outlines, while he leads to those who may give fuller information

Where the annual registers contracted the view of passing events given in the monthly press, the *Annals* should "enlarge upon the minuter points, less adapted to a general journal, and include accounts which scarcely come within the province of the former".[63]

Garnett's 448–page volume was published for 10*s*. 6*d*. by Cadell and Davies, who had published his *Observations on a Tour through the Highlands* the previous year. The basis of their agreement remains unclear, but when Garnett died of typhus in June 1802, the "proprietors" announced

[62] Anon, "Garnett's Annals of Philosophy and Arts," *British Critic* 20 (1802): 157 and 164–65.

[63] Anon, "Annals of Philosophy," *Critical Review* 2nd ser. 33 (1801): 255–56.

that his place had been taken by Charles and Arthur Aikin—whose father, John, was now one of Cadell and Davies's readers—working alongside the anonymous editor who had previously compiled the natural history section.[64] Making some structural modifications, the Aikins produced two further volumes of the *Annals* at the lower price of 9*s*. That the journal was then abandoned suggests an inadequate return. However, Arthur Aikin had also projected the *Annual Review* with Longmans in 1802—probably to an extent inspired by his experience of the *Annals*—and editing two such journals was probably impracticable.[65] Nevertheless, the involvement of two of John Aikin's sons in succession to Garnett again confirms the close connection between the new journals and the Metropolitan world of practically oriented and commercially engaged science.

Three "Archives" and "Retrospects"

The inception of three relatively long-lived journals of science and the arts in the years 1794 to 1798 clearly established the commercial viability of the form, as is evinced by the interest of Longmans, Phillips, Murray, and Constable in acquiring the *Philosophical Magazine*. That none of these leading publishers sought to introduce a competing journal, however, suggests that they now considered the market relatively well served by journals that were well respected. Indeed, after the initial flurry of activity, few further attempts were made to found new journals in the succeeding decade, and those few were unsuccessful. In May 1805, for instance, H. D. Symonds (a well-known periodical wholesaler of Paternoster Row) issued the first monthly number of the *Archives of Philosophical Knowledge*, priced at 2*s*. 6*d*., but it seems to have foundered after only two numbers.[66] The editor, B. Lambert, acknowledged that the "general merit" of the existing

64 Theodore Besterman, ed., *The Publishing Firm of Cadell & Davies: Select Correspondence and Accounts, 1793–1836* (Oxford: Oxford University Press and London: Humphrey Milford, 1938), 172; Anon, "Preface," *Annals of Philosophy* 2 (1802), p. v; and Anon, "Annals of Philosophy," *British Critic* 22 (1803), 307.

65 The initial prospectus (in the John Johnson Collection, Bodleian Library, Oxford) suggested that, in providing a systematic survey, the review would present the reader with "an uninterrupted history of the annual advancement of science". [Arthur Aikin], *Prospectus of a New Work, to be Entitled "The Annual Review, or Register of Literature"* ([London]: printed by T. Gillet, [1802]), 2–3. Somewhat confusingly, Arthur Aikin also assisted Thomas Thomson in editing his *Annals of Philosophy* in 1818–19.

66 *Morning Chronicle*, 18 May 1805, p. 2c; Thomas Rees, *Reminiscences of Literary London from 1779 to 1853, with Interesting Anecdotes of Publishers, Authors, and Book Auctioneers of that Period, &c., &c.* (London: Suckling and Galloway, 1896), 62.

philosophical journals might "seem to render any addition to their number unnecessary", but argued that there still existed "a material chasm in the speedy diffusion of experimental and practical knowledge".[67] Lambert's new journal, subtitled "manual of foreign discoveries and improvements in the experimental, mechanical, and physical sciences", was to consist "wholly of Translations, Abridgments, or Extracts from Foreign Memoirs and larger Works, in the Experimental, Mechanical and Physical Sciences, in order to enable the British Philosopher, Manufacturer, and Artisan to obtain an early acquaintance with the discoveries and improvements daily making in the arts and sciences in other countries".[68] Lambert, who had recently produced a translation of Berthollet's *Essay on Chemical Statics* (1804) and other French works, was quite probably a French émigré. In any case, all except five of the twenty-six translated memoirs were from French sources. However, his translations were not well-regarded and the journal made no progress against the established competition.[69]

A further attempt was made to found such a journal in January 1809, when the medical and scientific lecturer Alexander Walker issued the first quarterly number (priced 7*s.* 6*d.*) of his *Archives of Universal Science*.[70] The journal consisted largely of articles on the arts and sciences reprinted from other publications, British and foreign, interspersed with a significant number penned by the editor. Many of these related specifically to Walker's "Natural Theory of Universal Science", on which he had lectured to mixed audiences in Edinburgh, and the journal sought to reassert the value of an Enlightenment notion of universal science against the specialization of other journals. In addition, Walker offered prospective contributors ten guineas per sheet for their contributions.[71] The journal was initially advertised as being published by the publisher of the *Monthly Magazine*, Richard Phillips, in London, and by the small-time publisher Thomas Bryce in Edinburgh (where it was printed by the

[67] See the "Prospectus" on the paper wrapper of the first issue, present in the copy at the Senate House Library, London, which is also quoted in *Universal Magazine* 2nd ser. 3 (1805): 459.

[68] *Morning Chronicle*, 18 May 1805, p. 2c.

[69] See, for example, Anon, "Berthollet on Chemical Affinities, &c.," *Imperial Review* 4 (1805), 297 and, for the suggestion that he was a foreigner, Anon, "Clausen's Russian Anecdotes," *Critical Review* 3rd ser. 4 (1805), 181.

[70] *Caledonian Mercury*, 26 December 1808, p. 1d.

[71] [Alexander Walker], *[Prospectus for] Archives & Review of Universal Science* ([n.l.: n.p.], [1809?]). See the copy, dated November 1809, in the John Johnson Collection, Bodleian Library, Oxford, and the account of Walker's lectures in the *Caledonian Mercury*, 30 April 1810, p. 1a.

University printer Charles Stewart). In the event, the large periodical wholesalers Sherwood and Co. were the London agents, but the journal was not a success, surviving for only three numbers.

While these two "archives" failed, John Wyatt considered by 1806 that there were so many "scientific journals" that readers now needed "a digested index" of them.[72] He consequently began a quarterly, costing 3s. 6*d*. and entitled *A Retrospect of Philosophical, Mechanical, Chemical, and Agricultural Discoveries*, which was designed to provide "An Abridgement of the Periodical and other Publications, English and Foreign, relative to Arts, Chemistry, Manufactures, Agriculture, and Natural Philosophy", with occasional critical commentaries and suggestions for practical applications (fig. 4). The prospectus promised:

> This work will not resemble any other Periodical Publication, on the subjects comprehended in its Title; but, as the information on these heads is now scattered amidst a variety of Monthly Journals (the same articles, particularly those of Foreign origin, often appearing in all), it is presumed a concentration of that knowledge in one Work, published less frequently, and thereby less expensively, will meet the approbation of a numerous class of Readers.

Even those who bought all the monthly journals, Wyatt suggested, would find the new magazine helpful as "a complete Compendium or Epitome of Scientific and Practical Knowledge".[73]

Commentators certainly valued the *Retrospect* as a critical "Supplement and Index to all other scientific Journals", and even Robert Woodhouse, who questioned the adequacy of the abridgements and the lack of plates, considered it "likely to be highly useful".[74] However, while its prefaces boasted of the assistance and approval of scientific men, the magazine did not sell as well as the *Repertory*.[75] Moreover, the *Retrospect* was hampered by the difficulty of obtaining Continental publications during the Napoleonic blockade against Britain. From the first preface, Wyatt observed that the irregularity in the arrival of foreign books might cause publication delays, and this deteriorated so that by 1809, it was reported that "the troubled state of Europe" had deprived the journal "almost entirely

72 *Retrospect of Philosophical, Mechanical, Chemical, and Agricultural Discoveries* 1 (1806): [iii].

73 [John Wyatt], [*Prospectus for "Retrospect of Philosophical Discoveries"*] ([London]: printed by C. Spilsbury, [1806]); a copy is in the John Johnson Collection, Bodleian Library, Oxford.

74 *Universal Magazine* 2nd ser. 21 (1814): 60; and [Robert Woodhouse], "[Review of *Retrospect of Philosophical Discoveries*]," *Monthly Review* 2nd ser. 51 (1806): 220.

75 *Retrospect of Philosophical Discoveries* 3 (1808): [i]; 9 (1809): iv.

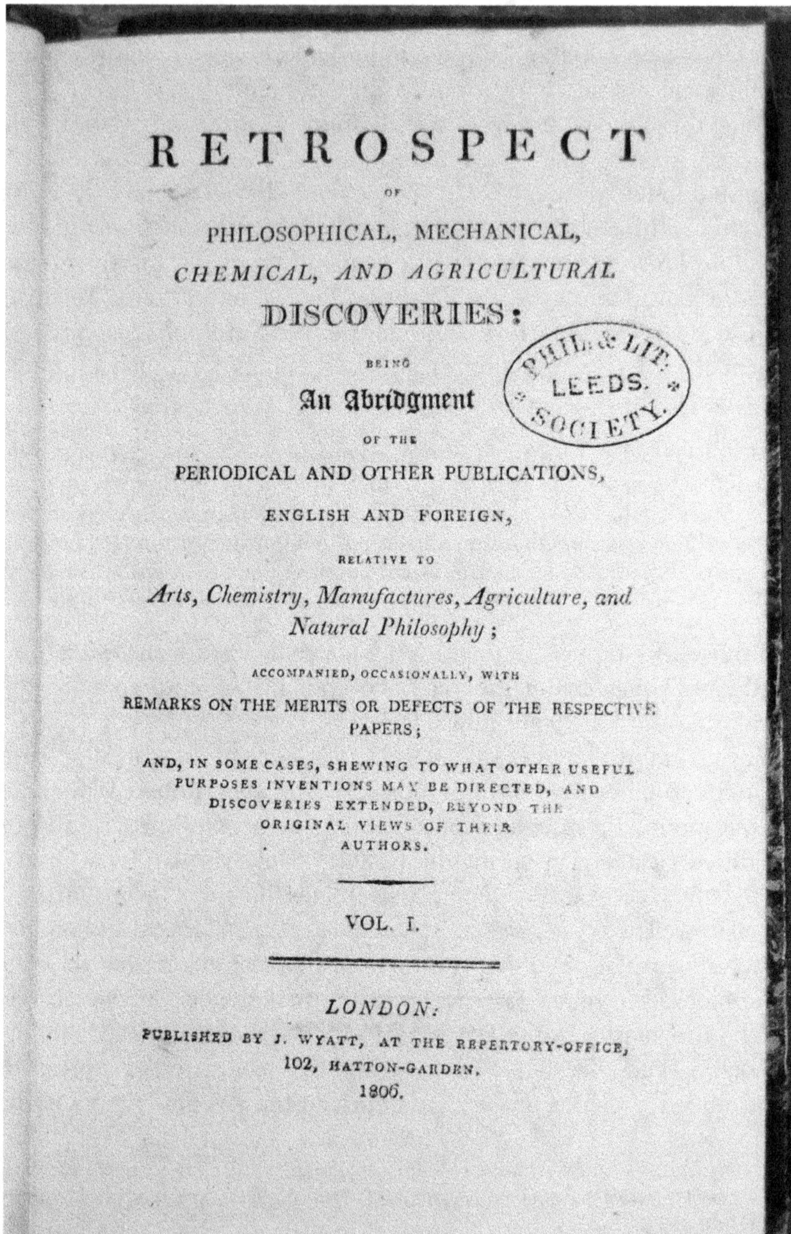

RETROSPECT

OF

PHILOSOPHICAL, MECHANICAL,

CHEMICAL, AND AGRICULTURAL

DISCOVERIES:

BEING

An Abridgment

OF THE

PERIODICAL AND OTHER PUBLICATIONS,

ENGLISH AND FOREIGN,

RELATIVE TO

Arts, Chemistry, Manufactures, Agriculture, and Natural Philosophy;

ACCOMPANIED, OCCASIONALLY, WITH

REMARKS ON THE MERITS OR DEFECTS OF THE RESPECTIVE PAPERS;

AND, IN SOME CASES, SHEWING TO WHAT OTHER USEFUL PURPOSES INVENTIONS MAY BE DIRECTED, AND DISCOVERIES EXTENDED, BEYOND THE ORIGINAL VIEWS OF THEIR AUTHORS.

VOL. I.

LONDON:

PUBLISHED BY J. WYATT, AT THE REPERTORY-OFFICE, 102, HATTON-GARDEN.

1806.

Fig. 4 The titlepage of the first volume of James Wyatt's *Retrospect of Philosophical, Mechanical, Chemical, and Agricultural Discoveries* (1806–15). Reproduced by kind permission of Leeds University Library.

of the materials for that portion of their work which they had hitherto received from the Continent".[76] However, it seems to have been the state of Wyatt's health that precipitated the end. Publication became infrequent and irregular, and while an announcement in the *Universal Magazine* in January 1814 promised a return to regular publication, the final numbers (for 1813) were issued in 1815.[77]

Conclusion

Reviewing Nicholson's *Journal* anonymously in the *Monthly Review* for July 1799, Robert Woodhouse reflected on the changing character of printed publications in relation to the sciences. Scientific journals like Nicholson's, he averred, were one aspect of a wider expansion of book production—including the production of small, cheap, and elementary books on the sciences—that had done much to increase the spread of knowledge in society. According to Woodhouse, the development that characterised the period was that "Epitomes, Abstracts, Synopses, Abridgements, Magazines, Journals, &c.", providing the "general outlines and principal features" of knowledge, had supplanted "profound and systematic treatises". Such publications, when conducted well, could prove valuable, most particularly in educating "the great majority of the people", and on the whole Woodhouse thought Nicholson's journal would be "useful".[78] It was by thus making recondite knowledge more generally available through modern techniques of book-making, rather than by providing prestigious places to publish authoritative and original scientific findings, that Woodhouse considered that the new scientific journals gained their value.

By seeing such journals from this contemporary perspective we are enabled to resituate the international circulation of scientific knowledge in this period within a particular commodity culture of print. Britain's earliest scientific journals emerged in a rapidly changing market in which literary entrepreneurs were able to earn a living by selecting and anthologizing. The founders of these early journals were chiefly men marginalized by their backgrounds in the practical arts, and in several cases by their involvement in rational dissent and political reform. They were also

76 *Retrospect of Philosophical Discoveries* 1 (1806): iv; 3 (1808): ii; 4 (1809): [ii]. See also Topham, "Science, Print, and Crossing Borders".

77 *Universal Magazine*, 2nd ser. 21 (1814): 60.

78 [Woodhouse], "Review of Journal", 301–03.

men who saw a dual opportunity in selectively anthologizing the latest scientific findings of Europe and North America: first, in securing a regular income and status from scientific work and, secondly, in making scientific knowledge more widely available. The journals they produced as a result significantly altered the means by which scientific knowledge circulated, both within Britain and beyond.

Of course, this study raises several new questions, the most important of which is how the anthologizing practices of these early journal editors impinged on scientific knowledge-making. Unfortunately, direct evidence of editorial practice and decision-making is very limited in the case of the early British journals.[79] Rather, practice must be inferred indirectly from the end product, with close scrutiny being given to what was being anthologized, how it was modified in the process, and how that changed over time. Detailed study of the relative extent to which work was reproduced from different countries, from different journals, from different authors, and on different subjects would be profoundly informative about the processes of knowledge circulation. As we have seen, editors prided themselves on being highly selective, and this is confirmed by even the briefest glance into the Royal Society's *Catalogue of Scientific Papers*, which lists the republication in other scientific journals of what were retrospectively deemed to be "scientific papers". For instance, whereas only two of Lamarck's thirty-one "papers" are listed as having been reprinted in the *Philosophical Magazine* or the *Journal of Natural Philosophy*, fifteen of the twenty-eight "papers" of the French industrial chemist Jean-Antoine Chaptal were reportedly reprinted.[80] Moreover, the process of republication was often highly transformative. For example, while the *Philosophical Magazine* translated and reproduced Georges Cuvier and Adolphe Brongniart's important joint paper on the mineral geography of the Paris region without comment or correction in January 1810, the next issue carried a lengthy review of the article from the land surveyor John Farey, accusing the authors of attempting to secure priority over William Smith's pioneering work on geological mapping. Farey also provided a fourteen-page glossary and index to place and species names and a tabular geological

[79] By contrast, a significant archive has permitted a detailed reconstruction of Marc-Auguste Pictet's practice as editor of the *Bibliothèque Britannique*. David Bickerton, *Marc-Auguste and Charles Pictet, the "Bibliothèque Britannique" (1796–1815), and the Dissemination of British Literature and Science on the Continent* (Geneva: Slatkine Reprints, 1986). See also Crosland, "In the Shadow".

[80] Royal Society, *Catalogue of Scientific Papers*.

section of the strata described, as well as a detailed discussion relating the French findings to British geology.[81] Such sophisticated reinterpretation of foreign science is suggestive of how far practices of knowledge circulation could contribute to knowledge making.

A further set of questions naturally arises concerning the process by which these early scientific journals came to be succeeded by (or in the case of the *Philosophical Magazine*, transformed into) something more recognizably modern. I have argued here that the early scientific journals were not solely or even primarily intended to provide authors with a means by which to secure credit or priority for original discoveries. Yet, as we have seen above, both Tilloch and Nicholson soon found themselves struggling to find space for abstracted materials alongside all the original papers they received. Nicholson's *Journal*, in particular, has been celebrated for the extent to which the pioneering work on electrochemistry was first published in its pages.[82] By offering an outlet for original researches that was more focused and authoritative than the *Gentleman's Magazine*, and at once more rapid and less socially restrictive than the *Philosophical Transactions*, the new journals contributed significantly to reconfiguring the process of scientific communication. Indeed, in doing so they presented a challenge to the ponderously slow-to-appear transactions of the learned societies, which by the 1820s and 1830s were responding with more timely volumes of "proceedings".[83] Yet, scientific journals at this later period continued to have a major role in reporting on and anthologizing findings published elsewhere, as the case of Thomas Thomson's *Annals of Philosophy* well illustrates.

Commencing his new journal against the established competition in 1813, Thomson claimed that its distinctiveness lay in the way it restricted itself on the one hand to "original papers" and on the other to "translations of foreign papers supposed to be unknown to the generality of the British public". The latter were still sufficiently important to his purpose that he felt obliged to apologize at length for the difficulty of securing

[81] Georges Cuvier and Alexandre Brongniart, "Memoir on the Mineralogical Geography of the Environs of Paris," *Philosophical Magazine* 35 (1810): 36–58; John Farey, "Geological Remarks and Queries on Messrs. Cuvier and Brogniart's Memoir on the Mineral Geography of the Environs of Paris," *Philosophical Magazine* 35 (1810): 113–39; and Martin J. S. Rudwick, *Bursting the Limits of Time: The Reconstruction of Geohistory in the Age of Revolution* (Chicago and London: University of Chicago Press, 2005), 494–95.

[82] Lilley, "Nicholson's Journal".

[83] See Secord, "Science, Technology and Mathematics," pp. 451–59; Csiszar, "Broken Pieces of Fact", esp. introduction and ch. 1.

"foreign papers" in wartime conditions.[84] Indeed, looking back at mid-century, one obituarist claimed that Thomson's journal had been valued because it "presented an epitome of all the transactions of physical science in an agreeable and readable form."[85] More generally, the anthologizing practices of scientific journals continued to be an important aspect of their operation through a significant part of the nineteenth century.

My purpose in this chapter has been to re-expose the extent to which the early scientific journals were conceived and justified as attempts to anthologize scientific discovery, drawing together for the harassed and means-limited reader claims to knowledge made across the Western world. Much of the analysis here has been devoted to manifesting how different such journals were from the scientific journals of twentieth-century science, and to explaining their origins in relation to the conditions of the late eighteenth-century book trade. Yet, as this conclusion shows, the bearing of these sources on our understanding of the transnational circulation of knowledge is weighty, and much remains to be done in interpreting and analysing the evidence they provide.

---

[84] [Thomas Thomson], "Preface," *Annals of Philosophy* 3 (1814): [iii]–iv.

[85] [Robert Thomson?], "Biographical Notice of the Late Thomas Thomson," *Glasgow Medical Journal* 5 (1857): 138–39.

BETWEEN CALCUTTA AND KEW: THE DIVERGENT CIRCULATION AND PRODUCTION OF *HORTUS BENGALENSIS* AND *FLORA INDICA*

Khyati Nagar*

This paper explores the divergent production and circulation of botanical knowledge between the East India Company's Calcutta Botanical Garden and its main centre for botanical data collection at the Royal Botanical Gardens at Kew in the early nineteenth century. Through the examples of two botanical catalogues, this paper argues that the networks of communication were not entirely controlled by powerful administrators in the Metropole and that the independent production of scientific knowledge in the colonies inhibited the global flow of scientific information.

As the director of Kew Gardens, Sir Joseph Banks (1743–1820) presided over a vast colonial network of botanical gardens from his private study in London. Banks was the driving force behind the founding of the Royal Institution and served as President of the Royal Society for forty-one years after being elected in 1778. But it was his involvement in Kew Gardens that led to his major role in colonial botany. The East India Company consulted Banks for all scientific matters concerning botanic gardens in the colonies. As a result Banks was able to influence the appointment of the Company's first Superintendents in the botanical gardens at Calcutta ensuring that he retained direct contact between Kew and Calcutta.[1] But Banks' control was not absolute. William Carey (1761–1834), a missionary, publisher, and amateur botanist stationed in India, pointed in his introduction to *Hortus Bengalensis*, published in 1813, to an aspect of the network of colonial gardens that was beyond Banks' control. He wrote,

* I am very grateful for the opportunity provided by the "Circulating Knowledge, East and West" conference organizers for presenting this paper at Halifax in July 2010. I can't thank Bernard Lightman enough, who patiently read numerous drafts of this paper and provided intellectual and moral support to pursue this research. Many thanks also to Larry Stewart, Gordon McQuat and other referees for their comments that greatly improved this paper. Suggestions from Ann Shteir and Jonathan Topham were invaluable.

[1] Richard Grove, *Green Imperialism: Colonial Expansion, Tropical Island Edens and the Origins of Environmentalism, 1600–1860* (Cambridge; New York: Cambridge University Press, 1995), 338.

> The labours of Tournefort, Linnæus, Smith, Jussieu, Cavanilles, D'Herritier, Gaertner, Bartonand many others deserving of notice, either never made their appearance, or would have appeared in a form very different, and probably much inferior to that in which they are now published, had such institutions never existed. But the Botanical Gardens and other noble collections of plants in Europe could never have been brought to the perfection in which they now appear, had not public or private repositories of plants been formed in different settlements in Asia, Africa and America.[2]

Carey attributed the progress in European botanical research to public and private gardens in the colonies. In his introduction he went on to say that it was not absolutely necessary that the gentlemen who form smaller repositories of plants be scientific botanists. Carey advised amateur enthusiasts about the mode of collection and documentation as well. He asked them to number plants instead of giving names and include the vernacular name of the plant opposite the number. He stressed on the fact that vernacular names must be written by a native so as to retain the original format of the vernacular language. Carey's intervention in the creation of a botanical catalogue thus refuted the exclusive reputation of Sir Joseph Banks.

Even though Sir Joseph Banks supervised the creation of all botanic gardens in the Colonies, the circulation of knowledge about plant species was mediated by numerous plant collectors who weren't necessarily scientifically trained botanists. While it was pertinent for Banks that the Superintendents of the gardens be trained Linnaean botanists like him, the Superintendents had the freedom to appoint plant collectors and amateur enthusiasts such as William Carey, who in this case also wrote the introductions to both *Hortus Bengalensis* and *Flora Indica* and published them without consent from Banks. These botanical catalogues were compiled by William Roxburgh (1751–1815) who worked on the manuscript for these catalogues during his appointment as the Superintendent of the East India Company's Calcutta Botanic Garden from 1793 to 1813. *Hortus Bengalensis* was a one hundred and thirty page catalogue of plants documenting three thousand five hundred plant species growing in the Botanic Garden at Calcutta published in 1814 at Carey's Serampore Mission press. *Flora Indica* was a catalogue of description of Indian plants collected by

[2] William Roxburgh, William Carey, and East India Company Botanic Garden, *Hortus Bengalensis, or, A Catalogue [by W. Roxburgh] of the Plants Growing in the…East India Company's Botanic Garden at Calcutta. [Followed by] A Catalogue of Plants Described by Dr. Roxburgh in His MSS. Flora Indica but not yet introduced into the Botanical Garden* (Serampore Mission Press: 1813; 1814), iii.

Roxburgh throughout his time in India, from 1773 to 1813. It was published in three volumes, starting in 1820 published by Carey again and right up to 1874 by different publishers. The catalogues were published in octavo format and contained no illustrations (Fig. 1).

The publication of these catalogues reveals the divergent production and circulation of scientific knowledge between the botanic gardens at Calcutta and Kew in the early nineteenth-century. Collecting plant specimens and documenting them was a process by which the metropole ensured the flow of knowledge from the colony to be classified in London. Printed catalogues were a means of documenting scientific knowledge and their circulation ensured a flow of knowledge usually from the metropole to the colony and back. A continuous publishing circuit hence ran from the author to the publisher, printer, distributors, readers, and back to the author.[3] In the case of *Hortus Bengalensis* and *Flora Indica*, the circuit was broken since the catalogues were published in Calcutta without waiting for approval from Sir Joseph Banks. They were edited and published by William Carey, Roxburgh's friend and owner of the Serampore Mission Press after Roxburgh had left Calcutta in 1813, due to his failing health. Both catalogues are unique because the botanic knowledge they contain was collected, created and modified in India with the help of numerous intermediaries, Indian and European. Even though Roxburgh used the Linnaean sexual mode of classification to map Indian Flora, he deviated from strict norms of classification to make these catalogues more accessible to the East India Company's officials as well as Indian gentlemen interested in the study of botany.

By focussing on Roxburgh's catalogues I will examine the complex circulation of knowledge between England and India. I will do so by exploring the intersection of the history of science with the history of the book as they relate to early nineteenth-century India. I want to show that combining approaches from these two areas helps us to better understand the role of printed texts in the dissemination and circulation of botanical knowledge and the larger social and cultural networks in which that knowledge circulated. In the last decade the scholarship in the history of imperial science has expanded the meaning of 'colonial science'. Scholars such as Kapil Raj, David Arnold, Dhruv Raina and Sujit Sivasundaram have

3 Jonathan Topham, "Beyond the "Common Context", the Production and Reading of Bridgewater Treatises," *Isis* 89 (1998): 233–262. Here Topham elaborates on the idea of the circulation circuit as proposed by Robert Darnton.

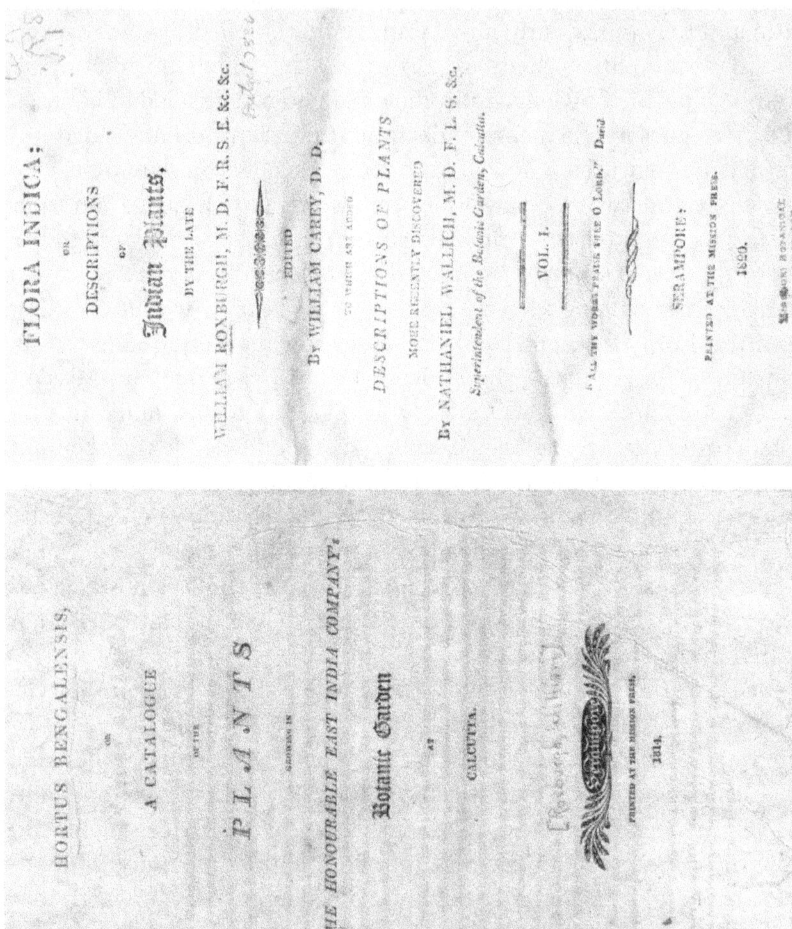
HORTUS BENGALENSIS, OR A CATALOGUE OF THE PLANTS GROWING IN THE HONOURABLE EAST INDIA COMPANY'S Botanic Garden AT CALCUTTA. SERAMPORE: PRINTED AT THE MISSION PRESS. 1814.

FLORA INDICA; OR DESCRIPTIONS OF Indian Plants, BY THE LATE WILLIAM ROXBURGH, M. D. F. R. S. E. &c. &c. &c. EDITED By WILLIAM CAREY, D. D. TO WHICH ARE ADDED DESCRIPTIONS OF PLANTS MORE RECENTLY DISCOVERED By NATHANIEL WALLICH, M. D. F. L. S. &c. *Superintendent of the Botanic Garden, Calcutta.* VOL. I. "ALL THY WORKS PRAISE THEE O LORD." *David* SERAMPORE: PRINTED AT THE MISSION PRESS. 1820.

Fig. 1 Title pages from *Hortus Bengalensis and Flora Indica* by William Roxburgh. Image courtesy of William Carey Institute and Biodiveristy Library.

argued that older diffusionist models, like the one proposed by George Basalla, not only essentialize indigenous knowledge, but also narrow the scope of studying the practice of science in the colonies.[4] My study analyses an instance of independent research in the colony where I build on the idea that colonial science practiced by European botanists in India was not merely an extension of 'metropolitan' science. A study of *Hortus Bengalensis* and *Flora Indica* reveals that the botanical knowledge produced in Calcutta did not always get relayed in its entirety to Kew because the production of textual and visual information was bound by geographical limitations, publication costs, material form of the catalogues as well as personal motivations such as those of Carey who wanted to make his mark on colonial botany. European botanists and botany enthusiasts such as Roxburgh and Carey acted as intermediaries between Kew, the East India Company, European botanical classification systems and Indian local uses of plants, thus creating and documenting new knowledge for the botanists at Kew. While they acknowledged metropolitan science, they also made autonomous decisions about publishing this new knowledge independently. This raises questions about the ways in which these catalogues have been discussed in earlier scholarship.

The Circulation of Knowledge and Historians

Scholarly work on *Hortus Bengalensis* and *Flora Indica* by historians of science has tended to ignore specific details of their production and circulation. Book historians mention the catalogues briefly as one of the many texts printed at the Serampore Mission Press but scholars in this area have concentrated on the revolution brought by the printing press and the emergence of the vernacular press in India. Book history focuses so much on the processes, production and form of the book and distribution networks that often other lesser-known genres outside the realm of literature, popular culture and rare books are left under-explored. Both fields tend to attribute successful revolutions to an individual. For example, in history

4 Basalla first proposed a model of diffusion of modern science in 1967 in his essay "The Spread of Western Science". According to his model, science originated in the nations of Western Europe and was diffused to the rest of the world through institutions and cognitive mechanisms resulting in the formation of independent scientific traditions. For a schema and discussion on the Eurocentrism of Basalla's model see Dhruv Raina, *Images and Contexts: The Historiography of Science and Modernity in India* (New Delhi; Oxford: Oxford University Press, 2003), 178–186.

of science, William Roxburgh is lauded as the father of Indian botany.[5] Similarly, in the scholarship on the history of the book, William Carey is celebrated as an important figure in the birth of vernacular publishing in Bengal. Yet, the role of Carey's press in the production of scientific knowledge remains under-acknowledged. Bringing Carey and Roxburgh into the discussion of the circulation and dissemination of botanic knowledge brings together ideas from both the history of science and the history of print culture.

This paper draws inspiration from James Secord's ideas in "Knowledge in Transit". Secord's approach is particularly advantageous when studying science and print culture in India because examining sites of intersection, like the catalogues in this paper, helps us to understand the wider settings of the creation of knowledge in the colonies. According to Secord:

> Attaining a global picture is not a question of transcending or erasing local practices but of giving more attention to practices of circulation on a wide variety of scales ... An approach grounded in communication opens up the possibility of integrating accounts of technical, specialist aspects of science with their wider uses ... So we need accounts of generic development of the field notebook, the experimental register, the museum catalogue, and other documents of practice , as bridging studies moving between specific passages of technical work and their wider settings.[6]

Secord's emphasis on an approach grounded in communication and the study of documents of scientific practice helps to create a sophisticated framework for studying *Hortus Bengalensis* and *Flora Indica* and their place in wider imperial networks.

In choosing the two catalogues as mediums of communicating scientific knowledge between Calcutta and Kew, I want to show that networks of communication were not entirely controlled by the metropolitan botanists and the creation of knowledge was in fact dependant on local intermediaries. These catalogues mark an important moment when 'colonial science' established its own course. They are proof that colonial science circulated in multiple communication networks, some of which were not controlled by powerful figures like Sir Joseph Banks who stood at the center of Britain's vast imperial network.[7]

---

[5] Tim Robinson, *William Roxburgh: The Founding Father of Indian Botany* (Chichester, West Sussex: Royal Botanic Garden Edinburgh, 2008).

[6] James Secord, "Knowledge in Transit," *Isis* 95 (2004): 666.

[7] Mark Harrison, "Science and the British Empire," *Isis* 96 (2005): 63.

Kew, the Calcutta Botanic Garden and William Roxburgh

In pre-Victorian and early Victorian England, natural theology continued to provide scientists with a framework in which to work. Nature was considered God's handiwork—searching for design led to understanding, and understanding allowed humans to make nature useful. Undiscovered and untended lands of the heathen could only be made useful through metropolitan and botanical intervention—tamed, cultivated and made productive by British intervention. This theme of improvement also influenced Sir Joseph Bank's thinking about Britain's overseas empire and shaped the conception of a network of botanical gardens in the colonies.[8] Moreover, the East India Company's administration in London and Calcutta realized the economic advantage of experimental crop plantations in the colonies, which meant that they gave much more importance to botanists and the science of botany.[9] A large part of British imperial wealth was based on plants—for example, teak, indigo, spices, opium and tea were used for trade. By 1790, 15 superintendents were appointed by Banks and the East India Company to run the network of botanical gardens in the tropics.[10] The first Superintendent of the Calcutta Botanic Gardens Captain Robert Kyd (1746–1793) was instrumental in the creation of satellite botanical sites in the colonies as he saw these as centres for preserving plants in transit.

Kyd established the botanic garden near Calcutta in 1787. In his proposal to the Company, Kyd suggested that the botanic garden be established not only to collect rare plants, but also to benefit trade, and to profit the local population by protecting them against famine. Roxburgh succeeded Kyd as the Superintendent of the Calcutta Botanic Garden in 1793. He had studied medicine at Edinburgh University where he was also introduced to Linnaean taxonomy by Professor John Hope (1725–1786).[11] Roxburgh had first joined the East India Company as a surgeon's mate on

---

8 Richard Drayton, *Nature's Government: Science, Imperial Britain, and the 'Improvement' of the World* (New Haven: Yale University Press, 2000), 98.

9 Richard Grove, *Green Imperialism: Colonial Expansion, Tropical Island Edens and the Origins of Environmentalism, 1600–1860* (Cambridge; New York: Cambridge University Press, 1995), 336.

10 Grove, *Green Imperialism*, 339.

11 John Hope greatly influenced Roxburgh's education as a botanist. It was in his lectures at Edinburgh that Roxburgh was introduced to minute botanical illustrations. See: Henry Noltie, *John Hope (1725–1786): Alan Morton's Memoir of a Scottish Botanist* (Edinburgh: Royal Botanical Garden Edinburgh Press, 2011), 58–81.

the Company's ships and then served as assistant surgeon at the Madras general hospital in 1776.[12] He was promoted to full surgeon in 1780 and spent many years in the Northern Circars, 200 miles north of Madras, where he was given permission to establish an experimental botanical garden.[13]

At the time, Johann Gerhard Koenig (1728–1785), student of Linnaeus, was already working there as a naturalist for the Nawab of Arcot. He was appointed the first Company naturalist in the Carnatics, followed by Scottish physician and naturalist Patrick Russell (1727–1805) in 1785 and then Roxburgh in 1789. While stationed in Samulcottah in 1781, Roxburgh frequently worked with Koenig and was allowed to grow pepper, breadfruit, coffee and mulberry among other plants. By 1787 he had more than 4000 pepper plants and had developed a great interest in documenting plant species along the Coromandel Coast.

Greatly impressed by the work of Koenig, Roxburgh had realised the importance of drawings as they were used as 'icontypes'. An icontype acts as a specimen when a dried herbarium species is not available and botanists use it to describe plants that are discovered for the first time.[14] Usually, plant types were preserved in herbaria but in the absence of specimens, drawings or even prints acted as types and are known as icontypes. Icontypes played an important role in establishing Linnaeus' nomenclature system because it relied heavily on the sexual morphology of flowers in mature plants. Since it is not possible to see all flower parts in dried herbarium specimens, Roxburgh knew that the drawings would be invaluable to the botanists in the Metropole. By the time Roxburgh was appointed the superintendent of the botanic garden at Samulcottah in 1789, he had more than 100, 000 plants and at his own expense employed two Indian artists who made 400 to 500 botanical drawings of his plant specimens growing in the Coromandel Coast.[15]

Roxburgh sent drawings to Kew starting in 1791 and continued until he left Calcutta in 1813. When Roxburgh had first sent the drawings to

---

[12] Ray Desmond, *The European Discovery of the Indian Flora* (Oxford; New York; Kew, England: Clarendon Press; Royal Botanic Gardens, 1992), 47.

[13] Desmond, *Flora*, 47.

[14] See, "William Roxburgh's Legacy," http://apps.kew.org/floraindica/htm/legacy.htm; and William T. Stearn, "Carl Linnaeus's Acquaintance with Tropical Plants." *Taxon* 37, no. 3 (1998): 776–781 (accessed December 5 2010).

[15] Ray Desmond, "Roxburgh, William (1751–1815)," in *Oxford Dictionary of National Biography*, online ed., ed. Lawrence Goldman, Oxford: OUP, http://www.oxforddnb.com.ezproxy.library.yorku.ca/view/article/24233 (accessed Dec 5, 2009).

Kew, Banks was not completely satisfied with them, but did not convey the defects he saw to Roxburgh. Roxburgh waited for Banks' approval for two years and in that period started viewing his illustrations more objectively and thought that they looked rigid. Revised drawings were sent again to Kew in the hope that Banks would notice that the drawings had improved. Banks surprised Roxburgh with a quick response and was impressed by the improved skill of the Indian artists, particularly with the precision of their botanical details.[16] He thought that these drawings would provide the European botanists with much superior renderings of Indian plants as compared to an earlier work on Indian botany, *Hortus Malabaricus.* [17] *Hortus Malabaricus* (1678) by Van Reed was one of the first textual and visual records showing a documentation of plants in the Malabar Coast of India. Drawings in *Hortus Malabaricus* did not follow one style of botanical illustration and sometimes the renderings were more stylised than scientific. Banks was also impressed by Roxburgh because he had funded the creation of the illustrations from his own pocket and requested that the East India Company fund the publication of *Plants of the Coast of Coromandel.*

The East India Company's response to Sir Joseph Banks' request for funding the publication of *Plants* was immediate and an amount of £300 was sanctioned for the first edition. *Plants of the Coast of Coromandel; Selected from Drawings and Descriptions Presented to the Hon. Court of Directors of the East India, Published, by their order, Under the Direction of Sir Joseph Banks* was published in three volumes: the first was published in 1795 followed by the second and third volumes in 1798 and 1819. All volumes were printed by W. Bulmer and Company in London for Mr. Nichol, his Majesty's bookseller, who had also assisted Banks earlier in publishing Captain Cook's voyage.[18] While the first two volumes were well received in terms of sales and reviews, the sales for the last volume were

---

[16] Ray Desmond, *The European Discovery of the Indian Flora* (Oxford; New York; Kew, England: Clarendon Press; Royal Botanic Gardens, 1992), 48.

[17] The first volume of Hortus Malabaricus was first published in 1678 under the leadership of Van Reed, commander of Malabar under the Dutch East India Company. See: Desmond, *Flora*, p. 22. and for a deeper analysis, see: Kapil Raj, *Relocating Modern Science: Circulation and the Construction of Scientific Knowledge in South Asia and Europe, Seventeenth-Nineteenth Centuries* (New York: Palgrave Macmillan, 2007), 44–52. It is also significant that Hortus provided Linnaeus "with most of his information about the Indian flora. He based many new names on illustrations and descriptions of plants in Rheede's work which would otherwise have been unknown to him..." see William T. Stearn, "Carl Linnaeus's Acquaintance with Tropical Plants," *Taxon* 37, no. 3 (1998): 779.

[18] *Plants of the Cost of Coromandel*, Vol. 1, preface by Col. Patrick Russell.

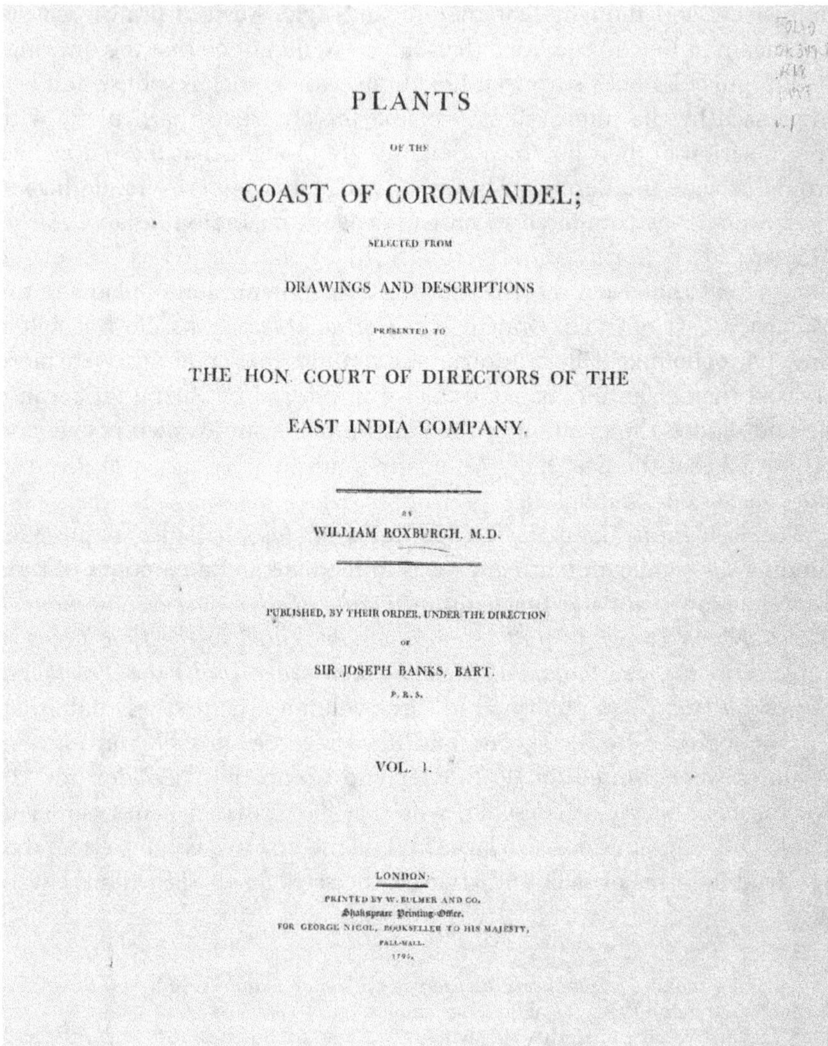

PLANTS

OF THE

COAST OF COROMANDEL;

SELECTED FROM

DRAWINGS AND DESCRIPTIONS

PRESENTED TO

THE HON. COURT OF DIRECTORS OF THE

EAST INDIA COMPANY.

BY

WILLIAM ROXBURGH, M.D.

PUBLISHED, BY THEIR ORDER, UNDER THE DIRECTION

OF

SIR JOSEPH BANKS, BART.

P. R. S.

VOL. I.

LONDON:

PRINTED BY W. BULMER AND CO.

Shakspeare Printing-Office,

FOR GEORGE NICOL, BOOKSELLER TO HIS MAJESTY,

PALL-MALL.

1795.

Fig. 2 Title page, Plants of the Coast of the Coromandel Vol. 1. Image courtesy: Biodiveristy Library: http://www.biodiversitylibrary.org/item/9711#

disappointing both in London as well as in India. Despite the interest in Europe of books pertaining to exotic oriental flora, *Plants* was not a huge commercial success.

However, the production and consumption of *Plants* of the Coast of Coromandel is an example of a complete publication circuit. The drawings and text were sent from India to Kew, regulated and certified at Kew, and then an improved version was published in book form in London, under the direct supervision of Banks. The books were printed, distributed and sold in London and also sent to India. Since they were expensive they solicited a limited readership among wealthy gentlemen of science. The knowledge produced in one part of the world was thereby communicated in its entirety to another part of the globe and then through print brought back to its place of origin. Throughout the nineteenth century, eminent botanists such as Robert Wight (1796–1872) and Joseph Hooker (1817–1911) were influenced by Roxburgh's work and used it in their own documentation of Indian Flora.

At the time of publication of *Plants of the Coast of Coromandel*, Roxburgh had already taken over the Superintendency of the Calcutta Botanic Garden. Soon after coming to Calcutta, Roxburgh had hired numerous European and Indian assistants, including gardeners, nurserymen and artists. He sent plant collectors to the Malay and Indonesian islands to collect rare and varied plant species. When plant specimens came to Roxburgh, he nearly always had a life-sized rendition of a plant or specimen by a native artist. He had them by his side when writing his descriptions. By the time Roxburgh left Calcutta in 1813, the garden had 2600 species out of which 2500 were documented in drawings.[19]

Thus, a very important aspect of Roxburgh's work in India was the botanical drawings created by his local artists. He called these the Flora Indica drawings. They influenced the production of *Hortus Bengalensis* and *Flora Indica* later, as they were used for describing plant species in both catalogues. The drawings are also important because they were part of the folio edition of *Plants of the Coast of Coromandel* but were later eliminated in octavo editions of *Hortus* and *Flora* to reduce production costs. The importance of illustrations in the production of botanical knowledge in India needs a brief discussion as these are an important aspect of Roxburgh's botanical legacy and local collaborations in India.

---

19 J. R. Sealy, "The Roxburgh Flora Indica Drawings at Kew," *Kew Bulletin* 11, no. 2 (1956): 297.

Indian Artists and the Uses of Local Knowledge

In the early nineteenth century, along with circulation of plant specimens and their textual documentation, visual documentation was also an important means of ensuring that communication of botanic knowledge to Kew was complete. Botanical illustrations reveal not only the intersection of art and science: they also disclose border crossings between two different cultures.[20] A closer study of the *Flora Indica* drawings reveals that scientific documentation in India relied heavily on the practices of Indian artists and they inherently brought in cultural and artistic aspects of art into their botanical renditions. Although there were accomplished botanical artists at Kew and in Europe, their numbers were few and bringing them to India would have proved to be a very expensive endeavour for the East India Company.

Roxburgh employed artists who were already trained in traditional Indian art-forms but very little is known about who they were. None of the illustrations were signed by the artists and the only writing seen on the illustrations is that of Roxburgh. However, the legacy of artistic skills employed by Indian miniature and textile artists is well known, along with a vast scholarship on Indian miniature paintings.[21] Miniature artists excelled at distinctive traditions of painting in different parts of India and were sought after because of their skill in painting minute details and their attention to form and colour.[22] Chief among these styles were the Rajput, Pahari, Deccani, and Mughal. Paintings from the royal courts in all these styles portray a rich style infused with vibrant colours and intricate painting techniques.

At the time of Roxburgh in the Carnatic, although the royal courts were entrenched in wars and debt, the legacy of older miniature paintings from the time of the Mughal emperors Akbar and Jahangir was strong. For the miniature artists the representation of a just and kind ruler was through the depiction of a paradise-like kingdom and the image of the Emperor's

---

20 Fa-ti Fan, *British Naturalists in Qing China: Science, Empire, and Cultural Encounter* (Cambridge, MA: Harvard University Press, 2004), 42.

21 See, Mildred Archer, *Natural History Drawings in the India Office Library* H.M.S.O, 1962; Pratapaditya Pal, *Court Paintings of India: 16th–19th Centuries* (New York: Navin Kumar for Kumar Gallery, New Delhi, 1983); Susan Stronge, *Painting for the Mughal Emperor: The Art of the Book 1560–1660* (London: V&A, 2002).

22 Tim Robinson, *William Roxburgh: The Founding Father of Indian Botany* (Chichester, West Sussex: Royal Botanic Garden Edinburgh, 2008), 95.

gardener resonated with the idea that the just king controls his kingdom to allow healthy growth. Their technique required them to paint in multiple layers and later burnish it to produce saturated flat colours.[23] Additionally, the clothing and borders of the paintings were decorated with minute gold rosettes, palms and other flora. This style focussed on the minute details and flat perspective-less technique which was very different than the flower painting techniques used by legendary European botanical painters of flowers such as Georg Dionysus Ehret (1708–1770) and Pierre Joseph Redoute (1759–1840). Roxburgh's artists were bound by the illustration techniques that they learned in the royal courts as well as by the Linnaean practices of botanical illustration developed over a period of time by artists like Ehret. Ehret is closely associated with popularizing Linnaeus' ideas through his illustrations and wielded power as a botanical illustrator. When Ehret first contributed his illustrations to *Hortus Cliffortianus* (1738), an account of the rare plants in the garden of banker George Clifford in Amsterdam, he refused to paint the pistils, stamina and other small parts because he thought it would spoil the drawings. Linnaeus, who was employed by Clifford for the project was not pleased and Ehret eventually gave in.[24] Artists such as Ehret defined the styles and techniques conducive to representing the Linnaean classification system. The 18th century saw numerous illustrated botanical books that attempted to explain and promote the Linnaean method. The illustrations were an improvement on visuals of flora in herbals and florilegia from preceding centuries and gave minute details of the flowering plant as well as the dissected parts of the flower.

When it came to botanical illustrations, Roxburgh's artists did not have the same advantage. They had to quickly learn to look at the plant specimens with the relevant view point of botanical illustrations promoted by European institutions of science. One cannot disregard the illustrations of Roxburgh's Indian artists as 'stiff' or 'flat' because the artists' viewpoint of nature in the courtly context was different. The royal courts they were initially employed by in fact functioned as intermediary institutions for promoting scientific knowledge because of the fact that they employed artists with skills that could be transferred for the use of documenting local flora

23 Susan Stronge, *Painting for the Mughal Emperor: The Art of the Book 1560–1660* (London: V&A, 2002), 88.

24 Gillian Saunders, and Victoria and Albert Museum, *Picturing Plants: An Analytical History of Botanical Illustrations* (London: Zwemmer, 1995), 89.

by European naturalists in India. The role of Roxburgh's artists in illustrating *Plants of the Coast of Coromandel* is central to the narrative of circulation of scientific illustration techniques. The naturalists employed in the colonies were intermediaries between institutions like the Royal Society, the Dutch East India Company and the British East India Company and local artists. The artists employed by Roxburgh remain unknown and present the same problem that Pamela Smith has discussed in the example of what she calls the Weiditz-Brunfels problem. While Otto Brunfels is recognized as the author of the famous *Herbarum vivae eicones*, the illustrator Hans Weiditz is not granted as much importance because the illustrations are supplementary to the text which constitutes the "more serious business of science".[25] In this case, Roxburgh was important because he was the company naturalist; his artists were not identified or recognized in the same capacity as him not only because they were Indian but because their art only supported Roxburgh's science. The artists themselves could not have claimed authorship over the paintings because they were made for an unfamiliar purpose using new techniques. It was realm and technique that they were beginning to get familiar with but it was not their own yet.

For centuries Hindu and Muslim royals had employed artists of the highest skill. Numerous other artists worked across the country in the flourishing textile and furniture industries. These artists were experts at capturing the beauty of Indian flora and fauna on cloth and elaborate inlay work in wood and marble. Many Europeans, including Governor Generals of India and prominent British artists such as Thomas and William Daniell, employed Indian artists for personal commissions and as assistants for their own work. It is possible that the artists did not belong to royal courts and were engaged in the textile industry or might have been painters of ethnic landscapes.[26] The cloth wall-hanging (Fig. 3), possibly from the Coromandel Coast, is an example of the technique used by Indian textile artists.

The wall hanging in this image from the Coromandel Coast shows the popular use of floral motifs in Indian textiles in the eighteenth and nineteenth centuries. These hand-made textiles were in high demand both in India and in Europe. The artists were first required to draw and then use

[25] Pamela H. Smith, "Art, Science, and Visual Culture in Early Modern Europe," *Isis* 97, no. 1 (March, 2006): 83–100.

[26] The Roxburgh Artists. Accessed on October 15 2010. http://apps.kew.org/floraindica/htm/artists.htm

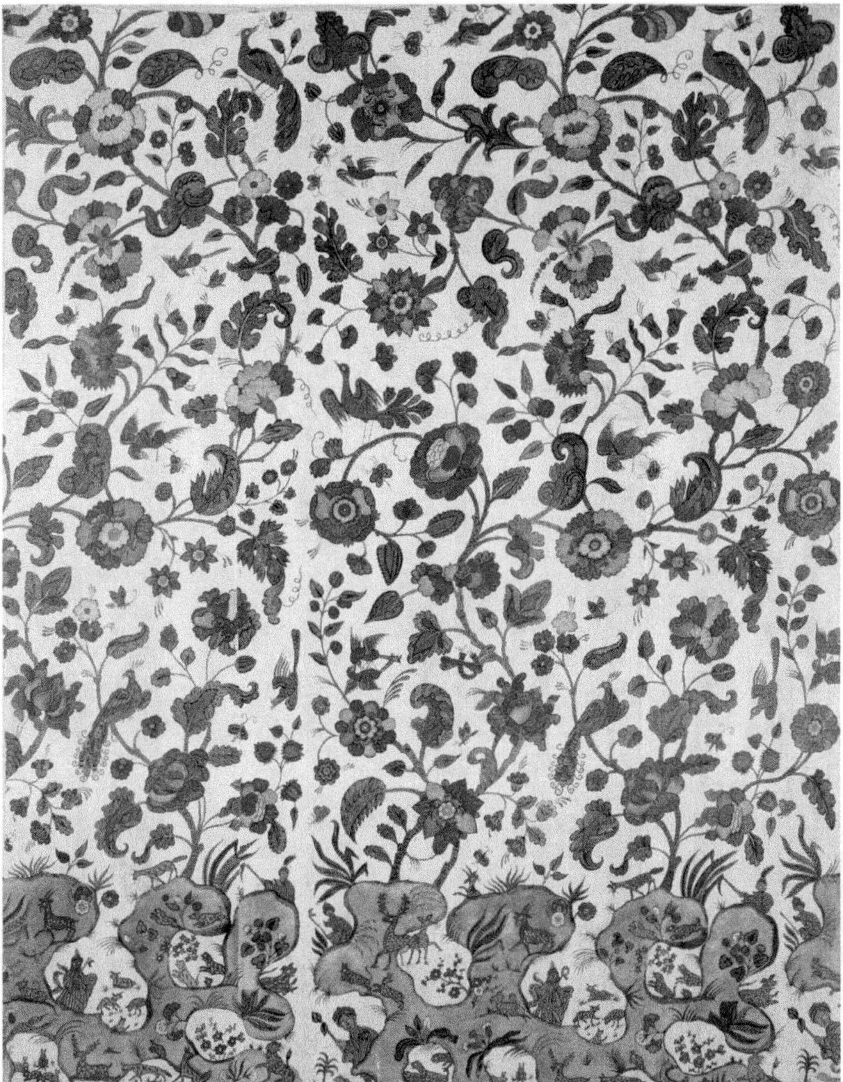

Fig. 3 Wall hanging, painted and dyed cotton c. 1700, possibly Coromandel Coast, India. Image courtesy: ©Victoria and Albert Museum, London.

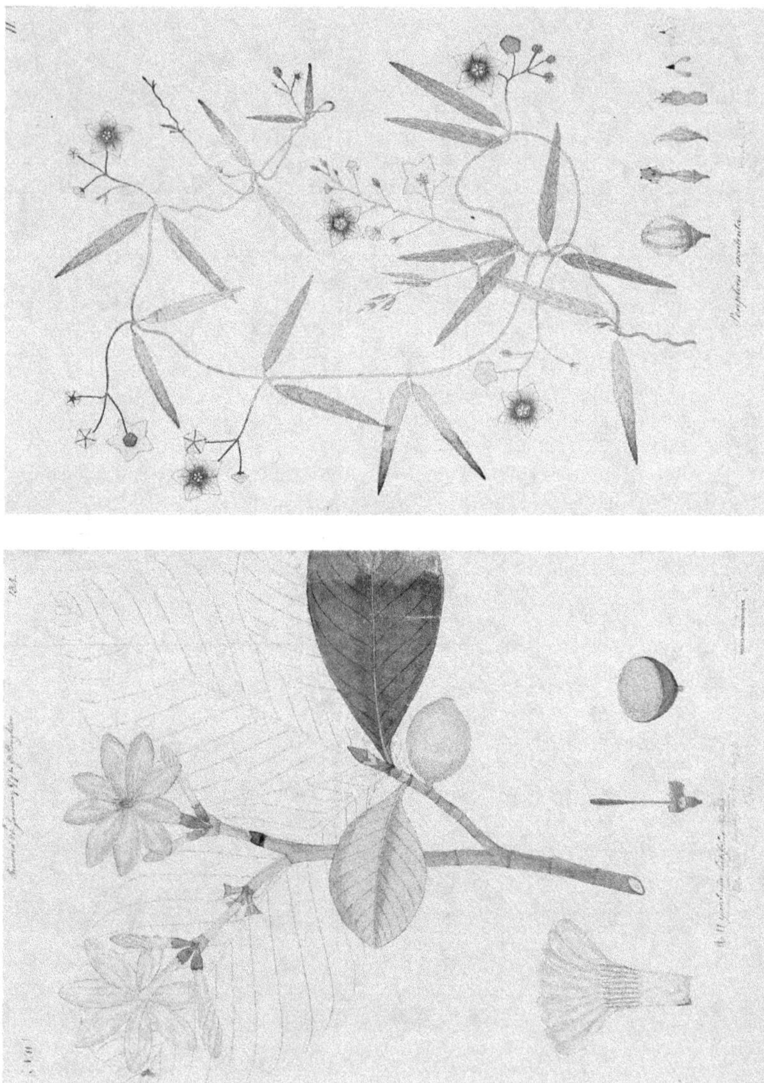

Fig. 4 Illustrations for Gardenia latifolia and Asclepias rosea. Image Courtesy of Kew Gardens, http://apps.kew.org/floraindica/displayImages

a complex set of dying techniques to achieve beautifully dyed textiles. This example also illustrates the flat rendition techniques that might have inspired Roxburgh's artists in their botanical drawings. The colour rendering of the flowers and leaves in illustrations from the *Plants of the Coast of Coromandel* are also two-dimensional and carry a pattern-like appearance much like the wall-hanging. The seeds and pistils look more decorative than scientific, as is obvious in the following examples (Fig. 4).

In both examples in Fig. 4, the illustrations cover the entire page in a pattern like style. Also, the parts of the flowering plant are displayed like borders for the illustration much like the borders in miniature paintings and textiles. Over a period of time Roxburgh's artists followed a more European botanical illustration style but the Flora Indica drawings remain unique in their visual quality. The drawings show the impact of available local resources on scientific practices much like the textual documentation in *Hortus Bengalensis* and *Flora Indica* shows local and global collaborations. This example proves that European scientific ideas found a place in Indian artistic collaborations but fell short of expectations at Kew when compared to European standards in realistic flower painting exemplified in the works of the famous flower painters and botanical artists such as Ehret and Redouté. Even though Roxburgh's artists never achieved the same standards as those of European flower painters, they excelled as botanical artists and their illustrations remained in global circulation and use by other botanists in the nineteenth century. For Roxburgh, the circulation of plant illustrations was as important to the construction of a global system of knowledge about plants as was the circulation of plants themselves. He ensured this by sending his drawings back to Kew for comments and simultaneously sending and collecting more plant species for documentation from all over the globe.

While Roxburgh's drawings provided relevant scientific information they did not appease the artistic sensibilities of those used to European flower paintings. From the perspective of European art, Roxburgh's artists proved to be good botanical artists but poor flower painters. The conflict in the convergence of Indian art styles and scientific illustration is evident in the drawings. However, this distinction fails to be recognized my most historians of art and science. These early drawings do not conform to the standards that came about through the confluence of European artistic styles in painting and the advancements in systems of classification or botanical illustrations. The drawings however do represent the confluence of Indian art and European scientific illustration styles thus making them unique and not just a poor representation of European botanical illustration styles in global circulation. The circulation of visual information thus remained as interesting as the circulation of plants species.

The Circulation of Plant Species, their Documentation and more Local Knowledge

Throughout his time in Calcutta, Roxburgh sent plants to London at least once every year, through Sir Joseph Banks, the botanist A.B. Lambert and the Linnaean Society, among other botanists and societies. Plants were also sent to other parts of the world, such as the West Indies and St. Helena. In addition to the specimens, seeds of teak, various sorts of hemp cultivated in Bengal, tobacco, indigo and coffee were also dispatched. This transfer of plants was not one-way. Plant species were constantly being sent to Calcutta from Kew and other parts of the world.[27] Roxburgh's *Hortus Bengalensis*, the catalogue that documented plants growing in the East India Company's garden at Calcutta, traces the range of origins of some of the plants to the West Indies, Peru, Virginia, Canary Islands, Europe, Arabia, Cape of Good Hope, Ceylon, Andaman Islands, Nepal, Tibet, Sumatra, New South Wales and China.[28] It is interesting to note that in *Flora Indica* (a description of all plant species collected in India), Roxburgh acknowledges his network of European colleagues who sent him plant specimens but does not mention the names of other local plant collectors who he might have employed to collect plants from other parts of India and the Indian Archipelago. His friends in Calcutta, including botanists, company surgeons and doctors, Orientalist scholars and Captains of trade ships, all sent him plants from wherever they were stationed in India and abroad. Some of these friends were William Carey, his missionary friend, Henry Thomas Colebrooke, Orientalist scholar and administrator, and Francis Buchanan, East India Company surgeon and botanist.

With the help of numerous plant collectors and through his own efforts, Roxburgh collected abundant new plant species, but the task of taxonomic classification was not an easy one. He decided to use the Linnaean sexual system of classification in both his catalogues, influenced by his earlier knowledge of the system that he had acquired from John

---

[27] Tim Robinson, *William Roxburgh: The Founding Father of Indian Botany* (Chichester, West Sussex: Royal Botanic Garden Edinburgh, 2008), 93.

[28] Robinson, *William Roxburgh*, 93.

Hope at Edinburgh and later with Koenig at Samulcottah. According to Roxburgh it was the perfect choice for the wide range of plant species that he found in India and its neighbouring countries: "The vegetable world is full of proofs that the sexual system of the immortal Linnaeus is founded on the soundest principles; and not a single plant have I ever found in India, that does not corroborate this fact."[29] Despite this adoration of Linnaeus, Roxburgh deviated from traditional rules of classification described in *Species plantarum* (1753) in the hopes that this would benefit a wider range of readers and familiarize them with local names of plants. He explains this idea to his audiences in the explanatory remarks at the beginning of *Hortus Bengalensis*:

> Botanists, particularly those who are not masters of the science, must derive equal advantages from a general knowledge of one universal nomenclature; I will therefore endeavour to contribute my mite, to render a more general communication of Botanic knowledge less difficult, by omitting, in this catalogue the usual column of English names and consign the few used, to that appropriated to the Asiatic names; such as the Sanskrit, Hindoostani, Bungali, Tamul, Telinga, &c. &c. provincial or vernacular names whenever they can be ascertained with any tolerable degree of certainty.[30]

Roxburgh's reference to "those who are not masters of the science" indicates that he is addressing non-specialists, including Indian gentlemen of science as well as European officers employed by East India Company. He writes that the vernacular names can be ascertained with only a tolerable degree of certainty because Indian gentlemen do not always agree on the names of plants until the plant has a peculiar quality, use or habit.[31] Moreover the plant had a different name in each Indian language even if the word differed to a small degree. As a result, Roxburgh had some reservations about using indigenous plant names. But he thought that they could be made useful through displaying them alongside European scientific names. According to C. A. Bayly, the noted scholar of Indian history, it is this mapping of Indian names on Linnaean nomenclature that later

[29] William Roxburgh, William Carey, and East India Company's Botanic Garden, *Hortus Bengalensis, or, A Catalogue of the Plants Growing in the Honourable East India Company's Botanic Garden at Calcutta. [Followed by] A Catalogue of Plants Described by Dr. Roxburgh in His MSS. Flora Indica but not yet introduced into the Botanical Garden* (Serampore Mission Press: 1813), in explanatory remarks on p. 1.

[30] Roxburgh and Carey, *Hortus Bengalensis*. See the explanatory remarks at the beginning.

[31] Roxburgh and Carey, *Hortus Bengalensis*.

became a catalyst for a systematic study of Indian botany.[32] In his remarks Roxburgh also adds that *Hortus* has more abbreviations and characters than usual in botanical catalogues to compress information and include local names in 15 Indian languages. He realized that problems of classification were compounded due to the aforementioned reasons. An instance of this problem can be seen through the example of numerous species of *Curcuma* collected by various people from different territories in and around India. Sometimes the donors sent more than one species and not all reached Roxburgh in good condition. He voiced his frustration about not being able to correctly recognize plant species because of their condition, in *Asiatick Researches*, Journal of the Asiatick Society, citing difficulty in classifying species of curcuma without having access to the colour of the flower.[33]

Roxburgh faced further difficulties because he did not have access to the new botanical research being conducted in London. His fears were not unfounded because he acknowledged an overlap of knowledge in *Flora Indica.* While describing the species *Curcuma Zedoaria*, he writes about a sample that he gave to Banks in 1805 who found out that it was the real *Zedoaria* of the British Materia Medica, and also ascertained that the root of Roxburgh's *Curcuma Zedoaria* was *Zedoaria Rotunda* of the shops in London.[34] The systemic name that Roxburgh assigned to the species was already given a different name by British botanists in London. It is for this reason that the *Flora Indica* drawings were so important. While working on *Flora Indica* Roxburgh numbered each plant description and he also gave the same number to the corresponding drawings with the intention of making more accurate observations. Without drawings in Flora Indica it must have been harder to ascertain the species with surety.

Other than meticulous scientific observations, documentation and classification, *Flora Indica* also offers unique accounts of social and cultural aspects of life in India found in the 'observations' section of different species. For *Curcuma zerumbet*, Roxburgh observed that the dry root when

---

32 C. A. Bayly, *Empire and Information: Intelligence Gathering and Social Communication in India, 1780–1870* (Cambridge; New York: Cambridge University Press, 1996), 272. Bayly very briefly mentions Roxburgh without elaborating on the reason why Roxburg's catalogues were a catalyst for a systemic study of Indian Botany.

33 Tim Robinson, *William Roxburgh: The Founding Father of Indian Botany* (Chichester, West Sussex: Royal Botanic Garden Edinburgh, 2008), 101.

34 William Roxburgh et al., *Flora Indica, Or, Descriptions of Indian Plants* (Serampore: Printed at the Mission Press, 1820; 1824), 23.

powdered and mixed with *Caeselpinia sappan*, produced the red powder *Abeer*. It was commonly used by Hindus for their festival of colours, *Holi*.[35] The wood of *Schrebera* was used by weavers to build many parts of their looms as it was less likely to bend or warp. The dried bark of *Asclepias pseudosarsa*, ground and fried in butter, was used by natives of the Coromandel Coast to treat thrush in children. Fresh buttermilk was often sold in the bazaars in India and the mucilaginous fresh leaves of *Pedalium Schreb* were used by vendors to thicken buttermilk that they had adulterated with water. They added water, and then simply stirred it a few times with the leafy branches of this tree. As a result their merchandise would seem rich and of the best quality and earn them higher profits.

These observations by Roxburgh, help to look beyond the specific scientific use of this catalogue. While it is common for Floras to have descriptions and native uses of plants, Roxburgh took extra effort to include names in different Indian languages, making the descriptions more accessible. Although when the mathematician and botanist C. B. Clarke (1832–1906) re-published *Flora Indica* again, in 1874, he was critical of Roxburgh for using so many names. According to him the Indian local names were perhaps more useful than the Sanskrit names that were used only to satiate the educated Bengali's need for enriching the language. However, observations of daily practices of natives, informs the reader of the indigenous association to plants. Indian plants then don't remain the sole property of Western science but are implicitly attributed cultural and indigenous scientific agency. The practices of metropolitan science are intertwined not just with practices of colonial science but also with indigenous science and culture. In the early nineteenth century these observations could have also been very useful to the officers of the East India Company who were trying to gather as much knowledge about India as part of improving its administration and gathering information about the natives of India. Roxburgh actively kept collecting plants and recorded all his findings regularly until 1813.

In his last few years in Calcutta, Roxburgh's health failed and he returned to England via the Cape of Good Hope in 1813. At this time Roxburgh left the final manuscript for *Hortus Bengalensis* with his dear friend William Carey as well as a manuscript for *Flora Indica*. Toward the end of his life,

35 To this day *Abir* or *Gulal*, the red powder, is used to colour fellow revellers of Holi or the festival of colours in India. Considered auspicious by the Hindus, the colour is also used playfully for celebrations relating to the Hindu Gods *Krishna* and *Ganesha*.

Roxburgh would have liked to see a complete *Flora Indica* but he did not live to see the printed version. This task was taken over by William Carey. *Flora Indica* and *Hortus Bengalensis* became collaborative projects that involved many people in Calcutta. Chief among the people who helped in the publication of *Flora*, were Nathaniel Wallich (1785–1854) superintendent of the Calcutta Botanic Garden after Roxburgh, who helped with the editing, and William Carey, who also helped edit and print the catalogues. Although Roxburgh's association with the metropole established him as a successful botanist, the publications of his work took on a life of its own, independent of Kew or Banks.

The Publication of Hortus Bengalensis *and* Flora Indica

The publication of both catalogues was the product of Roxburgh's term as Superintendent of the Calcutta Botanic Garden. In 1803 Roxburgh had already finished arranging his text for *Flora Indica* but was hesitant to publish it before consulting botanists in Britain. Roxburgh decided to exclude the drawings from this catalogue because many of the species were already depicted in *Plants of the Coast of Coromandel*. Possibly, Roxburgh also wanted to bring down the publication expense due to the failure of *Plants* to cover costs.[36] But the lack of illustrations was only one difficulty that stood in the way of conveying the appearance of plants to his readers. The biggest problem Roxburgh faced while describing plants in *Hortus* was the lack of access to botanic literature and ongoing research in Britain. By the time *Flora* was published, 185 species and their descriptions had already appeared in books by other British botanists in London. They had been assigned different systemic names than those used by Roxburgh, thus causing confusion about the species in his catalogues.

Roxburgh was not solely responsible for the printed version of his catalogues. Both catalogues were edited and printed by William Carey. Carey also wrote the introductions. While it is unclear when Carey might have first met Roxburgh, they both moved to Calcutta in 1793. They must have interacted soon after arriving in Calcutta because of Carey's interest in botany. In *Hortus*, Carey echoed Roxburgh's scepticism about using indigenous knowledge: 'But among the natives of the east, there are few who think about the productions of Nature ...', and added that native materia

[36] Ray Desmond, *The European Discovery of the Indian Flora* (Oxford; New York; Kew, England: Clarendon Press; Royal Botanic Gardens, 1992), 68.

medicas only documented around 400–500 plant species. In comparison Carey showed that *Hortus* had documented 3500 species, collected from all over the world. Carey had originally started the Serampore Mission press in 1800 in order to print bibles in Indian languages. Paper was imported from England and metal type was brought from the newly established type-foundry set up by Charles Wilkins (*bap.* 1749, *d.* 1836) in Calcutta.[37] The role of Carey's Serampore Mission Press in establishing modern missions in India is well-established because he used his press to print the bible in numerous Indian vernacular languages.[38] But Carey also used his press to produce botanical publications.

Before leaving for London in 1813, Roxburgh handed the manuscripts for both catalogues to William Carey. Roxburgh's health was failing and he realized that they needed to be printed quickly. William Carey was a perfect choice, because he was a knowledgeable amateur botanist, a dear friend and he owned a printing press. Carey edited and published *Hortus Bengalensis* in 1814. In a letter to the East India Company, Carey requested funds to publish it. Printing the catalogue would be expensive as he had calculated the cost at eight rupees. He requested the East India Company to subscribe for fifty copies to cover the costs. Carey's request was approved and eighteen months later Carey sent printed copies of the Catalogue to the East India Company. Although funded by the East India Company, the cost of printing *Hortus Bengalensis* was considerable. Eight rupees per copy can be translated to 4 pounds per copy, a considerable amount for that time. These copies were certainly only affordable by the East India Company or bigger institutions such as the botanic garden at Calcutta or educational institutions like Fort William College.

Right after leaving Calcutta in 1813, and before his death in 1815, Roxburgh also gave a copy of his manuscript of *Flora Indica* for editing to Robert Brown, clerk, librarian, and housekeeper for the Linnean Society of London, and an eminent botanist.[39] Brown took a long time to edit *Flora Indica* and was still slowly working on it at the time of Roxburgh's death. Frustrated by this delay, William Carey, with the help of Nathaniel

---

37 Anindita Ghosh, *Power in Print: Popular Publishing and the Politics of Language and Culture in a Colonial Society, 1778–1905* (New Delhi: Oxford University Press, 2006), 111. Charles Wilkins was a prominent printer and owned a press in Calcutta, before Carey.

38 Ghosh, *Power in Print*, 35.

39 D. J. Mabberley, "Brown, Robert (1773–1858)," in *Oxford Dictionary of National Biography*, ed. H. C. G. Matthew and Brian Harrison (Oxford: OUP, 2004); online ed., ed. Lawrence Goldman, October 2009, http://www.oxforddnb.com.ezproxy.library.yorku.ca/view/article/3645 (accessed January 18, 2010).

Wallich, decided to continue with the printing of the catalogue and the first edition of *Flora Indica* was produced at the Serampore Mission Press in 1820. Wallich helped edit this version of *Flora Indica* and also contributed to the description of species from his collection of plants. The first volume of *Flora* was first published in 1820 by Carey and Wallich. The second volume was published in 1824 and contained many additions by Wallich. Carey published it again in three volumes in 1832 and a final one-volume original version of Roxburgh's MS was published by C. B. Clarke in 1874. Clarke's motivation to re-publish *Flora Indica* was to prove that J. D. Hooker's *Flora Indica* was not superior in any way to Roxburgh's. In fact, he thought, it was more useful to the amateur botanist than Hooker's version.[40] In his introduction to the 1874 edition, Clarke informed readers, that Carey's version of Roxburgh's *Flora Indica* sold for 5 pounds in comparison to Piddington's index that sold for 1 pound while the 1874 edition was only 10 shillings.[41] He emphasized on the need for a cheaper and more accessible version of the *Flora* because he thought that in 1874 it was still the only handbook of the plants of India that was comprehensive enough for students and English denizens. Based on this publication history it is evident that *Flora* was not in high circulation until the late nineteenth century. Its importance was not recognized until long after it was first published.

The fact that earlier volumes of these catalogues were printed without sanction from Kew marks a moment in the story of colonial science where the production of knowledge was not entirely dependent on metropolitan science. People like William Carey wielded authority in the same capacity as leading colonial botanists in Calcutta. What gave Carey, who was neither trained as a surgeon nor botanist, the authority to become involved in the publication of a major botanical work? Carey's authority rested on his affiliation with an educational institution (the East India Company's Fort William College), his friendship with Roxburgh, and his status as a Baptist missionary with a commitment to natural theology. The parson naturalist was still a respected figure in early nineteenth century science. Carey's adherence to natural theology is evident on the title page of *Flora Indica* where he praises the work of God, "All thy works praise

---

40 Ray Desmond, "Clarke, Charles Baron (1832–1906)," in *Oxford Dictionary of National Biography*, online ed., ed. Lawrence Goldman, Oxford: OUP, http://www.oxforddnb.com.ezproxy.library.yorku.ca/view/article/32425 (accessed January 12, 2011).

41 Also known as Henry Piddington's English Index to the plants of India published at the Baptist Mission Press in Calcutta in 1832.

thee O Lord". Along with establishing his Serampore Mission for converting the natives of Bengal to Christianity, Carey saw his role as an amateur botanist also as work in the service of God. Carey was also interested in the involvement of native gentlemen in science and was instrumental in founding the Agricultural Society of India with an aim to recruit native gentlemen interested in farming and husbandry. Despite his contributions to the advancement of science in early nineteenth-century Calcutta, Carey is only better known as the shoemaker who turned into a missionary, the father of modern missions and the first person to print bibles in Indian vernacular languages. One sees a dichotomy in the acceptance of Carey's authority. He had authority as a missionary but not as a scientist. His role in the advancement of science is probably not acknowledged because he was not a professionally trained botanist. However, by publishing and editing Roxburgh's catalogues Carey ensured his place in the creation and circulation of botanical knowledge in India and Europe.

The Broken Circulation Circuit

The circulation of *Hortus Bengalensis* and *Flora Indica* was limited to the select elite in Calcutta in the first half of the nineteenth century. These elite groups consisted mainly of high officials of the East India Company, institutions such as the Fort William College and Orientalist scholars such as Sir William Jones. Unlike the *Plants of the Coast of Coromandel,* these catalogues were printed and circulated without certification or direct consent from Kew. This divergent and unique instance created a break in the publishing circuit because the circulation remained local instead of global. The separation of illustrations from plant descriptions at a time when it would have been very useful to have the two together makes the story of these catalogues even more interesting. Plant descriptions when sent to Britain could not always be deciphered until the botanical illustration was viewed. Two separate sets of drawings existed in Calcutta and Kew, creating a break in the flow of information to Kew. Moreover, after returning home in 1813, Roxburgh kept making changes to his manuscripts for *Flora Indica* and these were not accessible to Carey.

The timing of publication was a major factor in the disruption of circulation for two major reasons: the first being that the flow of information from the metropole to the colony was not timely due to geographical and financial constraints and, second, because newer species of plants were being constantly discovered and classified at Kew and any lag in the publishing affected the accuracy in taxonomic classification. By the time *Hortus Bengalensis* was published, 185 species had already been recognized and published under different systemic names than those used by Roxburgh in his catalogues. This caused obvious problems for scientists in London who were constantly classifying newer species of plants that they received from collectors and botanists the world over.

By bringing together the history of science and the history of the book in an examination of Roxburgh and Carey, a better understanding of the creation of colonial knowledge systems emerges. The fact that the printing of these catalogues could be done independently at Calcutta, with autonomous decision-making, undermines the assumption that everything had to be approved by the East India Company or the scientists in London. Roxburgh's botanical catalogues illustrate the significant power of print technology used by William Carey to promote and preserve his friend's work. Carey was not intimidated by the powerful Sir Joseph Banks or the Linnaean Society in deciding to edit and print *Hortus* and *Flora*. In fact he used these catalogues quite astutely to promote himself and to make his mark in the increasingly professional world of specialist botanists in the early nineteenth century. It did not matter if this botanical knowledge took a while to find its way into wider global knowledge networks because botanists remained interested in Roxburgh's work throughout the nineteenth century. While detailed histories of great individuals, their specialties and travels provide a rich resource for the history of science, it is the lesser-noticed accounts of the production and circulation of their works that can enhance the understanding of local and global knowledge networks in the early nineteenth century.

PART THREE

THE CIRCULATION OF EVOLUTION, GEOLOGY, AND ANTIQUITIES IN CHINA

KNOWLEDGE ACROSS BORDERS: THE EARLY COMMUNICATION OF EVOLUTION IN CHINA

Yang Haiyan*

A French daily newspaper, *Le Matin,* issued a remarkable challenge to its readers in January of 1907—"Is there anyone who will undertake to travel this summer from Paris to Peking by automobile?" It was to be the first transcontinental automobile race, held barely two decades after the motorcar was invented. The direction was reversed afterwards and the route ran from the Great Wall back to the Seine.[1] Coverage of the race by major newspapers in Paris (*Le Matin*), Milan (*Corriere della Sera*), London (*Daily Telegraph*) and many lesser titles was rapid and thorough, causing a sensation. For Western readers, it was another Western triumph. The swiftness and great potential of the automobile, like that of the steam locomotive, the telegraph and the mass-circulation press, symbolized and embodied the native supremacy of white civilization. For the Chinese people who were impressed by this automotive spectacle, another lesson was taught, one learnt again and again from the West ever since the First Opium War (1839–1842) at the start of the late Qing Dynasty. Using scientific and technological achievement as a measure to determine racial superiority had already been a familiar practice, which I will pay further attention to later in this chapter.

The encounter with the Western imperialist powers certainly had shaped Chinese history, but the nature and meaning of Western influence were open to question. In this article I try to get beyond the dichotomous "impact-response" model,[2] which makes China a passive recipient of

* I am grateful to Professors James Secord, James Moore, Bernard Lightman, Gordon McOuat and Larry Stewart, for their valuable suggestions and insightful comments in reading drafts of this essay. A previous version of this article was also presented in the international conference on "Darwin in Communication", held in Peking University from the 26th through the 28th of August 2010.

[1] Genevieve Obert, *Prince Borghese's Trail: 10,000 Miles over Two Continents, Four Deserts, and the Roof of the World in the Peking to Paris Motor Challenge* (San Francisco: Council Oak Books, 1999), 3.

[2] The typical examples of this model can be found in Ssu-yü Teng and John K. Fairbank, *China's Response to the West: A Documentary Survey, 1839–1923* (Cambridge, Mass.: Harvard University Press, 1954), 1–5, and in John K. Fairbank and Edwin O. Reischauer, *China: Tradition and Transformation* (Boston: Houghton Mifflin, 1978), 307–345.

stimulus from an active West. Equally, I hope to avoid the "China-centered" approach,[3] which privileges national history and local agencies of change while pushing the West into the background. Instead, I want to place the East-West encounter in a global context, showing how the unprecedented mobility brought about through the global communications revolution, together with Western imperial expansion based on so-called scientific notions of racial supremacy, shaped the introduction of a new form of knowledge, evolution, into China.[4] In doing so, I draw on the recent scholarship on science and communicative practices to discuss how the occurrence of a communications revolution in China became tied together with the dissemination of evolutionary ideas, and how intellectual concerns and technological functions were intimately related, the one sustaining the other.

The previous literature on the reception of evolution in China only rarely engages with these wider historiographical questions, but for the most part has focused on the history of ideas within the Chinese setting, and their relation to political and social changes.[5] For example, for James R. Pusey in *China and Charles Darwin*, the most important question is, "In fitting China for Marxism and the Thought of Mao Tse-tung, what did Darwin do to China?"[6] These works deal with very important issues, but particularly central to my concerns, instead, is "science as a form of

[3] For the detailed explanation of this approach, see Paul A. Cohen, *Discovering History in China: American Historical Writing on the Recent Chinese Past* (New York: Columbia University Press, 1984).

[4] For a comparative framework of global history applied to examine the pedagogy of imperialism and the nature of "colonialism" in China, see James L. Hevia, *English Lessons: The Pedagogy of Imperialism in Nineteen-century China* (Durham & London: Duke University Press, 2003). In this book, the combination of intellectual and material elements in historical discussion is also very instructive.

[5] The most important writings that belong to this tradition are: James R. Pusey, *China and Charles Darwin* (Cambridge, Mass.: Council on East Asian Studies, Harvard University, 1983); Benjamin I. Schwartz, *In Search of Wealth and Power: Yen Fu and the West* (Cambridge, Mass.: Harvard University Press, 1964), chapter 4; Wang Hui, *Xiandai Zhongguo sixiang de xingqi* (*The Rising of Modern Chinese Thoughts*) (Beijing: SDX Joint Publishing Company, 2004), chapter 8; Wang Zhongjiang, *Jinhua zhuyi zai Zhongguo* (*Evolutionism in China*) (Beijing: Capital Normal University Press, 2002). A partial exception to this rule is the work of Wang Tiangen, Tianyan lun *chuanbo yu Qing mo Min chu de shehui dongyuan* (*Communication of* Tianyan lun *and Social Mobilization during the End of Qing Dynasty and the Beginning of the Republic of China*) (Hefei: Hefei University of Technology Press, 2006), which tries to combine intellectual history and media studies, but which lacks a global perspective, and fails to pay enough attention to the interconnections between science, technology and culture.

[6] Pusey, *China and Darwin*, 452.

communication".[7] James A. Secord has effectively shown that the mechanization of printing promoted the dissemination of knowledge through cheap books and mass-circulation periodicals, particularly the public awareness and discussion of evolution, from the mid-nineteenth century in Britain and later on in urban centres globally.[8] In this chapter, I will first show that the independent publishing network in China up to the 1890s was weak and that constitutes one of the main reasons why Darwin and evolution were not discussed widely until the middle of the 1890s. Then with the first boom of Chinese journalism related to the defeat in the Sino-Japanese War and the Reform Movement of 1895–1898, evolutionary theories were actively introduced and appropriated in China. Associated with the New Culture Movement and the May Fourth Movement, a second boom of mass-circulation periodicals, mainly in vernacular language, occurred over the years 1915–1919. This played an important role in the more elaborate discussion and debate among different groups of people. All in all, through a flourishing and modernized journalism, Chinese intellectuals seized on western resources for their own purposes while building up their sense of cultural and national identity.

New communications technologies developed for the most part in Britain, such as mechanized papermaking (1803), steam-powered printing (1814), electric telegraphy (1830s), the "penny post" (1840), networked railways (1840s), undersea cables (1850s), wireless telegraphy (1901) and so on—accompanied and abetted Britain and other Western powers' imperial expansion and eventually resulted in profound transformations in communications worldwide.[9] During this process, new communication technologies served as a cross-cultural interface for the transmission of scientific knowledge, and the material forms of communication, particularly weekly and monthly journals, helped considerably in making evolution

[7] James A. Secord, "Knowledge in Transit," *Isis* 95 (2004): 654–672.

[8] James A. Secord, *Victorian Sensation: The Extraordinary Publication, Reception, and Secret Authorship of* Vestiges of the Natural History of Creation (Chicago & London: The University of Chicago Press, 2000). On Darwin's global transmission, see James A. Secord, "Global Darwin," in *Darwin*, ed. William Brown and Andrew C. Fabian (Cambridge: Cambridge University Press, 2010), 44–52.

[9] The rise of "global media system" from the 1860s and the relation between empire and communication are discussed in detail in Dwayne R. Winseck and Robert M. Pike, *Communication and Empire: Media, Markets, and Globalization, 1860–1930* (Durham & London: Duke University Press, 2007). For the Chinese case, the authors mainly focus on the close connection between the building of the telegraph and cable communication networks from 1870s and the whole idea of building a "new China", though the related boom of "new journalism" came two decades later, in the 1890s.

an integral component of global knowledge. This globalization could only happen when other countries around the world also experienced a communications revolution.

For a number of technological, economic, and social reasons, however, the press and journalism dealing with new knowledge had a very limited impact in China until the middle of the 1890s.[10] Printing and publishing were of course pervasive in the Qing period. Chinese traditional woodblock printing originated in the seventh century, and movable type methods were in use as early as the eleventh century, all long before Gutenberg.[11] The civil service examination system that reached its heyday in the Ming Dynasty (1368–1644) with expanding educational opportunities, promoted a widely based market selling books affordable to a growing, literate population.[12] Techniques of paper- and ink-making, engraving, binding and multi-colour printing had steadily evolved in traditional ways into the nineteenth century, while the early efforts, made mainly by missionaries, to domesticate modern printing techniques using metal alloy type casting, lithography and steam-driven cylindrical printing, had no substantial influence until the late nineteenth century. The Chinese traditional woodblock still prevailed for its economic and operational advantages.[13] Publications on science and technology were scarce; translations on related subjects provided by Kiangnan Arsenal (the foremost imperial military establishment) and missionary organizations were expensive and poorly distributed.[14] Missionaries and foreign merchants founded Chinese periodicals adapted to the interests of local readers,[15] but journalism conducted by Chinese writers with their own domestic concerns (apart from the imperial official gazettes that broadcast news from above) only began to appear in the 1870s. These new ones were very

[10] Certainly, the development of print technology in China is a vast topic, deserving of further detailed study. Due to limitations of scope I only touch on it lightly here.

[11] Zhang Xiumin, *Zhongguo yinshua shi* (*The History of Chinese Printing*) (Hangzhou: Zhejiang Ancient Books Publishing House, 2006), 8–16, 530. See also Tsien Tsuen-Hsuin, *Science and Civilisation in China, Vol. 5, Part I: Paper and Printing* (Cambridge: Cambridge University Press, 1985), 148–149, 201–203.

[12] Kai-wing Chow, *Publishing, Culture, and Power in Early Modern China* (Stanford: Stanford University Press, 2004), 38–56.

[13] Chow, *Publishing*, 59–67. See also Joseph P. McDermott, *A Social History of the Chinese Book: Books and Literati Culture in Late Imperial China* (Hong Kong: Hong Kong University Press, 2006), 21–39.

[14] Ferdinand Dagenais, ed., *The John Fryer Papers*, 3 vols. (Guilin: Guangxi Normal University Press, 2010), Vol. 2. 151–153.

[15] Barbara Mittler, *A Newspaper for China? Power, Identity, and Change in Shanghai's News Media, 1872–1912* (Cambridge, Mass.: Harvard University Asia Center, 2004), 3–7.

few in number and mainly based in Hong Kong, which had become a British colony in 1842.[16] As for the literacy rate, according to Evelyn Rawski, in the mid-to-late nineteenth century "30 to 45 percent of the men and from 2 to 10 percent of the women in China knew how to read and write."[17] However, the literate ranged from sophisticated scholars who knew classical Chinese to ordinary people who knew the few hundred characters needed in daily life. On the whole, by the middle of the 1890s, the press and journalism were mainly controlled either by missionaries and foreign merchants or by the Manchu government.

From the early 1870's till the middle of the 1890s there are only brief, scattered references to Darwin and evolution in the press. In 1871 the Translation Department of Kiangnan Arsenal began publishing translations of Western works about practical knowledge only. One translation work was based on the sixth edition of Charles Lyell's *Elements of Geology* (1865), which came out in 1871 with the Chinese title of *Dixue qianshi* (*A Brief Introduction to Geology*).[18] In the Chinese version, Darwin's theory is alleged to claim that "species could select the suitable places to live, and their characters are changeable", and it is said that the validity of the theory "is not established."[19] The publication in 1872 of Darwin's

---

16 The most successful and important one is *Xunhuan ribao* (*Xunhuan Daily*) founded by Wang Tao in 1874 in Hong Kong, famous for its editorials with concerns about current affairs and suggestions for political-institutional reform. See Paul A. Cohen, *Between Tradition and Modernity: Wang T'ao and Reform in Late Ch'ing China* (Cambridge, Mass.: Harvard University Press, 1974), 73–81. See also Zhuo Nansheng, *Zhongguo jindai baoye fazhan shi: 1815–1874* (*The History of Modern Chinese Journalism Development: 1815–1874*) (Taibei: Zhengzhong Book Company, 1998), 212–241.

17 Evelyn S. Rawski, *Education and Popular Literacy in Ch'ing China* (Ann Arbor: University of Michigan Press, 1979), 140.

18 The author adopted 1871, rather than widely believed 1873, as the year when the Chinese translation came out for the first time. One supporting evidence can be found in an article from John Fryer, "Jiangnan zhizao zongju fanyi xi shu shilüe (The Summary of the Translated Western Books by Kiangnan Arsenal," *Gezhi huibian* (*The Chinese Scientific and Industrial Magazine*) (Autumn, 1880), 10. It clearly indicated *Dixue qianshi* was published in 1871. The other evidence is the preface written by Hua Hengfang, one of the translators, in 1873. At the every beginning of it, he says, "the 38 volumes of *Dixue qianshi* have been engraved and proofread, and several hundred printed copies have spread out already." See Hua Hengfang, preface to *Dixue qianshi*, by Charles Lyell (Shanghai: Jiangnan Jiqi Zhizao Zongju, 1873).

19 Charles Lyell, *Dixue qianshi*, trans. Hua Hengfang and Ma Gaowen (Daniel Jerome Macgowan) (Shanghai: Jiangnan Jiqi Zhizao Zongju, 1873), vol. 13. In the original text of a note in chapter 13, Lyell notes, "Mr. Darwin's theory of natural selection removes many of the principal difficulties which stood in the way of Lamarck's doctrine of transmutation," and "The main proposition which seemed so startling in 1833, namely, that species in general may be older than the lands and seas they inhabit, is now the creed of almost every geologist, whether he adopts or rejects the theory that species may be indefinitely modified

The Expression of the Emotions in Man and Animals was reported in a *Shen bao* (*Shanghai Journal*) article in 1873.[20] *Shanghai Journal* was founded by an English businessman, Ernest Major (1841–1908), and would prove to be one of the long-lived newspapers in China, running from 1872 to 1949. According to the short article, this recent work from "the Western learned scholar" investigates whether or not, judging by their expressions, different human races in the world have one origin. Darwin's method of "enquiring from scholars visiting all corners of the world to obtain first-hand information" is praised as a manifestation of the Westerner's concentration on the hard work of real learning.[21] In 1877, "Hundun shuo (The Theory of Chaos)" appeared in the autumn issue of *Gezhi huibian* (*The Chinese Scientific and Industrial Magazine*), edited by Fu Lanya (John Fryer, 1839–1928), a "missionary of science", and published by the Chinese Polytechnic Institution and Reading Rooms. This article endorsed mankind's ape-ancestry, but eighty percent of it discussed the ten ways that humans might become extinct through changes in the earth's environment, assuming the end of the human race to be a more important consideration than its origin.[22] A version of theistic evolution appeared in Ding Weiliang's *Xixue kaolüe* (*Introduction to Western Learning*) in 1883. Ding Weiliang was in fact William A. P. Martin (1826–1916), and he stated that Darwin offered explanations for "transformation of species", one reason being that those with attributes fit for environments are "strong and then can survive", but he claimed that the overall process of "all organisms get formed one after another" involved an "intelligent and wise Governor".[23] In the spring issue of 1891, "Bowu xinwen (Scientific News Items)" in *The Chinese Scientific and Industrial Magazine* printed a news story

in their organization under the influence of new conditions in the animate and inanimate world." See Charles Lyell, *Elements of Geology* (London: John Murray, 1865), 194–195.

20 From all sources I have seen, the book mentioned in this short article is mistaken as Darwin's other book, *The Descent of Man, and Selection in Relation to Sex*, which was published in 1871.

21 "Xi boshi xin zuo ren ben yi shu (A Western Scholar's New Publication: Ren Ben)," *Shen bao* (*Shanghai Journal*), issue 404, 2, 21 August, 1873.

22 "Hundun shuo (The Theory of Chaos)," *Gezhi huibian* (*The Chinese Scientific and Industrial Magazine*), the autumn issue (1877), 6–7. Among those ways mentioned, the tenth one holds a circular view that man degenerates into the worm and then starts all over to become man again. The endless circulation in fact dissolves the doomsday of the human race.

23 Ding Weiliang, "Xixue kaolüe (Introduction to Western Learning)," in *Xuxiu si ku quan shu* (*The Continuation Books of Complete Library of the Four Treasuries*), ed. the Editing Committee (Shanghai: Shanghai Ancient Books Publishing House, 1883/2002), the 1299th *ce*, 739–740.

provided by Ai Yuese (Joseph Edkins, 1823–1905) about the English physicist John Tyndall's retirement in London. Tyndall is said to regard Darwin's theory—"thousands upon thousands of present kinds of animals and plants in fact derived from a few of kinds"—as a sign of scientific progress during the previous fifty years.

It is easy to understand why all these references to Darwin and evolution did not lead to heated discussions. There was no religious tradition opposing the transmutation of species and the animal origin of mankind in Chinese culture, and Darwin's provocative concept of "struggle for existence" or "survival of the fittest" was not greatly emphasized. Some Protestant translators, keen to import Western knowledge to China, deliberately avoided the notion. For example, Joseph Edkins, the missionary and sinologist who translated Macmillan's *Science Primers* into Chinese, promoted natural theology while omitting the "fittest survive" in his translation of Darwin's friend Joseph Hooker's *Botany* textbook in 1886.[24] As a new theoretical discovery about the descent of living organisms, evolution without struggle was not particularly controversial. From its debut, it had nothing to do with the "wealth and power" that concerned the first generation of reformers confronted with the West, such as Feng Guifen (1809–1874), Wang Tao (1828–1897) and Zheng Guanying (1842–1923). Urgent practical problems took precedence over any consideration of Darwin as a pure scientist or of evolution as a new theoretical discovery in the world of plants and animals.

Yet there were exceptions. For the purpose of inducing "the Chinese literati to investigate the various departments of Western knowledge with the view to their applications in the Middle Kingdom,"[25] John Fryer and Wang Tao devised Gezhi shuyuan keyi (the China Prize Essay Scheme), which ran parallel to the civil service examination. Li Hongzhang (1823–1901), then Viceroy of Chihli, once offered the special theme question in spring 1889. One of the questions is: "With respect to the 'Science' referred to in the 'Great Learning' from Ching-kiang-ching downwards, there have been several tens of scholars who have written on the subject. Do any of them happen to agree with Western scientists? Western science began with Aristotle in Greece; then came Bacon in England who changed the previous system and made it more complete. In later years, Darwin's and

24 Benjamin A. Elman, *On Their Own Terms: Science in China, 1550–1900* (Cambridge, Mass.: Harvard University Press, 2005), 327–330.

25 Dagenais, *John Fryer Papers*, Vol. 2, 125.

Spencer's writings have made it still more comprehensive. Give a full sketch of the history and bearing of this whole subject."[26] In answer to it, Zhong Tianwei (1840–1900), a middle-aged alternate county governor in Guangdong Province, in an account of Darwin's theory of evolution, clearly described the principle that "the strong survive and the weak perish" and he commended it as the "natural principle of the heavenly way." Yet, according to Wang Tao, a Christian convert who supervised the contest, the principle of struggle "has no basis in fact."[27] Wang's remark might explain why Zhong came fourth in the competition, though he demonstrated a better knowledge of Darwin and Herbert Spencer than the other three contestants. Zhong probably picked up Darwin and Spencer's ideas on his trip to Europe in 1880–1882, and his approach to evolution would become commonplace in the coming decades.

Perhaps the most immediate reason why the earliest introductions of evolution provoked little reaction was that access to this new knowledge was as problematic as its content. By the middle of the 1890s, the independent press and journalism dealing with new knowledge were very weak in China. On the whole, Chinese intellectuals didn't discuss Darwin and evolution because, quite simply, they didn't read him or much about him. They had very few local channels for acquiring new knowledge from the West and, more importantly, for making their own knowledge through creative discussion in print.

The real evolution-sensation began with the shock and humiliation of defeat in the Sino-Japanese War and the following Reform Movement of 1895–1898. In March 1895, "Yuan qiang (Whence Strength)" written by Yan Fu (1854–1921), was published in *Zhi bao* (*Chih pao*, or *Tianjin Newspaper*). He was among the first Chinese to be educated in England, where he had studied naval science from 1877–79, mainly in the Royal Naval College, Greenwich, and had eagerly absorbed Western philosophy and social science. In the article, Yan Fu introduced the concepts of "struggle for existence" and "natural selection" from *The Origin of Species*. Drawing on the yellow-white-brown-black race categories prevalent in Western writings, he argued that the previous neighboring warlike "barbarians" who eventually

[26] Dagenais, *John Fryer Papers*, Vol. 2, 137. Ching-kiang-ching (Zheng Kangcheng) is a prestigious Confucianist living in the Han Dynasty (206 BC–220 AD).

[27] Zhong Tianwei, "Zhong xi gezhi zhi xueyi tonglun (Comparison on Gezhi of China and of the West)," In *Jindai kexue zai Zhongguo de chuanbo: wenxian yu shiliao xuanbian* (*The Introduction of Modern Science into Late 19th and Early 20th Century China: Selected Works and Documents*), ed. Wang Yangzong (Jinan: Shandong Education Press, 1889/2009), 342. See also Elman, *On Their Own Terms*, 345–351.

Fig. 1 Yan Fu, a photo from *Zongsheng huizhi (The Tsung Sheng Review)*, Vol. 1 No. 4 (1913).

became assimilated into the Chinese culture were not an alien race at all; instead, people from the West who threatened the survival of Chinese people were really a different and superior race. He said, "Western people are not only taller, stronger and braver, their virtue, intelligence, techniques and learning are also advanced than my people." "The reason, I guess, is that they take freedom as the fundamental principle and democracy as the practical measure." He also mentioned that the national competition within Europe resulted in mutual and continual progress. But in China, the constraints, enslavement and deprivation exerted on the people by the imperial government led to stagnating intelligence and impoverished physical strength. "If thus, then is it far from ceasing existence and survival?" Brown and black people were in danger, and so were the yellow population of 400 million. Overall, he traced the origin of Western wealth and power; lamented China's weakness, poverty and backwardness; and clearly stated his ultimate solution. This was to transform the Chinese people mentally, physically and morally: to open their minds, strengthen their bodies, harmonize their virtues. In this way, individuals' energy could be liberated in order to make communal bonds and thus a stronger "qun" (collective, or Spencer's social organism), which could better survive in the struggling world.[28]

Although the *Tianjin Newspaper* was founded by a German noble, Constantin von Hanneken (1855–1925), Yan Fu used it as a platform to advocate his reformist ideas, publishing five important polemical essays from February to June 1895. Yan Fu finally joined other reformers to set up their own periodicals—*Guowen bao* (*National News Daily*), and *Guowen huibian* (*Collection of National News* or *The Light Seeker*)—on 26 October and 8 December 1897 respectively. The latter came out every ten days and carried important news, reviews and academic articles, mostly from foreign sources.[29] It was in the *Collection of National News* that part of Yan Fu's paraphrased translation of Huxley's "Evolution and Ethics: Prolegomena"[30] first appeared, in the second, fourth, fifth and sixth issues, between

---

[28] Yan Fu, "Yuan qiang (Whence Strength)," in *Yan Fu ji* (*Collected Works of Yan Fu*), ed. Wang Shi (Beijing: Zhonghua Book Company, 1895/1986), Vol. 1, 10–12, 14.

[29] Fang Hanqi, ed., *Zhongguo xinwen shiye tongshi* (*A General History of Chinese Journalism*) (Beijing: Renmin University Press, 1992), Vol. 1, 604–619.

[30] The Prolegomena was written as a preface to the published edition of Huxley's Romanes lecture "Evolution and Ethics", and the two made their first appearance in the ninth volume of Huxley's collected essays. See Thomas H. Huxley, "Evolution and Ethics: Prolegomena," in *Collected Essays, Vol. IX: Evolution and Ethics, and Other Essays* (London: Macmillan, 1894), 1–45.

18 December 1897 and 15 February 1898. The publication of Yan Fu's translation was halted because the *Collection of National News* itself ceased publication after six issues, each having sold only some 500 or 600 copies, partly because the content and language were difficult and abstruse for a wider audience. The circulation of the *National News Daily* was about 1500—better, but not large.[31]

After several revisions, Yan Fu's paraphrased translation of Huxley's Romanes lecture "Evolution and Ethics" and "Prolegomena", with extensive commentaries, was published in 1898 with the Chinese title *Tianyan lun* (*The Theory of Evolution*, literally *The Theory of Heavenly Evolution*).[32] It came out first in Mianyang (woodblock printing) and then in Tianjin (lithographical printing) in 1898, and several dozen editions appeared in subsequent years.

Huxley's horticultural metaphor for colonization[33] struck Yan Fu sharply, warning him of the doom of the Chinese people in the international struggle for existence. At the same time, Huxley's emphasis on human efforts to create favourable conditions of existence, restrain ruthless self-assertion and strengthen the social bond[34] appealed to him greatly, though the opposition between the "cosmic process" and the "ethical process", stated in Huxley's original argument, was disagreeable in his opinion. For Yan Fu, and for Wu Rulun, each of whom wrote a preface to *Tianyan lun*, tianyan (evolution) is the underlying principle of an all-encompassing process, taking in the biological, intellectual, moral, social and political realms.[35] As long as one followed the universal rule, "Everyone has freedom, but this shall be bound by the freedom of others" (i.e. self-assertion plus self-restraint equals "enlightened self-assertion"), cosmic

---

[31] Fang, *Zhongguo xinwen*, Vol. 1, 617, 606.

[32] Tian, the Chinese concept with complicated meanings, appears in the title of the Chinese edition. According to one of Yan Fu' commentaries in a later translation of Herbert Spencer's *The Study of Sociology*, tian in "tianyan" means neither God, nor sky, but something that things can develop from according to causality, though it is itself unchangeable. See Yan Fu, trans., *Qunxue yiyan* (*The Study of Sociology*) (Beijing: Commercial Company, 1903/1981), 298, note 5. Tianyan, coined by Yan Fu and functioned as the core concept of his worldview, is a universal way for continuous change in cosmos instinct with ethical significance.

[33] Huxley, *Collected Essays, Vol. IX*, 16–17.

[34] Huxley, *Collected Essays, Vol. IX*, 43, 35–36, 81–82.

[35] Yan Fu, trans., *Tianyan lun* (*The Theory of Evolution*) (Zhengzhou: Zhongzhou Ancient Books Publishing House, 1898/1998), 57, 60. See also Wu Rulun, preface to *Tianyan lun* (*The Theory of Evolution*), translated by Yan Fu (Zhengzhou: Zhongzhou Ancient Books Publishing House, 1898/1998), 1.

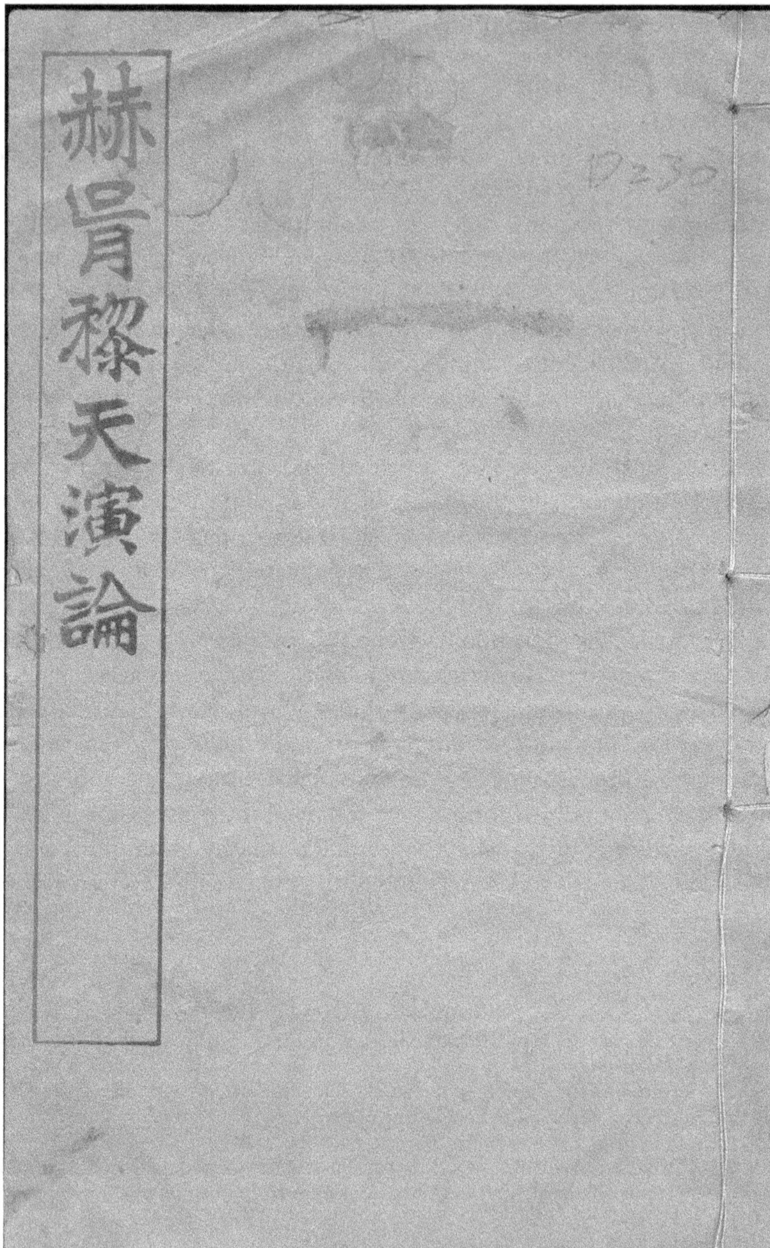

Fig. 2 The book cover of *Tianyan lun*, Mianyang: Lushi Shenshi Jizhai, 1898.

nature and ethical nature could be reconciled.[36] In an extensive commentary occupying one-third of the book, Yan Fu introduced Spencer's progressive "social organism" and his argument that functionally induced brain improvement defeats Malthusian over-population and natural inequality. On this basis, for Yan Fu, a sort of optimism replaced Huxley's pessimism: as evil is checked, so good increases day by day, and a well-organized, united and strong society emerges.[37]

Thus *Tianyan lun* was a confection of Huxley's and Spencer's ideas concocted by Yan Fu. Huxley's advocacy of human action served for Yan Fu as a wedge to keep the door of reform wide open, allowing him to assimilate Confucianism into evolutionary theory as a basis for urging government and people to act immediately to eliminate artificial checks to progress. *Tianyan lun* is filled with stirring assurances of social and political progress, always contingent on right actions. A reformed China need not just survive international competition; it could in fact prosper by bending the evolutionary process to nationalist ends.

Knowledge from the West was appropriated by Yan Fu to constitute a unique cosmology with a complex structure suited to his own reformist needs.[38] At the same time, he aimed to use *Tianyan lun* to open up people's minds and mobilize them, but his classical writing style did not help much. The book was difficult to read,[39] except for those with a high level of classical education, who found it to be really fascinating. One of them was Liang Qichao (1873–1929), an important reform movement leader and pioneer of Chinese journalism. Over his long career, from 1895 to 1920, he was the editor-in-chief of more than ten periodicals.[40] On 15 October 1896, after finishing the preface to *Tianyan lun*, Yan Fu posted the book manuscript to Liang,[41] who in turn mined the translation for a series of articles in *Shiwu bao* (*Current Affairs*), founded that August in Shanghai as a political vehicle for reformists.[42] More of the gist of Yan Fu's Huxley paraphrase appeared in Liang's journals *Qinyi bao* (*Honest Criticism*) and

---

36 Yan, *Tianyan lun*, 187, 433.

37 Yan, *Tianyan lun*, 422, 196–197.

38 Wang, *Xiandai Zhongguo*, chapter 8. See also Wang, *Jinhua zhuyi*, chapter 3.

39 He Lin, "Yan Fu de fanyi (Yan Fu's Translation)," *Dongfang zazhi* (*The Eastern Miscellany*) Vol. 22, No. 21 (1925): 80–81.

40 Fang, *Zhongguo xinwen*, Vol. 1, 557.

41 Sun Yingxiang, *Yan Fu nianpu* (*Chronicle of Yan Fu*) (Fuzhou: Fujian Renmen Press, 2003), 83–84.

42 Pusey, *China and Darwin*, 84–126, and passim. See also Wang, Tianyan lun *Chuanbo*, 128–139.

Xinmin congbao (*Journal of a New People*), and thus the survival of China, competition among nations, the strong versus the weak, history of the races, and the idea of progress emerged as popular motifs. A lecture on the general idea of evolution by a Japanese scholar, Oka Asajirō (1868–1944), was translated and published, which outlined Darwin's theories, and presented the embryo images of eight kinds of animals to show the vestiges of animal evolution.[43] Furthermore, *Journal of a New People* published photos of Darwin, Huxley and Spencer for the first time in the Chinese press.[44] For Liang's own part, among many other articles, he wrote, "The Theory of Darwin, the Founder of Evolution, and His Brief Biography".[45] In this article, using simple and plain language, Liang described Darwin's life, his natural and artificial selection analogy, and the wider implications of his theories. Compared to Yan's writings, it is much easier to read and understand, though the underlying spirit was the same. *Honest Criticism* and *Journal of a New People* were published in Japan where Liang lived in exile after the abortive 100 Days Reforms in the summer of 1898. With the help of the Japanese authorities, the issues were smuggled into China by merchants and immigrants. Liang had a tremendous following among Chinese readers: the circulation of *Current Affairs* reached 17,000 and *Journal of a New People* 10,000. Reprinting took place on the mainland despite imperial prohibitions, and copies were dispatched from over 90 consignment points inside and outside of China, some even reaching parts of the hinterlands.[46]

One important reason for Liang Qichao's great success was his writing style, famous for its clarity and smoothness, in contrast to the traditional writing style. Liang observed that "60 percent of Chinese people are illiterate, and 75 percent of the literate persons can't understand grammar", so "teaching school children and unenlightened people", through simple language and flexible writing style, "is the first thing to do nowadays in saving China".[47] Liang and his colleagues understood that: they saw how first-rate journalism would communicate far more widely and effectively than

---

[43] Oka Asajirō, "Jinhua lun da lüe (A General Introduction to Evolution)," *Xinmin congbao* (*Journal of a New People*) No. 46–48 (1903): 175–192.

[44] For these photos, see "Yingguo shuo ru yixiang (The Portraits of Famous British Scholars)," *Xinmin congbao* (*Journal of a New People*) No. 6 (1902).

[45] Liang Qichao, "Tianyan xue chuzu Da Er Wen zhi xueshuo jiqi lüezhuan (The Theory of Darwin, the Founder of Evolution, and His Brief Biography)," *Xinmin congbao* (*Journal of a New People*) No. 3 (1902): 9–18.

[46] Fang, *Zhongguo xinwen*, Vol. 1, 559, 651.

[47] Fang, *Zhongguo xinwen*, Vol. 1, 635.

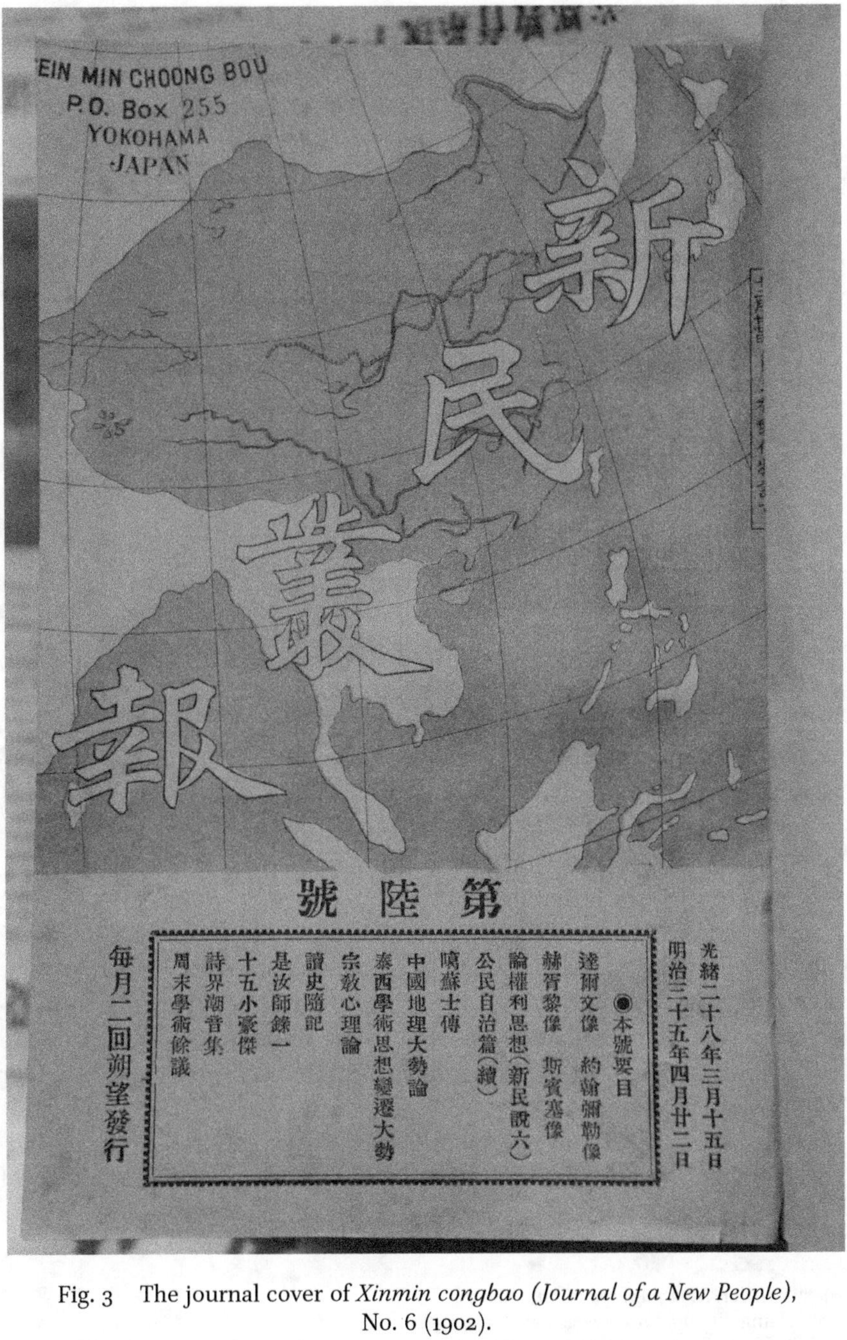

Fig. 3 The journal cover of *Xinmin congbao (Journal of a New People)*, No. 6 (1902).

Fig. 4 The portraits of famous British scholars, *Xinmin congbao* (*Journal of a New People*), No. 6 (1902).

school teaching. Indeed, Liang believed, "the more the people read the newspapers, the more educated these people become; the more newspaper companies are established, the stronger the country".[48] The message got through. During 1895–1898, about 150 newspapers and journals sprang up; over three-fourths founded by Chinese people themselves.[49] They constituted the first boom of Chinese journalism. Most of the titles were run by reformists, and these new journals contributed tremendously to the Reform Movement of 1895–1898. Official publications, aimed at privileged groups, now began to give way to the public vehicles to promote social and political reform at the time of national crisis, through the inner dynamism of Confucianism.[50]

Free communication was said to strengthen the nation, and the periodicals founded by Chinese intellectuals and flourished from the middle of

[48] Mittler, *Newspaper for China*, 22.

[49] Fang, *Zhongguo xinwen*, Vol. 1, 539.

[50] Zhang Xiaotao, *The Origins of the Modern Chinese Press: The Influence of the Protestant missionary press in late Qing China* (London & New York: Routledge, 2007), 114–126.

the 1890s, were themselves seen as constitutive parts of a technologically progressive, evolving and improving society. With a circulation of 10,000 or more, with 30,000 words per issue, and appearing every ten days, Liang Qichao's *Current Affairs* could not depend on printing by xylography and wood movable type. Lithographical printing and letterpress printing were adopted instead. Of the 36 journals examined, published between 1895 and 1898, 21 used stone-based lithography or letterpress.[51] Rapid bulk printing as required by mass-circulation periodicals promoted many technological innovations at the end of the nineteenth century. These were introduced into China and adapted to local needs, showing that the mechanization and industrialization of communication had advantages over traditional ways. The material communications infrastructure in modern China grew up intertwined with periodical discussion and journalistic debate over evolution. For people who had interests in evolution, their first access to it was side by side with the latest war news and world affairs on, for example, the Turkey-Greece draft peace treaty, the new commercial regulations of various countries, Britain's coastal interests and rights, and the possibility of a Russian revolution, the tendency of competition among nations, and the evolutionary history of Chinese despotism. Readers took evolution not as an abstract and philosophical theory, but as a new way of thinking from the West, actually as an instrument to be put into use readily to make their own arguments.

Boosted by Liang's eloquent and emotional essays, the main ideas of *Tianyan lun* caused an immediate sensation among the literati and young students. A scientific concept and related theories were exploited to promote the survival of the Chinese people and social progress. The tipping-point came in the years around 1900. Darwin's name now entered into the vocabulary of every educated household. The Chinese equivalents of "struggle for existence", "natural selection" and "survival of the fittest" became popular slogans; evolutionary cosmology grew so familiar that key words from Yan Fu's translation were even adopted for people's names. In the 1970s, Cao Juren, a scholar of modern Chinese intellectual history, reported after reading over 500 autobiographical memoirs from this period that almost all the authors were influenced by Yan's *Tianyan lun*, including leading intellectuals such as Lu Xun and Hu Shi.[52] In coming

---

51 Su Jing, *Ma Lixun yu zhongwen yinshua chuban* (*Robert Morrison and Chinese Printing and Publication*) (Taipei: Student Press, 2000), 285.

52 Cao Juren, *Zhongguo xueshu sixiang shi suibi* (*Essays on Chinese History of Academic Thinking*) (Beijing: SDX Joint Publishing Company, 2003), 371.

decades, Darwin's name was invoked by almost all sides: reformers or revolutionists; nationalists or communists. The ideologues showed little interest in biological knowledge *per se*; political, social and moral concerns were paramount, Darwin's painstaking methods and the scientific details of his arguments all but irrelevant. Evolutionary cosmology and the "natural" law of progress were exploited to clinch political arguments and create public opinions. Newspapers and journalism used the latest innovations to disseminate these doctrines of historical progress. Following evolution, whatever that meant, would keep a reformed China on the right track.

Not all joined in this kind of optimism. Some Chinese observers chose to see the dark side of evolution. For example, Zhang Taiyan, in an article in *Min bao* (*People's Journal*), acknowledged with Huxley that good and evil evolved together.[53] Zhou Zuoren, a younger brother of Lu Xun, found "the theory of evolution is great, but too cruel". Or as he put it, "If we use strong and weak as the standards, and regard competition as the vital link, how can the world be in peace?"[54] At the same time, through the translation of the Russian anarchist Peter Kropotkin's works, advancing by "mutual aid" rather than competitive struggle became a popular discourse. To be fair, in the devastating aftermath of World War I, the early advocates of evolution published doubts too. Yan Fu wrote in 1918, "I feel that the European race's last three hundred years of evolutionary progress have all come down to nothing but four words: selfishness, slaughter, shamelessness, and corruption." Liang Qichao chimed in, offering his own calculation based on a year's visit to Europe: *The Origin of Species*' "thousands of words all came down to ten: 'the struggle for survival and the survival of the fittest'." Under the spell of these words, "for individuals, the worship of influence and money became the sacred rule, and, for countries, militarism and imperialism became the most stylish political policies. Surely it was out of this that came this great international World War."[55] A different view of the West was rising in China, according to Chinese people's own judgments.

---

53 Zhang Taiyan, "Ju fen jinhua lun (Ju Fen Evolution)," *Min bao* (*People's Journal*), No. 7 (1906): 1–14.

54 Zhou Zuoren, "Gu er ji (Legend of an Orphan)," in *Wan Qing wenxue congchao* (*Late Qing Literature Series*), ed. A Ying (Beijing: Zhonghua Book Company, 1906/1961), "Novel" Vol. IV, the second *ce*, 497.

55 The English translation of Yan Fu and Liang Qichao's remarks are from Pusey, *China and Darwin*, 439–440.

Fig. 5 Ma Junwu, a photo from *Daxia zhoubao (The Great China University Weekly)*, Vol. 24 No. 1 (1947).

What happened to *The Origin of Species*? Arguably, this is a separate and different story. Darwin's "Historical Sketch of the Progress of Opinion on the Origin of Species" (added to *Origin*'s third edition in 1861) was translated by Ma Junwu (1881–1940), a follower of reformists and an overseas student in Japan then, and published in 1902 in *Journal of a New People*.[56] The influence of *Tianyan lun* can be revealed clearly in a poem composed by Ma in 1900, which starts as: "A vast expanse of past to present, viewing

56 Ma Junwu, "Xinpai shengwu xue (ji tianyan xue) jia xiao shi (The Brief History of New School of Biologists, or Specialists of Evolution)," *Xinmin congbao* (*Journal of a New People*) No. 8 (1902): 9–18.

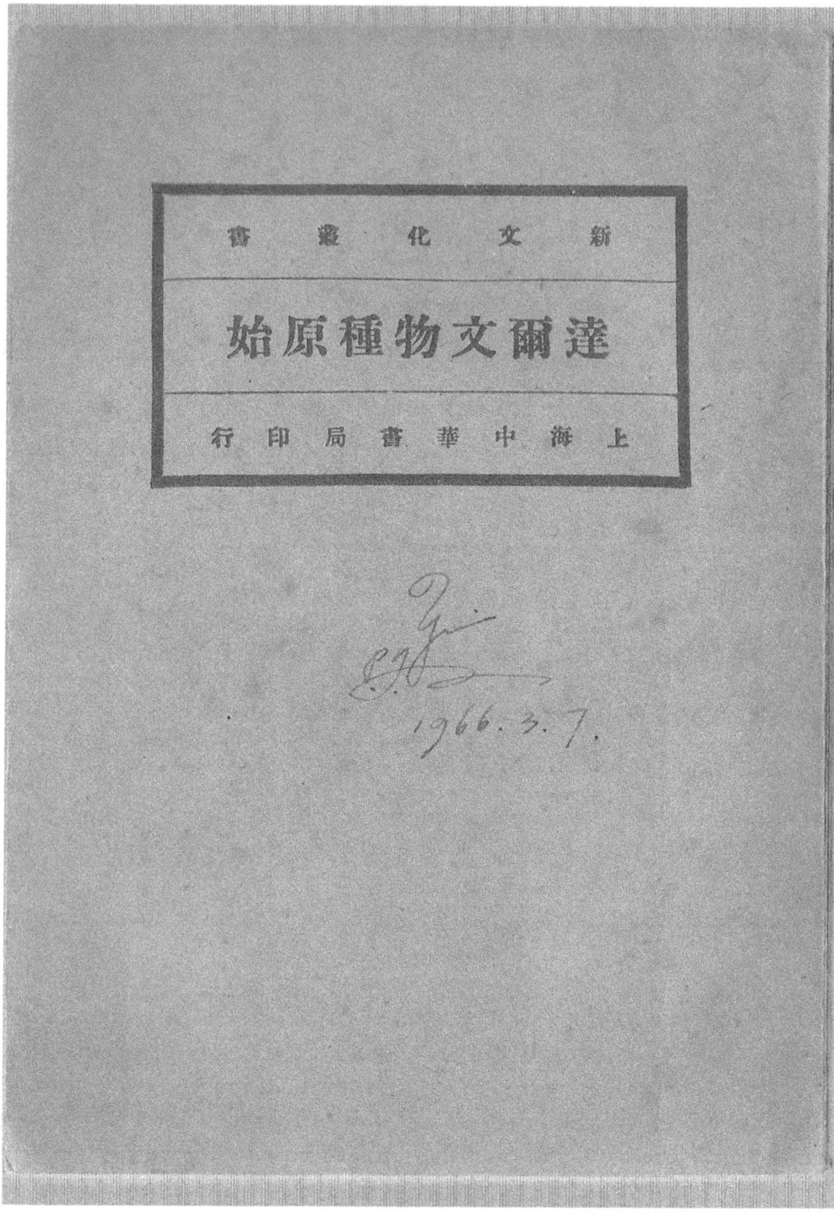

Fig. 6 The book cover of *Wuzhong yuanshi*, Shanghai: Zhonghua Book Company, 1920.

tianyan in action; the fierce struggle for existence prevails on earth".[57] His translation of chapters III and IV, 'Struggle for Existence' and 'Natural Selection', came out later in 1902, and the first five chapters, with Darwin's "Sketch" were published together as the first volume of the whole book in spring 1904. The complete English text in Chinese by the same translator, titled as *Wuzhong yuanshi*, did not appear until 1920.

The Origin of Species took a back seat to *Tianyan lun* for several reasons, Ma's doubts about the facts in the later chapters revealing one.[58] The validity of Darwin's natural selection was doubted especially when Mendel's laws were rediscovered and elaborated, which came to the attention of Chinese intellectuals. Their critical concern marked a new interest in the factual and theoretical basis of evolution, an interest especially true for those young biologists who were trained overseas and came back. Even while at Cornell University pursuing doctoral studies in entomology, Bing Zhi (1886–1965) joined other Chinese students in 1914 to found the Science Society of China, with its own journal, *Kexue* (*Science*). Bing's translation of the canine section of chapter one of Darwin's *Variation of Animals and Plants under Domestication* (1868) was published in 1915 in the second, third and sixth issues of *Science*, and many further translations of evolutionary texts, articles and news items about Darwin appeared in its pages. The Chinese mass periodical press was experiencing the second boom from 1915, thanks to the vernacular language movement and the May Fourth Movement. The golden days of Chinese journalism had arrived, or so it would seem. In 1919 alone, about 400 periodicals in vernacular language were being published;[59] popular and academic science periodicals sprang up alongside growing scientific institutions. These journals now furnished the platform for public debates about the meaning of evolution.

In 1921, the fourth issue of *Bowu zazhi* (*The Magazine of Natural History*), founded by Society of Natural History in Peking Normal College, published four articles to commemorate the 112th anniversary of Darwin's birth. A year later, the third volume of *Min duo* (*People's Bell*) devoted its

---

57 Ma Junwu, *Ma Junwu shi zhu* (*The Annotated Collected Poems of Ma Junwu*), ed. Tan Xing (Nanning: Guangxi Nationalities Publishing House, 1985), 13.

58 Zhou Jianhou, "*Ren ji dongwu zhi biaoqing* yizhe bianyan (Translator's Preface for *The Expression of the Emotions in Man and Animals*)," in *Jinhua lun de yingxiang li* (*The Influence of Evolution*), ed. Shen Yongbao & Cai Xingshui (Nanchang: Jiangxi Universities Press, 1937/2009), 24.

59 Chow Tse-tsung, *The May Fourth Movement: Intellectual Revolution in Modern China* (Cambridge, Mass.: Harvard University Press, 1960), 176–182. See also Fang, *Zhongguo xinwen*, Vol. 2, 1, 114–117.

fourth and fifth numbers to evolution, printing 16 articles in total, including a detailed chronicle of Darwin's life and a reading list of evolutionary works (almost 200 items). Ordinary Chinese readers, though urban and still few in number, had their access to detailed expositions of Darwin's theories. However, appearing beside Darwin were Gregor Mendel, William Bateson, Hugo de Vries and other proponents of the new genetics. This complicated the picture. Thomas Hunt Morgan was dubbed "the first person to successfully oppose Darwin's theory".[60] The mechanism of natural selection faced skepticism as other evolutionary mechanisms were popularized. Writers began to insist that the fact of evolution and the theory of evolution should be carefully distinguished. One should know "*Origin of Species* is insufficient to represent all evolutionary theories".[61] Although working biologists focused on their own specialties to unearth the potential of evolutionary theory, they barely touched the question of evolutionary mechanism. Ideologically, Darwinian slogans like "struggle for existence" back up their aspiration of saving China through science; practically, the general concepts and narratives of evolution were localized through their own investigating, collecting and excavating. Their work was viewed as an effort to nationalise science and build up their identity as Chinese scientists. Darwin's name was still shining, yet his most original contribution was somehow out of focus, except later on in a few yet important researches made by Tan Jiazhen (C. C. Tan, 1909–2008), who obtained his Ph.D. in 1936 from Morgan's lab at Caltech, under the direction of Theodosius Dobzhansky. Tan's work on the chromosome structure of *Drosophila* was shaped heavily by Dobzhansky's interest of combining genetics and evolution; in turn his work was incorporated in Dobzhansky's *Genetics and the Origin of Species*, a widely read classic for modern evolutionary synthesis. Tan's later work on the geographical and seasonal variations of ladybird beetles in China offered experimental evidence for the newly founded study of population genetics.[62] As for theoretical population genetics, Li Jingjun (C. C. Li, 1912–2003), a 1940 Cornell Ph.D. in plant breeding and genetics and then a postdoc in mathematics and statistics, had his textbook *An Introduction to Population Genetics*

---

[60] Qian Tianhe, "Tianyan xinshuo (New Evolutionary Theory)," *Kexue* (*Science*) Vol. 4, No. 12 (1919): 1211. It is a translation of an article from an American genetics journal.

[61] Chen Jianshan, "Da Er Wen yihou zhi jinhua lun (Evolutionary theories after Darwin)," *Min duo* (*People's Bell*) Vol. 3, No. 5 (1922): 2.

[62] Tan Jiazhen, *Tan Jiazhen lunwen ji* (*Collected Scientific Articles of Tan Jiazhen*) (Beijing: Science Press, 1987), 202, 243. Laurence Schneider, *Biology and Revolution in Twentieth-Century China* (Lanham: Rowman & Littlefield Publishers, Inc., 2003), 75–78.

Fig. 7 The commemorative photo for the 112th anniversary of Darwin's birth, *Bowu zazhi (The Magazine of Natural History)*, No. 4 (1921).

published in English by National Peking University Press in 1948. However, it was never on sale publicly because of the chaotic situation at that time.[63] Consequently it almost had no influence in China, except among his immediate students.

During the early decades of the Republican China, knowledge-transmission and opinion-formation were characteristically prompt. In 1925, *Dongfang zazhi* (*The Eastern Miscellany*), an influential comprehensive journal founded in 1904 by the Shanghai Commercial Press, published three articles about the so-called Monkey Trial, which took place from 10 to 21 July in the U. S. state of Tennessee. In a short news piece published on 10 August, "Penalty for Teaching Evolution", the author expressed great surprise at the proceedings, for "evolution has been common sense even for women and children", not least in Thomas Huxley's centenary year.[64] On 10 October, "Monkey Incident in America" gave a detailed and vivid

63 Partha P. Majumder, "C. C. Li (1912–2003): His Science and His Spirit," *Journal of Genetics* Vol. 83, No. 1 (2004): 103.

64 Cong Yu, "Jiaoshou jinhua lun zhi bei fa (Penalty for Teaching Evolution)," *Dongfang zazhi* (*The Eastern Miscellany*) Vol. 22, No. 15 (1925): 70.

AN INTRODUCTION TO

POPULATION GENETICS

BY

CHING CHUN LI, Ph. D.

College of Agriculture, National Peking University

NATIONAL PEKING UNIVERSITY PRESS
PEIPING, CHINA.
1948

Fig. 8 The title page of *An Introduction to Population Genetics*, Peiping: National Peking University Press, 1948.

description of the courtroom encounters. The writer affirmed the descent of man from monkeys (not apes) and thought the trial exposed the shallowness of the American people, which must "incur humiliation from civilized countries".[65] One month later, "Bryan who is against Evolution" reported the prosecutor's final speech, with its four arguments against evolution. William Jennings Bryan had resigned in protest as U.S. Secretary of State when the country entered the World War; now he wound up before the court, declaring struggle, competition and selection to be inhuman, leading to the worship of power rather than to justice. He blamed the War on the poison of evolution. Only moral power replacing evolutionary power would preserve world peace and upgrade the spiritual life of humanity.[66] Although the writer reminded readers that Bryan was an anomaly in competitive materialistic America, one who stood for "spiritual civilization" and "spiritual life", he also thought Bryan's cause shallow and extreme. In China, students of the West were growing up. Condemning "ludicrous and pitiful" Americans for acting "fussily" showed the new Chinese sense of superiority.[67] Learning from the West had turned into judging the West, backed by a belief in evolution.

Let us return to the first transcontinental automobile race from Peking to Paris in 1907. During the imperial expansion of the late nineteenth century, railroads, steamships, motorcars and the telegraph combined with existing postal systems and the mass periodical press to create a highly mobile, fast-paced global information network, crossing borders, breaking geographical boundaries and penetrating deep into hinterlands. At the same time, however, anthropologic barriers replaced the geographical barriers. They set the white race apart from other races, basing their superiority on "scientific" notions of human nature.

According to the Italian journalist, Luigi Barzini (1874–1947), reporting directly from the winner's Itala car, as the diplomatic quarter, crowded with cheering "sons of Europe and America," wished the racers *bon voyage*, "a triumph of the West" was being "solemnized in the heart of Pekin". The

65 Bai Ben, "Meiguo de houzi shijian (Monkey Incident in America)," *Dongfang zazhi* (*The Eastern Miscellany*) Vol. 22, No. 19 (1925): 59–62.

66 Paolo E. Coletta, *William Jennings Bryan III. Political Puritan, 1915–1925* (Lincoln: University of Nebraska Press, 1969), 240–271. See also James Moore, "The Creationist Cosmos of Protestant Fundamentalism," in *Fundamentalisms and Society: Reclaiming the Sciences, the Family and Education*, ed. Martin E. Marty and R. Scott Appleby (Chicago & London: The University of Chicago Press, 1993), 45–47.

67 Bai Ben, "Fandui jinhua lun de Bai Li An (Bryan who is against Evolution)," *Dongfang zazhi* (*The Eastern Miscellany*) Vol. 22, No. 21 (1925): 71–74.

car passed the disciplined and obedient everyday Chinese people in the market place, but the journalist saw only "utter indifference" in the "profoundly ignorant" crowd. He even made "curiosity" a typical "Tartar" trait and "indifference" Chinese.[68] As a pair of telegraph wires came in sight, crossing the road and over the Great Wall, Barzini said,

> it was like the sight of familiar faces; *these* were friends to us, and would carry our news to the world outside. Poor ancient wall, the labor and pride of three royal dynasties and of some millions of men—it is not by the cannon alone that thou art rendered useless; a thread of wire suffices. The most distant peoples can quietly commune with one another above thy head, ignoring thy very existence![69]

Using science and technology as the measure, he and his fellow travelers felt

> the pride of a civilization and a race, and are conscious of representing something more than ourselves. The civilization of Europe overshadows us; it is resumed and symbolized by the speed of our flight. The great longings of the Western soul, its strength, the true secret of all its progress, is resumed in the short word-*faster!* Our life is pursued by this violent desire, this painful insatiability, and this sublime obsession-Faster! Here in the midst of Chinese immobility we truly carry with us the essence of our feverish advance.[70]

Barzini was neither original nor unusual. Scientific thought and technological innovation as indexes of racial superiority shaped Europeans' attitudes to non-Western peoples. Indeed, the opinions of John Barrow (1764–1848), a member of the 1793 Macartney embassy to China, and his contemporaries "formed the nucleus of a cluster of ideas about China that informed virtually all nineteenth-century accounts of the Qing Empire".[71] Restated in the language of evolutionary theory the backwardness of China in scientific and technological achievement was attributed to innate racial deficiencies. Evolution was used variously to mark out racial boundaries and make up racial hierarchy, notably during the era of imperial expansion.

---

68 Luigi Barzini, *Pekin to Paris: An Account of Prince Borghese's Journey across Two Continents in a Motor-car*, trans. L. P. de Castelvecchio (London: E. Grant Richards, 1907), 64, 69–70, 109, 110.

69 Barzini, *Pekin to Paris*, 98.

70 Barzini, *Pekin to Paris*, 109.

71 Michael Adas, *Machines as the Measure of Men: Science, Technology, and Ideologies of Western Dominance* (Ithaca: Cornell University Press, 1989), 4, 183.

Here then is the wider context in which Darwin and evolution were encountered in China. Yan Fu, Liang Qichao and their followers domesticated the new Western knowledge by using large-circulation newspapers and journals to create the conditions for lively introduction, assimilation and debate. New communication technologies and material forms of knowledge-exchange, particularly mass-circulation periodicals, have far-reaching consequences on human perception and practice.[72] In the case here, they contributed substantially to making Darwin a global phenomenon and evolution a global knowledge. What happened during the period from the end of Qing Dynasty to the first decades of the Republican China can be seen as an outgrowth of this process. Those people active in the newspaper industry of the new era also played a huge part in the formation of the matrix of the active introduction, appropriation and discussion of Darwin and evolution. Knowledge-transfer across borders must be understood within the context of the communications revolution that took hold at different times and in different places. For this to take effect in China, it is clear that the "Chinese made a series of choices from the broad range of Western printing technologies available to them".[73] One study of Shanghai's news media, for instance, shows that there was a strong tendency to domesticate the Western newspaper "for Chinese use and Chinese understanding, for only thus—so it must have appeared to China's newspaper makers—could it be an effective agent of change".[74] And the author concludes, "The Chinese experience with the newspaper is 'China's response to the West' in reverse. The foreign medium was not so much the stimulant but the object of a stimulus".[75]

If my argument in this chapter is right, then, it follows that the texts carried by the press medium and served by its technologies—here, evolutionary theories and their interpretations—were themselves in China "not so much the stimulant but the object of a stimulus" and, no less than newspapers, serviceable to social and political concerns, anxieties and aspirations. The encounter with Darwin and evolution compelled Chinese people to confront their own traditions, to compare and to judge and to reshape their own identity. Reciprocal influences of material and

---

72 James Moore, "Communications," in *Science, Technology and Everyday Life, 1870–1950*, ed. Colin Chant (London: Routledge, 1989), 200–249.

73 Christopher A. Reed, *Gutenberg in Shanghai: Chinese Print Capitalism, 1876–1937* (Vancouver: The University of British Columbia Press, 2004), 258.

74 Mittler, *Newspaper for China*, 31.

75 Mittler, *Newspaper for China*, 415.

intellectual culture have always characterized Chinese engagement with the West both inside and outside of the scientific realm. The case of Darwin and evolution demonstrates the imperfection of the notion of a static China as the passive recipient of an impact from the dynamic West. But it also reminds us that Chinese agencies of change have to be understood within a global context.

CIRCULATING MATERIAL OBJECTS: THE INTERNATIONAL CONTROVERSY OVER ANTIQUITIES AND FOSSILS IN TWENTIETH-CENTURY CHINA

Fa-ti Fan*

In recent years, scholars have been increasingly interested in the circulation of knowledge in science. We now recognize that knowledge production and scientific practice were not confined to the familiar sites of museums and laboratories. Ideas and information transmitted by networks and often across cultures played a major role in the making of modern science. Equally important, though less discussed, is the circulation of material objects. The circulation of specimens and other objects of nature in maritime trade, for example, contributed to the development of natural history in the early modern world. Such objects were also instrumental to cultural exchange in global encounters. The circulation of knowledge and material objects grew as the globalizing process expanded in the later centuries. The practice and institutions of science cannot be properly understood outside this global context.[1]

The image of circulation, however, can be misleading. It tends to suggest that people, information, and material objects flowed smoothly along

* Earlier drafts of this paper were presented at the Modern China Seminar at Columbia University, the Institute of Modern History at Academia Sinica, the School of Historical Studies at the Institute for Advanced Study, and the "Circulating Knowledge, East and West" conference at the University of King's College, Halifax. I am grateful to the organizers and participants of these events for their valuable comments and suggestions. I also deeply thank the editors of this volume for their careful reading of an earlier version of this paper.

[1] Harold Cook, *Matters of Exchange: Commerce, Medicine, and Science in the Dutch Golden Age* (New Haven: Yale University Press, 2008); Lissa Roberts, "Situating Science in Global history: Local Exchanges and Networks of Circulation," *Itinerario* 33, no. 1 (2009): 9–30; Lissa Roberts, Simon Schaffer, Kapil Raj and James Delbourgo, eds., *The Brokered World: Go-betweens and Global Intelligence, 1770–1820* (Cambridge: Science History Publications, 2009); Lydia Liu, *Tokens of Exchange: The Problem of Translation in Global Circulations* (Durham: Duke University Press, 1999); Fa-ti Fan, *British Naturalists in Qing China: Science, Empire, and Cultural Encounter* (Cambridge: Harvard University Press, 2004), chaps 1 and 2; idem, "Science in Cultural Borderlands: Methodological Reflections on Science, European Imperialism, and Cultural Encounter," *East Asian Science, Technology and Society: An International Journal*, 1, no. 2 (2007); 213–231. For a critique of the model of circulation in the history of science, see Fa-ti Fan, "The Global Turn in the History of Science," *East Asian Science, Technology, and Society: An International Journal* 6, no. 2 (2012), 249–258.

networks and channels. Circulation appears to be a "natural" or default condition. Yet, not only did the movement of knowledge and material objects require work—consider the efforts put in to transport plants and animals, live or dead, across oceans—but its trajectory may not have been as teleological as circulation would imply. Besides, not all things circulated the same way. Some traveled far and fast; others stalled. Therefore, what is called "circulation" may have been really a series of negotiations, pushes and pulls, struggles, and stops and starts. These caveats aside, one can still recognize the importance of knowledge and material objects in motion in modern science. Its importance is reflected in part in the measures set up to regulate, monitor, and control the flow. To underline this point, we can use the regulation of human migration as an analogy. It has been convincingly argued that the effort to tighten border control and identity documentation, represented by the modern passport system, cannot be separated from the rise of global mobility as well as of nation-state sovereignty in the 19th century.[2] On the one hand, global networks facilitated the migration of people; on the other, political entities asserted their authority and imposed restrictions on such movements. Just as the circulation of material objects, such as specimens, was crucial to science, the regulation, monitoring, inspection, and control of the same had consequences in science. It is, therefore, imperative to investigate the historical reasons and circumstances that enabled or hindered the movement of knowledge or material objects; that is, to find out how and why certain mechanisms were introduced to control the coming and going of people and things. This is an unusual, but necessary, perspective on the topic of the circulation of knowledge and material objects in science.

To illustrate this point, I shall examine an international controversy over antiquities, including fossils, in Republican China. The historical episode shows how certain intellectual, cultural, and political trends converged and challenged the ways in which foreign (mainly Western) scientific activities were conducted in China. The primary contention was about the ownership of antiquities and fossils, and the Chinese deployed cultural and political apparatus to assert their legal authority over the objects at

[2] Adam McKeown, *Melancholy Order: Asian Migration and the Globalization of Borders* (New York: Columbia University Press, 2008). In not unrelated manners, there were new sentiments and/or measures to erect border control against diseases and biological species. Mark Harrison, "Disease, Diplomacy and International Commerce: the Origins of International Sanitary Regulation in the Nineteenth Century," *Journal of Global History* 1, no. 2 (2006): 197–217; Philip J. Pauly, "The Beauty and Menace of the Japanese Cherry Trees: Conflicting Visions of American Ecological Independence," *Isis* 87 (1996): 51–73.

issue. It must be noted that the controversy was not an isolated incident. The particular instance not only shared crucial resemblances with other similar cases of international disputes at the time, but it also foreshadowed many controversies over antiquities today—to name only the better known, the lawsuits against the Metropolitan Museum of Art and the Getty Museum, the tug of war between Greece and Britain over the Elgin Marbles, the Egyptian demand for the return of the bust of Nefertiti, and the Chinese delegation to search for "lost national treasures" in Western and Japanese museums.[3] In all of these cases, antiquities seem to have embodied the complex historical relations among imperialism, nationalism, and internationalism. What is striking about these recent controversies is perhaps not the disagreements between the parties. Rather, it is the fact that there are generally shared definitions of cultural property and heritage. Interestingly, too, there exists a set of national and international legal systems to deal with such cultural property. How did this come about? How did the national and international sentiments and laws regulating the movements of antiquities originate? How did national sovereignty extend to these objects? These are important questions in the history of art, science, and law.

To answer these questions demands a large canvass. In this essay, I can only sketch out the general historical background and tell the story of the international controversy over antiquities and fossils in Republican China in few details. My purpose is to draw attention to an important topic and make suggestions for future research. Since the essay will deal with diverse issues that have rarely been brought together in a history of science paper, it may be helpful to spell out its structure and argument at the outset. The narrative goes like this. In the early 20th century, there was an increase of foreign scientific, especially archaeological, explorations

[3] The literature on this topic is growing fast and is extremely contentious. E.g. James Cuno, *Who Owns Antiquity: Museums and the Battle over Ancient Heritage* (Princeton: Princeton University Press, 2008); James Cuno, ed., *Whose Culture?: The Promise of Museums and the Debate over Antiquities* (Princeton: Princeton University Press, 2009); Kate Fitz Gibbon, ed., *Who Owns the Past?: Cultural Policy, Cultural Property, and the Law* (New Brunswick: Rutgers University Press, 2005); Sharon Waxman, *Loot: The Battle over the Stolen Treasures of the Ancient World* (New York: Times Books, 2009); John Henry Merryman, ed., *Thinking About the Elgin Marbles: Critical Essays on Cultural Property, Art and Law* (London: Kluwer Law, 2009); Peter Watson and Cecilia Todeschini, *The Medici Conspiracy: The Illicit Journey of Looted Antiquities—From Italy's Tomb Raiders to the World's Greatest Museums* (Public Affairs, 2007); Magnus Fiskesjö, "The Politics of Cultural Heritage," in *Reclaiming Chinese Society: The New Social Activism*, eds. Ching Kwan Lee and You-tien Hsing (London: Routledge, 2010), 225–45.

in Inner Asia, or as commonly referred to at the time, "Chinese Central Asia." Archaeology was an expanding scientific discipline, and many archaeologists considered Chinese Central Asia an important region for fieldwork. This period also witnessed the transformation of antiquities into cultural heritage and national property. This transformation was influenced by the international circulation of cultural assumptions and legal practices, including nationalism and internationalism. These broad movements similarly influenced the institutions and practice of science at the time. And, finally, the controversy invites us to consider the issue of science in relation to international cosmopolitanism and national sovereignty. The issue is central to understanding the ideas and actions of the Chinese as well as of the foreign scholars and scientists involved in the controversy.

By examining the historical episode, this essay will demonstrate that the intersections of several historical developments in China and beyond are crucial to a better understanding of science in a global context at the time; and that the controversy provides a useful opportunity to interrogate the circulation of knowledge and practice in science by focusing on material objects, their movements, and the historical actors, institutions, and cultural values that facilitated or regulated their movements. I hope this investigation will complicate the increasingly popular perspective of "circulating knowledge" on the history of science.

China and Central Asia

The introduction of Western learning in the late nineteenth and the early twentieth century provided Chinese intellectuals with new perspectives on Central Asia.[4] During this period, Western orientalists and sinologists paid more attention than Chinese scholars to the historical connections between China proper and central/western Eurasia. Western scholars in philology, history, and archaeology actively pursued research in Mesopotamia and Central Asia.[5] Their work had significant impact on the search

[4] E.g. Joanna Waley-Cohen, *Exile in Ming-Qing China: Banishment to Xinjiang, 1758–1820* (New Haven: Yale University Press, 1991); James Millward, *Eurasian Crossroads: A History of Xinjiang* (New York: Columbia University Press, 2007); Pamela Kyle Crossley, Helen F. Siu, Donald S. Sutton, eds., *Empire at the Margins: Culture, Ethnicity, and Frontier in Early Modern China* (Berkeley: University of California Press, 2006).

[5] E.g. Svetlana Gorshenina, « De la Tartarie à l'Asie centrale. Le coeur d'un continent dans l'histoire des idées entre la cartographie et la géopolitique, » PhD diss., Université de

for the origins of Chinese civilization and on Chinese national narratives in the first decades of the 20th century.[6] Westerners also mounted scientific expeditions into Central Asia, one of the less explored areas in the world, whose history, geography, peoples, and potential political significance attracted scholarly attention. For example, in 1902, the explorer Aurel Stein presented a lecture on his expedition to Chinese Central Asia at the International Congress of Orientalists and made a splash.[7] Representatives of major Western powers divided up Chinese Central Asia among themselves, and Britain, France, Germany, and the United States all prepared to launch archaeological expeditions into their designated areas.

The European zeal in the search for the origins of races and civilizations, which had spurred a series of archaeological excavations in Asia Minor, now moved eastwards to Central Asia. These European scholars saw Central Asia as a historical corridor of migration and cultural diffusion, transmission, and contact between East and West. They asked: What were the relationships among ancient peoples and civilizations in Mesopotamia, in the Steppes, and in the region lying between Turkestan and China proper? What were the historical connections between Europe and China? Their Central Asia was a historical landscape of the "Aryans" and their bronze technology, steppe horse riders thundering across the grasslands, the Silk Road, the Chinese Empire of the Han and of the Tang dynasty, Buddhist monks on pilgrimages to India, Genghis Khan and the Mongol warriors, Marco Polo, and other topics that linked Europe and

---

Lausanne, 2007; Suzanne Marchand, *Down From Olympus: Archaeology and Philhellenism in German, 1750–1970* (Princeton: Princeton University Press, 1996); Bruce Kuklick, *Puritans in Babylon* (Princeton University Press, 1996); Peter Hopkirk, *Foreign Devils on the Silk Road* (London: John Murray, 1980); Jeanette Mirsky, *Sir Aurel Stein: Archaeological Explorer* (Chicago: University of Chicago Press, 1998).

6 Fa-ti Fan, "How Did the Chinese Become Native?: Science and the Search for National Origins in the May Fourth Era," in *Beyond the May Fourth Paradigm: In Search of Chinese Modernity*, eds. Kai-wing Chow et al. (Lanham, MD: Lexington Books, 2008), 183–208. On archaeology and nationalism, see Lynne Meskell, *Archaeology Under Fire: Nationalism, Politics and Heritage in the Eastern Mediterranean and Middle East* (London: Routledge, 1998); Philip L. Kohl, Mara Kozelsky, and Nachman Ben-Yehuda, eds., *Selective Remembrances: Archaeology in the Construction, Commemoration, and Consecration of National Pasts* (Chicago: University Of Chicago Press, 2008); Philip L. Kohl and Clare Fawcett, *Nationalism, Politics and the Practice of Archaeology* (Cambridge: Cambridge University Press 1996).

7 M. A. Stein, "A Journey of Archaeological Exploration in Chinese Turkestan," *International Congress of Orientalists, 13th, Hamburg, 1902* (Leiden: Brill, 1904), 85–87. "Proceedings (Extract) of XII International Congress of Orientalists, Rome, October 1899," accessed July 6, 2011, http://idp.bl.uk/4DCGI/education/orientalists/index.a4d.

Asia.[8] In the meantime, for paleontologists and zoologists, Central Asia was also an important place; many speculated that the region was where mammals and humans had first originated. Henry Fairfield Osborn, Director of the American Museum of Natural History, declared Central Asia to have been "the chief theater of evolution of both animal and human life."[9] In fact, the express purpose of the American explorer Roy Chapman Andrews's expeditions to the Gobi desert in the 1920s was to find evidence of human origins.

Now let us turn to the Chinese. For the Chinese intellectuals in the Republican period, the region was primarily a site of contestation over national origins, national sovereignty, and national identity. In the late 19th century, the region felt the threat from Britain and particularly Russia as the two imperial powers engaged in the Great Game of imperial competition.[10] From the political standpoint of the Qing Empire, the region constituted its northern and western frontiers against Western imperial powers.[11] Republican China faced a new challenge. It was struggling to hold on to a disintegrating empire, much of which was beyond its political control, and to forge it into a modern nation-state. To pursue that end, Chinese intellectuals had to deal with the activities and discourses produced by the scientific/archaeological explorations conducted by foreigners. Since the ultimate goal was to integrate the region into the geo-body of the Chinese nation as they envisioned it, history, archaeology, anthropology, and geography were crucial to this enterprise.[12] Because of the lack of experience and resources, however, they could hardly launch their

---

8 On 19th-century comparative philology, see, e.g., Maurice Olender, *The Languages of Paradise: Aryans and Semites A Match Made in Heaven* (New York: Other Press, 2003); Suzanne L. Marchand, *German Orientalism in the Age of Empire: Religion, Race, and Scholarship* (New York: Cambridge University Press, 2009). Many of the noted sinologists of the late 19th and the early 20th century wrote about the connections between China and central/western Asia for example, Henry Yule, Henri Cordier, Joseph Edkins, Emil Bretschneider, Édouard Chavannes, Friedrich Hirth, and Berthold Laufer.

9 Quoted in Brian Regal, *Henry Fairfield Osborn: Race and the Search for the Origins of Man* (Burlington: Ashgate, 2002), 98.

10 Peter Hopkirk, *The Great Game: the Struggle for Empire in Central Asia* (New York: Kodansha, 1992). Cf. Malcolm Yapp, "The Legend of the Great Game," *Proceedings of the British Academy* no. 111 (2001): 179–198.

11 But of course, for the Tibetans, the Uighurs and the Mongols, these areas were not so much "frontiers" as "homelands at the time, none of these collective identities existed in their modern forms.

12 E.g. Fan, "How Did the Chinese Become Native?" 183–208; Clayton Brown, "Making the Majority: Defining Han Identity in Chinese Ethnology and Archaeology," PhD diss., University of Pittsburgh, 2008; James Leibo, *Reconfiguring Chinese Nationalism: How the Qing Frontier and its Indigenes Became Chinese* (New York: Palgrave Macmillan, 2007).

own scientific explorations into Central Asia. As we'll see, this embarrassment would become piercingly poignant when the combined threats of foreign scientific exploration and the loss of antiquities reached a new height in the late 1920s.

Antiquities and Cultural Heritage

For the bookish Chinese literati in the late Qing, the loss of cultural heritage meant first of all the loss of books. In the first decade of the 20th century, they were shocked to see that some of the best library collections in China had been sold to Japan.[13] Before the era of large public libraries, these private collections, containing many unique rare editions, were regarded as the jewels of Chinese civilization. More than any other objects, books excited strong passions; many a Chinese intellectual shed tears for the losses, and they sensed a crisis. On the eve of the Republican Revolution in 1911, the revolutionary nationalists contributed greatly to the formation and popularization of the discourse of national treasures. For example, the National Essence circle, a group of noted intellectuals, not only called for the preservation of national learning, but also championed the preservation of books, paintings, and other cultural objects that, in their view, embodied national essence.[14]

In the meantime, the discovery of the Dunhuang manuscripts—a trove of religious works, official documents, and literary texts from about one thousand years before—in Chinese Central Asia in the first decade of the 20th century ushered in new concerns. It was Western explorers and scholars who brought the manuscripts to light. This forcefully reminded Chinese intellectuals that they were losing control of their history, territory, and cultural objects to foreigners. When in 1909 the sinologist Paul Pelliot brought cases of Dunhuang manuscripts to Beijing and invited Chinese scholars to examine them, many Chinese scholars were distressed to see these "cultural treasures" had fallen into foreign hands.[15] Their

---

[13] Xu Zhenji, *Qianyuan yishi* (Shanghai: Sanlian shudian, 1996); Song Luxia, *Bainian shoucang* (Taipei: Lianjing, 2005).

[14] Fa-ti Fan, "Nature and Nation in Chinese Political Thought: the National Essence Circle in Early Twentieth-Century China," in *The Moral Authority of Nature*, eds. Lorraine Daston and Fernando Vidal (Chicago: University of Chicago Press, 2004), 409–37; Tze-ki Hon, *Revolution as Restoration: Guocui xuebao and Chinese Nationalist Modernity* (Brill, forthcoming).

[15] Shana Brown, *Pastimes: From Art and Antiquarianism to Modern Chinese Historiography* (Honolulu: University of Hawai'i Press, 2011); Sang Bing, *Guoxue yu hanxue* (Beijing: Zhongguo renmin daxue chubanshe, 2010).

concept of ownership was different from the modern notion of national property, however, and they didn't ask Pelliot to return the manuscripts. They lamented the fact that these historical documents were to leave China, but did not challenge Pelliot's possession of the manuscripts. Other similar incidents caused much pain and anxiety. Foreign scientific expeditions penetrated into Manchuria and Chinese Central Asia dozens of times and carried away numerous archaeological collections. This situation forced Chinese intellectuals to pay close attention to foreign expeditions to Chinese Central Asia. Even in those troubled times, when civil wars burned across the land, newspapers still found space for reports on these expeditions as if they were major public events.[16]

The concept of the preservation of antiquities and historical sites gathered full momentum in the 1920s. Major universities, notably Peking University and Qinghua University, established institutes for national studies, part of whose training and research included *kaogu* (which did not correspond exactly to modern archaeology because of its distinctive concerns and techniques derived from traditional Chinese antiquarianism, but which similarly required research into antiquities).[17] Established in 1928, Academia Sinica, the highest national research institute, regarded history and archaeology as one of its main areas of research. Individual scholars also took it upon themselves to document historical sites. In the 1930s, the famous couple Liang Sicheng and Lin Huiying started their lifelong commitment to the study and preservation of historical buildings.[18] By then, there had already been significant interest at the local level in preserving ancient tombs, monuments, temples, architectures, and other historical sites.[19] Many provincial and city governments had started preservation projects.[20] In addition, there was also the opening of the National

---

[16] E.g. *Chenbao*, March 6 and 10, 1927. *Shuntian shibao*, March 12 and 15, 1927.

[17] Sang Bing, *Wan Qing Minguo de guoxue yanjiu* (Shanghai: Guji chubanshe, 2001). We cannot go into the details here, but it should be noted that the Chinese had a long tradition of antiquarianism and collecting. The significance of the new development lies in (1) a (variegated) shift from the "traditional" concepts of *guwu, kaogu, jinshi, etc.* to the "modern" concept of antiquities, archaeology, etc. and (2) the scholarship and techniques that came with the shift.

[18] Wilma Fairbank, *Liang and Lin: Partners in Exploring China's Architectural Past* (1994). Lin Zhu, *Jianzhushi Liang Sicheng* (Beijing: Zhongguo jianzhu gongye chubanshe, 1996); Wang Jun, *Cheng ji* (Beijing: Sanlian chubanshe, 2003).

[19] E.g. Minguo Neiwubu, ed., *Minguo Jing Lu Jin Yu gu qi wu diao cha ming lu* (Beijing Shi: Beijing tu shu guan chu ban she, 2004 [1918]), which contains catalogs of antiquities and historical monuments complied by the Chinese government in the late 1910s.

[20] E.g. Peter Carroll, *Between Heaven and Modernity: Reconstructing Suzhou, 1895–1937* (Stanford: Stanford University Press, 2006).

Palace Museum in 1924.[21] After the Emperor Puyi was expelled from the Palace earlier that year, the Beijing government decided to make it into a national museum.[22] Scholars were summoned to examine the collections, and after months of inventory work, the museum was opened to the public. National museums across the world served the function of canonizing and propagating the ideas of "national treasures," and the Palace Museum was no exception.[23] The museum proved popular until its holdings were shipped south on the eve of the Sino-Japanese War.

There was another dimension of the urgent concern about antiquities—the draining of antiquities into the international market. The collapse of the Qing dynasty left behind thousands of Manchu aristocrats who lost their privileges and fell on hard times. Many of them had no alternatives but to sell family treasures, often cheaply, to make ends meet.[24] Jewelry, books, paintings, stationery, objects of decoration, and other valuables flooded the markets.[25] Similarly, many Han and other non-Manchu officials also lost their social standings and family wealth and had to sell their collections of treasures. The antique markets were overflowing. Letters and memoirs of Chinese intellectuals documented the booming antique markets in Beijing and other large cities. Antique collectors and book hounds of all kinds—professors, merchants, government officials, writers, and Japanese and Western visitors—swarmed the shops. Major antique dealers in Beijing could obtain on order prized items, including ancient bronzes and the heads of Buddhist statues from the Longmen Grottoes.[26]

21 Na Zhilian, *Gugong bowuyuan sanshi nian zhi jingguo* (Taibei, Zhonghua cong shu wei yuan hui, 1957).

22 The details of the expulsion of Puyi should not concern us here. For more information, see Peter Zarrow, "The Expulsion of Puyi from the Palace: Culture and Historical Memory in 1920s China," (in Chinese, "Puyi be zhuchu gong ji: yijiuerling niandai de Zhongguo wenhua yu lishi jiyi"), *Yijiuerling niandai de Zhongguo* (Taipei: Zhonghua minguo shiliao yanjiu zhongxin bianyin, 2002) 1–32.

23 David Boswell and Jessica Evans, *Representing the Nation* (London: Routledge, 1999). Wen Fong and James Watt, eds., *Possessing the Past Treasures from the National Palace Museum, Taipei* (New York: Metropolitan Museum of Art, 1997).

24 The family of Duan Fang, a Manchu aristocrat and major collector of antiques, had to sell many of his prized collections, including a collection of ancient bronzes to John Ferguson (the Metropolitan Museum), after his death and the collapse of the Qing dynasty.

25 E.g. Sun Dianqi, *Liulichang xiaozhi* (Beijing: Beijing gu ji chu ban she, 1982); Zhou Zhaoxiang, *Liu li chang za ji* (Beijing Shi: Beijing Yan shan chu ban she, 1995).

26 Satoko Fujita, "Okakura Kakuzo (1862–1913) and Boston Brahmins," Ph.D. diss., University of Michigan, 1986; Daijo Tokiwa and Tadashi Sekino, *The Buddhist Monuments in China* (Tokyo: Bukkyo-Shiseki Kenkyu-kwai, 1926); Stanley Abe, "Rockefeller Home Decorating and Objects from China," in *Collecting China: The World, China, and a Short History of Collecting*, ed. Vimalin Rujivacharakul (University of Delaware, 2010): 107–23;

By this time, the Chinese antiques had been well integrated into the international market through the networks of Chinese, Japanese, and Western antique dealers. The famous dealer, C. T. Loo, who had moved to Paris when young, installed his agencies in Shanghai, Beijing, Paris, and New York.[27] He became an international celebrity in the antique world, hobnobbing with the rich and powerful. He dined with European aristocrats and American tycoons, exchanged notes with major scholars of Chinese art, and received visits from curators of the grand museums in Europe and America. The Rockefellers were only one of his better-known customers.

With the refinement of connoisseurship, the institutionalization of Asian art in the museum world, and the market interest in Chinese antiques came the formation of the academic discipline of East Asian art. There were no clear lines drawn between these intersecting worlds. The American art historian John Ferguson started out as a missionary and teacher before taking up various advisory positions with a succession of Chinese governments.[28] Like many Westerners in China and Japan at the time, he became interested in the art of his resident country.[29] He traded in the international antique market and served as a buyer of Chinese art for the Metropolitan Museum of Art. His work helped define Chinese art as an academic discipline in the United States. Similarly, his rough contemporaries Stephen Bushell and Walter Percival Yetts, both physicians to the British Legation in Beijing, collected Chinese art and antiques

idem, "Collecting Chinese Sculpture: Paris, New York, Boston," in *Journeys East: Isabella Stewart Gardner and Asia*, eds. Alan Chong and Noriko Murai. (Boston: Isabella Stewart Gardner Museum, 2009), 432–442; idem, "China, The Buddha, and Modern Aestheticism," in *Re-Imagining Asia: A Thousand Years of Separation*, ed. Shaheen Merali (Berlin: Haus der Kulturen der Weit, 2008), 124–133; idem, "From Stone to Sculpture: The Alchemy of the Modern," *Treasures Rediscovered: Chinese Stone Sculpture from the Sackler Collections at Columbia University* (New York: Miriam and Ida D. Wallach Art Gallery, Columbia University, 2008), 7–16; Cynthia B. Altman, "A Family's Love of Asian Art," in *A Passion for Asia: The Rockefeller Legacy* (New York: Asia Society, Hudson Hills Press, 2006), 51–105; Futian Sheng. *Jin dai ri ben de zhong guo yi shu ping liu zhuan yu jian* shang (Shanghai: shi ji chu ban ji tuan and Shanghai gu ji chu ban she, 2005); Lawton, Thomas. "Yamanaka Sadajiro: Advocate for Asian Art," *Orientations*, 26 (January 1995): 80–93.

[27] Yiyou Wang, "The Loouvre from China: A Critical Study of C. T. Loo and the Framing of Chinese Art in the United States, 1915–1950," PhD diss., Ohio University, 2007.

[28] Lara Netting, "Acquiring Chinese Art and Culture: the Collections and Scholarship of John C. Ferguson (1866–1945)," PhD diss., Princeton University, 2009.

[29] E.g. Ernest Fenollosa and Edward Morse in Japan. Warren Cohen, *East Asian Art and American Culture* (New York: Columbia University Press, 1992); Christopher Benfey, *The Great Wave: Gilded Age Misfits, Japanese Eccentrics, and the Opening of Old Japan* (New York: Random House, 2003).

during their sojourns in China in the late 19th and the early 20th century.[30] Through their collaboration with the British Museum and their scholarly work, they shaped the field of East Asian art in Britain. Yetts would eventually become Professor of Chinese art and archaeology at the University of London.

Thus, the institutionalization of Asian art, the making of the academic discipline, the availability and the continual discoveries of Chinese antiquities, the thriving international market, the increasing interest in Asian art among private collectors, and the growth of public and university museums in Europe and the United States all contributed to the massive outflow of Chinese antiquities in the early decades of the 20th century. Many Chinese intellectuals watched the rapid loss of antiquities with anger and pain. They despised the Chinese dealers who exploited the opportunity for personal gain, and they resented what they saw as foreign imperialism in disguise. They yearned to be able to challenge the complex of academia, museum, market, and politics that was imposed on China.

International Controversy and Antiquities Law

In the second half of the 1920s and early 1930s, a number of historical factors converged to create the confrontation between some Chinese intellectuals and the foreign scientific expeditions in China. One of the immediate causes was that foreign exploration of China reached a new height. In Beijing alone, there were several foreign expeditions, including those of the American Museum of Natural History, the Fogg Museum, the Freer Gallery, Sven Hedin from Sweden, and Haardt-Citroën from France. Their research ranged from art history to archaeology, from anthropology to paleontology, and their activities were reported in newspapers, magazines, and scholarly publications. The Chinese intellectuals were very much aware of their presence. More broadly, there was also the changing political climate that came with the ascendance to power of the KMT (Guomindang) in 1928. The KMT forces had marched under the banner of nationalism and anti-imperialism, and they adopted, for a while anyway, a firmer stance against "imperialist" activities in China than the defeated warlord governments. This political atmosphere fueled the rising nationalism and anti-imperialism among Chinese intellectuals since the May

30 Judith Tybil Green, "Britain's Chinese Collections, 1842–1943: Private Collecting and the Invention of Chinese Art," DPhil diss., University of Sussex, 2002.

Fourth Movement in 1919. Finally, there was also the evolving attitude of the Chinese scientific community. As the scientific community grew, the Chinese became more confident and assertive in dealing with foreign scientists. The founding of Academia Sinica, a national research institute, in 1928 boosted the morale of many Chinese scholars.[31] Take archaeology, for example. Academia Sinica pushed aside Carl Bishop and the Freer Gallery expedition and pursued its own archaeological projects, notably the Anyang excavation that explored ancient Chinese history. Based on their discoveries and interpretations of pre-historical sites, scholars at Academia Sinica also championed their own theory of the origins of Chinese civilization.[32]

To the Chinese intellectuals, Langdon Warner of the Fogg expeditions exemplified the arrogance and deceit of foreign imperialism in action.[33] On his first expedition on behalf of the Fogg Museum at Harvard University, in 1923–24, Warner peeled wall paintings of Buddhist art off the walls of the Dunhuang Caves and absconded with them. In tears, his Chinese interpreter, a student selected from the Yanjing University, reported later to his Chinese professor what had happened. On his next expedition to Dunhuang, in 1925, Warner planned to do the same again, but his action caused a disturbance among the locals. Angered by Warner's destruction of the portraits and statues of deities, the locals confronted him. He tried to bargain with them, but the news went back to Beijing. It caused an outcry among the faculty of Peking University, which had been collaborating with Warner on the expedition. They terminated the cooperation and ordered the Chinese representative, Chen Wanli, a young lecturer, to return.[34] Besieged, Warner aborted his expedition and retreated. It should be noted that the local villagers didn't confront Warner out of their

---

31 Shiwei Chen, "Government and Academy in Republican China: History of Academia Sinica, 1927–1949," (PhD diss, Committee on History and East Asian Languages, Harvard University, 1998.

32 We cannot go into the details here. Suffice it to say that led by Li Ji and Fu Sinian, the Chinese archaeologists began to develop their own theory and perspective on the origins of Chinese civilization in the 1920s and 1930s. They rejected the "Western origins" theory, which emphasized Central/Western Asian influences on Chinese civilization, and replaced it with the theory that the Chinese people and Chinese civilization had originated in the territory of modern northern China. This narrative of national origins would become the standard narrative of national history for many Chinese till today. For more, see Fan, "How Did the Chinese Become Native?".

33 Sanchita Balachandran, "Object Lessons: The Politics of Preservation and Museum Building in Western China in the Early Twentieth Century," *International Journal of Cultural Property* 14 (2007): 1–32.

34 Chen Wanli, *Xixing riji* (Lanzhou Shi: Gansu renmin chubanshe, 2002).

nationalistic feelings; they were especially infuriated by the removal of their religious figures.[35] The intellectuals, on the other hand, saw Warner's behavior as imperialistic and took action to challenge him. In this way, this incident energized the Chinese intellectuals, whose attitude toward similar situations stiffened.

The international controversy over antiquities in the late 1920s led to the nationalization of antiquities (*guwu* or ancient objects), including archaeological remains and fossils. The objects became Chinese national property by law in 1930 and had to be preserved. It may seem surprising that artifacts and fossils were lumped together as *guwu* in China's antiquities law. There were three main reasons. First, the scientific expeditions in China included both archaeological and paleontological teams; in fact, the same expedition often collected both kinds of objects. Since one of the immediate purposes of the law was to stop the collections of scientific objects from leaving China before thorough examination, it was practical to include a variety of objects. Second, Chinese intellectuals pushing for the law regarded foreign collecting activities in art history, archaeology, and paleontology in a similar light. In their view, these activities were all attempts to remove valuable objects—national treasures—from the proprietary of China. Indeed, we must not try to draw a demarcation between science and art here. Chinese antiquarianism based partly on traditional scholarship had played a role in the development of modern archaeology in China. Besides, scientific explorations at the time frequently included research in geography, meteorology, geology, natural history, and archaeology, with different priorities and emphases. It is no coincidence that the Chinese intellectuals who first organized themselves and objected to foreign scientific expeditions in the late 1920s were mostly antiquarians, textual scholars, and archaeologists. Their group, called the Chinese Association of Scholarly Organizations (Zhongguo Xueshu Tuanti Lianxihui, established in 1927), set the tone for the Chinese stance. And, third, the law reflected how totalizing nationalism had become that even nature was integrated into the body of the nation.[36] Western scientists

---

35 Chen Wanli, *Xixing riji*, 42–43.

36 In the late 19th century, major preservation movements in the United States actually started with the preservation of nature (not cultural heritage). In Germany and Britain, movements for the preservation of nature were also active. Due to space constraint, I cannot elaborate on the historical relationship between the preservation of nature and of historical/cultural heritage. Suffice to say that the notion of national/historical monuments played a notable role. (cf. the monument acts in 19c Europe and US). Françoise Choay, *The Invention of the Historic Monument* (New York: Cambridge University Press,

had roamed in China investigating geology, geography, and natural history for decades, and they had collected numerous specimens and deposited them in European and American museums. Several developments in China in the late 19th and the early 20th century changed this attitude. Most relevant was probably the introduction of Western science. Chinese intellectuals now also appreciated the scientific value of such specimens. They would not let specimens be taken away so freely. This helps explain why fossils were included in the category of *guwu* that were protected by antiquities law. However, this new attitude toward objects of nature can be properly understood only in a broader context, one that includes cultural products. The changing notions of nationhood and national property expanded from cultural heritage to the domain of nature. In this nationalist discourse, objects of nature were integral parts of the nation: Nature underlined the permanence of nation, and nation defined the meaning of nature.[37]

Founded in 1928, the Committee for the Preservation of Ancient Objects (a more formal organization based on the Chinese Association of Scholarly Organizations) consisted of about twenty prominent intellectuals who were familiar with Western cultures and societies, as many of them had received education abroad. The organization confronted the foreign scientific teams in China; they obtained the government's support and seized the foreigners' archaeological and fossil collections. The scientific expeditions caught in this nationalist agitation included those led by the American naturalist Roy Chapman Andrews, the Swedish geographer Sven Hedin, and the British archaeologist Aurel Stein, all of whom had conducted research in Chinese Central Asia before. The president of the Committee was Zhang Ji, a KMT dignitary. His election to the presidency

2001); Rudy Koshar, *Germany's Transient Pasts: Preservation and National Memory in the Twentieth Century* (Chapel Hill: University of North Carolina Press, 1998); Hal Rothman, *America's National Monuments: The Politics of Preservation* (Lawrence: University of Kansas Press, 1994).

[37] Fan, "Nature and Nation in Chinese Political Thought." This is not to say that the Chinese intellectuals could not or did not make any distinction between the two. For instance, Ren Hongjun, an indefatigable promoter of science education and scientific research in China, indicated the differences between the goals of preserving artifacts and objects of nature. In his view, the former was historical material and the latter sources of scientific research. The very need to underline the distinction, however, proves how strongly the two were linked in the minds of many Chinese intellectuals at the time. Ren Hongjun, "Meiguo tianran lishi bowuguan Zhong Yasiya caiji shiye zhi jiaoshe," in Ren Hongjun, *Kexue jiuguo zhi meng: Ren Hongjun wencun* (Shanghai: Shanghai keji jiaoyu chubanshe, 2002), 418–425.

was due in part to his position in the KMT government, which had only recently taken over Beijing, but there was no question about his dedication to the cause of the preservation of antiquities. If anything, his view came close to that of the National Essence circle. Most of the members had been based in Beijing before the arrival of the KMT forces and held no significant political posts. They were mainly academics. The role of the newly established Academia Sinica in Nanjing was also significant, as it supplied several members, including its president Cai Yuanpei, the historian Fu Sinian, and the archaeologist Li Ji. Members of the Committee for the Preservation of Antiquities were neither the likes of the Boxers nor the conservative cultural traditionalists of the late Qing. Rather, they were mainly highly-educated intellectuals of the new generation; many of them had studied abroad and were promising scholars in their respective fields. They had been drilled in traditional Chinese learning when young, but they also had first-hand experience and knowledge of Western society and culture. Moreover, they knew the global condition of imperialism, colonialism, and nationalism.

Although the Committee was at first a non-government organization, it maintained connections to the KMT government. Some of the members held quasi-government positions, such as those employed by Academia Sinica in Nanjing. In a statement against Aurel Stein's projected expedition in 1930, the group articulated several points that underpinned the general direction of their thinking and argument. First, they emphasized the importance the Chinese placed on their history. Taking away their historical objects would be violating their sense of history. Second, they laid out the conditions under which archaeological remains could be removed from the host country: that the acquisition of archaeological objects from the rightful owner had to be completely legal, that the removal of any archaeological objects would not damage the completeness or wholeness of the collection to which they belong, and that there would be no one in the host country who had the interest or ability to study or preserve the archaeological objects. The Chinese argued that so far, these conditions had not been met. If foreigners performed excavations and took away archaeological objects without meeting these conditions, then their action would not count as scientific archaeology. It would be seen as vandalism.[38]

---

[38] Liu Shiping and Meng Xianshi's popular account, *Dunhuang bainian* (Guangzhou: Guangdong jiaoyu chubanshe, 2000), 288–295, provides a succinct summary.

Citing the example of the Dunhuang manuscripts, they chided Stein and other foreigners who carried away the manuscripts. The statement insisted that the temple keeper who sold away the manuscripts was not the rightful owner of the collection. Stein and others shamelessly took advantage of his ignorance and paid only a pittance for the invaluable manuscripts. The unfortunate result was that the Dunhuang collection was torn apart and scattered throughout London, Paris, and Tokyo. Moreover, collections in European cities had been neglected and had hardly been studied. The Committee lamented that as a result the Chinese were deprived of both the rightful ownership of the manuscripts and the opportunity to study them. Chinese scientists and scholars did not intend to bar foreign scholars from doing work in China. Bona fide scholars would be welcome to participate in collaborative research projects. The statement also appealed to the international scientific community for sympathy and justice. The authors were aware of the international scene, and they cited the precedents of Egypt and Russian Turkestan to fortify their position and the emerging authority of antiquities law.[39]

While such scientific controversies belonged largely in the realm of cultural affairs and remained marginal in the official work of major diplomatic organizations, they sometimes forced themselves upon the embassies and foreign offices. The scientific institutes involved in the controversies—such as, the American Museum of Natural History, the Smithsonian Institution, Harvard University, the British Museum—all had official or personal connections with their respective national governments. They did not hesitate to pull strings, and subsequently the British and American diplomats in China exerted pressure on the Chinese government. Although the British and American diplomats were not willing to go too far to intervene in a matter largely outside of their regular affairs, they tried to satisfy the scientists and their scientific institutes in their home countries. The Chinese government also had to do something to maintain its image and credentials among the Chinese intellectuals, but it didn't want to risk diplomatic relations with the Western nations. Things dragged on. Some foreign explorers, such as Sven Hedin, agreed to the terms set by the Chinese and they co-organized the Swedish-Chinese

[39] In the early 1930s, the Central Government Commission for the Preservation of Antiquities, which was the official organization that had succeeded the Committee, published a volume on the laws of the preservation of antiquities in a number of different nations—Turkey, Egypt, Italy, and the Soviet Union. It seems that they had studied some of these laws before they designed the Chinese laws.

expedition in 1930. Others, such as Aurel Stein, refused. Roy Chapman Andrews and Carl Bishop were worn out. They folded their projects and left.[40]

For some Western explorers, this situation resembled a confrontation between noble, impartial scientists and fanatic mobs of natives (albeit clothed in the modern garments of nationalism). It was a tale of ignorance obstructing the progress of human knowledge. Andrews's expedition reports, characteristically entitled *The New Conquest of Central Asia* (1932), were written largely in the tradition of a Victorian exploration narrative that celebrated the heroic courage and fortitude of the scientific traveler and cast the natives as the naïve or conceited Other; or else, the natives simply did not exist—hence the empty uninhabited space, the unknown, to be conquered.[41] The natural environment of Chinese Central Asia—deserts, grasslands, high mountains, etc.—evoked in the American explorer the geographic imagination of the American west (which was perceived to be disappearing in the United States itself). Perhaps Andrews and his crew likened themselves to the earlier explorers of the American west, standing on top of the mountain, rifle in hand, and surveying the rugged landscapes reminiscent of the badlands back home (Figure 1).

The Chinese nationalists presented a counter-narrative, one that called for a sense of urgency and pitted patriotism against imperialism. On one level, it was just another nationalist movement in those turbulent years. They saw the scientific expeditions as the last example of decades of ruthless pillaging of Chinese treasures. Memories of the looting during the Boxer Uprising remained bitterly fresh.[42] The discovery of Dunhuang now seemed to be a brutal act of theft, robbery, and imperialism. Aurel Stein, Langdon Warner, Albert von le Coq, and the like were but educated tomb

---

40 For the controversy over Anderson's expedition, see Guwu baoguan weiyuan hui, *Guwu baoguan weiyuanhui gongzuo huibao* in Zhang Yan and Sunyanjing, eds., *Minguo shiliao congkan* vol. 1122 (Daxiang chubanshe, 2009), 123–154; Luo Guihuan, *Zhongguo xibei kexue kaochatuan: Zong lun* (Beijing: Zhongguo kexue jishu chubanshe, 2009). For a view of the diplomatic negotiations, see, e.g., Huo Yunfeng and Liu Jinbao, "Sitanyin di sici Zhongguo kaocha huzhao de jiaoshe shimo," *Nanjing Shida xuebao* (*shehui kexue ban*), no. 5 (2008): 61–67.

41 Roy Chapman Andrew, *The New Conquest of Central Asia: a Narrative of the Explorations of the Central Asiatic Expeditions in Mongolia and China, 1921–1930* (New York: American Museum of Natural History, 1932).

42 On the Yuanmingyuan, see, e.g., James Hevia, *English Lessons* (Durham: Duke University Press, 2003); Haiyan Lee, "The Ruins of Yuanmingyuan or, how to Enjoy a National Woun," *Modern China* 35, no. 2 (2009): 155–190; Young-tsu Wong, *A Paradise Lost: The Imperial Garden Yuanming Yuan* (Honolulu: University of Hawaii Press, 2001).

Fig. 1 Roy Chapman Andrews overlooking the Gobi landscape from up high. (Image #411083. American Museum of Natural History Library.).

raiders.[43] Historical objects were transformed into invaluable treasures that embodied national heritage and national pride. The imperialists, they argued, simply wanted Chinese land and Chinese treasures. Even objects of nature like fossils became national. They were symbols of sovereignty; they were national properties. Thus, the explorers were the invaders into the body of the nation.

Although many Chinese intellectuals shared this sentiment and perspective, some of them, especially scientists who had already had close working relations with their Western colleagues, found themselves in a delicate position. These scientists (e.g. the geologist Weng Wenhao) were

[43] To be sure, Chinese intellectuals could and did distinguish Western explorers who were more cooperative from those who were not. They disliked Stein, Warner, and Andrews, but were relatively friendly toward Hedin and Pelliot. But even Hedin and Pelliot were frequently criticized. In an open letter signed by prominent intellectuals in Beijing in 1935, both Stein and Pelliot were again condemned for their "looting" of the Dunhuang manuscripts, and Fu Sinian felt the need to separate the two and defend the latter. Sang Bing, *Guoxue yu hanxue*, chap. 4.

staunch nationalists, but they also knew well the necessity and advantages of collaborating with Western scientists. The success of Chinese science depended on its international participation. They supported the idea and action to institute regulations and antiquities law. However, they had also to display scientific cosmopolitanism to preserve ties with their Western colleagues. Finally, the Chinese government appointed the Central Government Commission for the Preservation of Antiquities to study the topic of antiquities law. The Commission's proposal specified that the staff of an international expedition must include a Chinese co-director and Chinese scientists and that the fossils and archaeological objects must remain in China or may be shared by the parties if there are duplicates. The law was passed in 1930 to regulate the excavation, preservation, and export of antiquities (including fossils).[44] The Commission also sponsored investigations related to plans for historical preservation. For example, under its aegis, Zhu Xizu, a professor at Peking University, and his son Zhu Qi, an economist and antiquarian, visited dozens of historical sites, including many ancient tombs, and documented them in the mid-1930s.[45]

The Preservation of Antiquities and Cultural Heritage

The story of how the circulation of material objects inside and outside China altered over time may sound like an example of native resistance against imperial domination, a perspective popularized by postcolonial studies. This perspective is helpful, for it both underlines power relations and attributes some agency to the natives. However, the perspective still oversimplifies historical categories and dynamics. It tends to reduce Western action to oppression and non-western action to resistance—a response or counteraction rather than initiation. In the case of antiquities law, modern legal concepts and practices had a complex origin. They did not arise from Europe and spread to other parts of the world. It is perhaps true, as has been argued, that modern border control systems had been developed by Western nations to regulate Asian migrations. Yet, the same cannot be said about how the conception and practice of antiquities law developed in late 19th and early 20th centuries. In this case, major

---

44 Luo Guihuan, "Shilun ershi shiji xianqi 'Zhongyang guwu baoguan weiyuanhui cheng li ji yiyi," *Ziran kexue zazhi* (2006), 137–144.

45 Zhu Xizu, *Liuchao lingmu diacha baogao* (Nanjing: Zhong yang gu wu bao guan wei yuan hui, 1935).

Western powers did not always lead the way. Nevertheless, there is a crucial similarity—both were products of the same process of the globalization of borders. Hence, one can see the case of Chinese antiquities law as part of this larger historical development rather than a later imitation of Western legal concepts and institutions. For even in European nations, laws regulating the export of antiquities were relatively new. France, for example, did not have strong laws preventing the sale of moveable antiquities to foreigners (as opposed to laws protecting historical monuments) until the 1910s. Large numbers of its early medieval artifacts thus went to American collectors, such as J. P. Morgan, and museums, such as the Metropolitan Museum. In fact, Britain couldn't prevent some of its prized private antique objects from being sold abroad.[46]

In Europe, Mediterranean countries like Italy and Greece, which had only recently joined the ranks of modern nation-states, found themselves in the ironical position of being at the same time the origins of European civilization and the periphery of "European modernity." These countries had long been the hunting grounds of antiquarians, archaeologists, and art collectors from the European metropoles in the north. The Elgin Marbles in the British Museum were only one of the better-known examples. In the late nineteenth and the early twentieth century, German philhellenism propelled teams of archaeologists to Greece and Asia Minor to excavate ancient ruins. Both Greece and Italy introduced regulations for excavating and exporting objects of antiquity by 1900, though antiquities

[46] This can be explained in part by the different approaches of the European nations regarding how to deal with antiquities in private hands. Britain was more cautious in infringing on private property rights, whereas Italy and France officially adopted a stronger policy regarding the national ownership of such cultural objects. All of the nations were compiling inventories of antiquities (including those in private possession), but not all of them had the same policy or legal terms as to how much the private owner can exercise his or her ownership or property rights over such objects. Other factors also played a role. For example, there was an exodus of antiquities and objects of art from Europe to the United States before and after World War One. The decline of the British aristocracy in the late 19th century and early 20th century was another factor.

Although this paper doesn't deal directly with the issues of war and the preservation of cultural heritage, it should noted that war was central to many efforts as well as controversies about antiquities and the preservation of cultural heritage. Charters were signed for the protection of antiquities/historical sites during the war. War and the plunders of art/antiquities is another contentious topic (most notably, the controversies over the collections of art and antiquities taken by the Nazi and the Soviet Union during War World Two.) All of these historical incidents contributed to the international legal regimes regulating repatriation of antiquities, etc. The focus of this paper, however, is on the outflow of antiquities from antiquities-rich countries to foreign destinations through trade and archaeology.

laws in the comprehensive form were not in place until the early 20th century (Greece in 1932 and Italy in 1909 and again 1939).[47]

Farther afield, in the Middle East, British, French, German, and later American teams searched for traits of ancient civilizations and for Biblical sites.[48] Egypt had established an agency for the protection of antiquities as early as 1883, but since Egypt was basically under British and French colonial supervision at the time, the officials in charge of the protection of antiquities were actually European scholars.[49] They were against random export of archaeological finds because of their scholarly interests and their work for the semi-colonial state. More significantly, only a year later, the Ottoman Empire also adopted antiquities law to control the export of items of cultural heritage out of its borders.[50] In Latin America, to protect their pre-Columbian monuments, El Salvador passed an antiquities law as early as 1903.[51] Similarly, the United States passed its first national antiquities act in 1906, though the law intended mainly to protect Indian sites on public lands.[52]

---

[47] Aggi Dimacopoulou and Andreas Lapourtas, "The Legal Protection of Archaeological Heritage in Greece in view of the European Union Legislation: A Review," *International Journal of Cultural Property* 4 (1995): 311–324; Elisabetta Povoledo, "Italy Defends Treasures (and Laws) With a Show," *New York Times*, October 8, 2008.

[48] William Stiebing, *Uncovering the Past: A History of Archaeology* (Oxford: Oxford University Press, 1993); Bruce Kuklick, *Puritans in Babylonia* (Princeton: Princeton University Press, 1996).

[49] Donald Reid, *Whose Pharaohs: Archaeology, Museums, and Egyptian National Identity from Napoleon to World War I* (Berkeley: University of California Press, 2003); Norman Bentwich, "The Antiquities Law of Palestine," *Journal of Comparative Legislation and International Law*, vol. 6, no. 4 (1924): 251–254. Egypt issued its first antiquities law in 1835, although it was an isolated act.

[50] Wendy M. K. Shaw, *Possessors and Possessed: Museums, Archaeology, and the Visualization of History in Late Ottoman Empire* (Berkeley: University of California, 2003); Debbie Challis, *From the Harpy Tomb to the Wonders of Ephesus: British Archaeologists in the Ottoman Empire 1840–1880* (London: Duckworth, 2008); Izabella Donkow, "The Ephesus Excavations 1863–1874: in the Light of the Ottoman Legislation on Antiquities," *Anatolian Studies* 54 (2004): 109–117.

[51] See "International Antiquities Law since 1900," published on line by the Archaeological Institute of America. http://www.archaeology.org/online/features/schultz/intllaw.html. Close to East Asia, Japan issued its colonial antiquities law in Korea 1916. Hyung Il Pai, "The Creation of National Treasures and Monuments: The 1916 Japanese Laws on the Preservation of Korean Remains and Relics and Their Colonial Legacies," *Korean Studies* 25, no. 1 (2001): 72–95.

[52] The 1906 antiquities act of the United States was very limited. In the US, the preservation of cultural heritage really started with the preservation of national monuments, referring mainly to natural monuments and national parks, in the second half of the 19th century. So, the nature/culture divide was a little blurred even in the 1906 antiquities act, in the sense that its immediate intent was to protect Indian antiquities in national parks.

How do we best explain the formation of antiquities laws in the world during this period? It seems that the concepts and practices of antiquities law did not originate in one particular nation or continent. Instead, they had sprouted from the tensions and interactions between what have been called the "colony" and the "metropole." The law was both a defensive measure and a bargaining tool of the countries with ancient remains within their boundaries. It was also a means to assert their sovereignty. As the prominent British art historian, G. Baldwin Brown explained in 1912: "There is, first of all, the case of historic lands where the soil enshrines remains of the ancient civilizations of which they were the seats. Many of these lands are under more or less despotic rule, a fact which makes distinctive regulations more feasible than in lands constitutionally more advanced. In such regions, when there is no formal Monumental Act applying to the case, it is usual for the Government to claim by edict proprietorship over all objects of antiquity that may come to light accidentally or in excavations, as well as over all standing monuments already in evidence.... This is the case in Bosnia, Egypt, India, Tunis, Turkey." Baldwin went on to discuss nations with less comprehensive antiquities laws and, finally, commented that "In the case of remaining countries, Prussia, Great Britain, and the United States have no law or regulation bearing on this particular subject of the freedom of sale of works of art. ..."[53] Leaving aside his orientalist assumptions of Eastern despotism and Western liberalism, this statement offers a useful observation on the formation of legal regimes regarding the circulation of antiquities in the early 20th century. Influenced by the ideology of nationalism, these countries often claimed to be the descendants of the ancient civilizations and therefore rightful inheritors of the cultural heritage. The antiquities law served to authenticate their national histories and identities and to assert their stance vis-à-vis the metropole.[54]

53 G. Baldwin Brown, "Recent Monument Legislation and the Care of Artistic Treasures," *Journal of the Society of Comparative Legislation, New Series* 12, no. 2 (1912): 266–279. See also, idem, *The Care of Ancient Monuments* (Cambridge: Cambridge University Press, 1905).

54 Iraq provides a good example: "Since its birth as a nation in 1923 [sic. 1932?], Iraq has had a Department of Antiquities with a model antiquities law and a National Museum. In order to build the museum's collections quickly, foreign expeditions were induced to work in the country for a share in the excavated objects. The choice of which half of the duplicate objects would remain in Iraq was left to the Director of Antiquities, with all unique items going to the Museum. The long-term excavations at Kish and Ur set the pattern and showed how effective the policy was in forming a collection of Mesopotamian artefacts that is second to none." McGuire Gibson, "The loss of archaeological context and the illegal

This development had to do with several concurrent global discourses and practices: nationalism and the writing of national history; a change in the taste and concept of art; new directions in archaeological research; the expansion of museums and other institutions for preserving items of national memory; and the expansion of global and international legal institutions and cultural communities. From this broader perspective, then, China actually participated in and contributed to a process of defining national properties and rights in international scientific activities. The fossil episode, then, meant something more than the challenge of a surging Chinese nationalism to an unreflective, unchanging Western imperialism. We may instead see it as part of a process of reconfiguring legal and scientific practices in a global context.

Cultural Internationalism and World Order[55]

Parallel to the circulation and transmission of antiquities law globally, there was a proliferation of international organizations of science: the International Congress of Orientalists was initiated in 1873, the International Geological Congress in 1878, the International Congress of Zoology in 1889, the International Botanical Congress in 1900, and the Pan-Pacific Scientific Congress in 1920, to name only a few. The two phenomena were not unrelated. The rise of nationalism in Europe, Latin America, and Asia

---

trade in Mesopotamian antiquities," in *Culture Without Context*, issue 1 (1997), accessed July 6, 2011, http://www.mcdonald.cam.ac.uk/projects/iarc/culturewithoutcontext/issue1/gibson.htm.

The law was first drafted by Gertrude Bell when Iraq was still under British mandate, and it was similar to that of the Ottoman Empire and other similar laws. The Ottoman Empire tightened its regulations over the export of antiquities. The finds of foreign archaeological excavations had to be submitted to the Ottoman national museum for inspection and the museum had the right to judge and keep the best of the finds. Therefore, the establishment of the national museum was directly connected to the Ottoman antiquities law. After the legislation, on the one hand, Western archaeologists complained vociferously about the trouble this law caused them in their scientific work, and on the other hand, they cynically said how ineffectual the law was because the sultan gave permissions when he liked to. Whether their observation was true or not is not the main issue. The more relevant point is the relationship between the establishment of the national museum and the legislation. Donald Reid, *Whose Pharoahs?: Archaeology, Museums, and Egyptian National Identity from Napoleon to World War I* (Berkeley: University of California Press, 2003); Elliot Colla, *Conflicted Antiquities: Egyptology, Egyptomania, Egyptian Modernity* (Durham: Duke University Press, 2007); Janet Wallach, *Desert Queen: The Extraordinary Life of Gertrude Bell* (New York: Anchor Books, 1999), 349, 355, 357–78.

[55] I've borrowed the section title from Akira Iriye, *Cultural Internationalism and World Order* (Baltimore: Johns Hopkins University Press, 2000).

evolved along with the formation of scientific internationalism. This phenomenon cannot be separated from Western scientific imperialism, but it also reached into a different dimension. International scientific congresses and organizations blossomed in the last quarter of the 19th century, and the enthusiasm continued into the 20th century. In their ideal vision, the scientific congresses represented a new incarnation of the Republic of Letters, an imagined community of disinterested scholars who spoke for reason and truth in the public sphere. They provided convenient forums for the circulation and exchange of knowledge and ideas. In order to be part of the international scientific communities, however, Chinese scientists had to fulfill the intellectual and institutional criteria shared by the members. The transnational discourse of science derived its authority from its claim to objectivity, modernity, practical utility, universal truth, and global validity. Admittedly this claim owed much of its force and persuasion to Western imperialism, but its appeal to enlightenment universality also, however grudgingly, permitted the participation of "qualified" non-Westerners.

Global networks of science required rules that would govern scientific practices. The multiplication of institutions for botanical research over the world, for example, led in 1900 to the establishment of the International Botanic Congress, which met every six years to discuss, among other things, rules of taxonomy and the international code of botanic nomenclature. China participated in these events. For example, it attended the Botanic Congress in 1926 and was an active member of the Pan-Pacific Congress of Science throughout the 1920s and 30s. Ironically, scientific internationalism was actually a byproduct of the rise of nationalism in much of the world. International congresses of science found themselves in the company of world expositions, the Olympics, and other organized international events that simultaneously promoted national competition and internationalism.[56] Of course, just like in any other international

[56] Akira Iriye, *Cultural Internationalism and World Order* (Baltimore: Johns Hopkins University Press, 2000); Elizabeth Crawford, *Nationalism and Internationalism in Science, 1880–1939* (Cambridge: Cambridge University Press, 2002); idem, *Global Community: The Role of International Organizations in the Making of the Contemporary World* (Berkeley: University of California Press, 2004); Eckardt Fuchs, "The Politics of the Republic of Learning: International Scientific Congresses in Europe, the Pacific Rim, and Latin America," *Across Cultural Borders: Historiography in Global Perspective*, eds. Eckardt Fuchs and Benedikt Stuchtey (Lanham, MD: Rowman & Littlefield, 2002), 205–244; Mark Mazower, *No Enchanted Palace: The End of Empire and the Ideological Origins of the United Nations* (Princeton: Princeton University Press, 2009).

organization, the reality of power politics and imperialism frequently overshadowed the ideas of common spirit. Chinese delegations sometimes left the meetings feeling humiliated, as they did at the Tokyo meeting of the Pan-Pacific Congress of Science in 1926.[57]

Nevertheless, China did manage to capitalize on these forums and opportunities. The Chinese entered the international community of science primarily through research in natural history, geology, paleontology, and archaeology. One of their strategies was to use the natural riches of China to bargain for participation in international scientific explorations of China. Foreign scientists had the knowledge, equipment, international prestige, institutional support, money, and other resources that Chinese scientists badly needed. Cooperation seemed to be a good solution for both parties. Chinese scientists involved in the fossil controversy therefore were walking a fine line between their compatriots and their international colleagues, between nationalism and scientific internationalism, and between scientific idealism and political reality. They had scant resources to mount scientific expeditions to Chinese Central Asia, which had attracted much scientific attention over the past decades and which was, on paper anyway, within the political dominion of China. The result of the controversy over fossils enabled them to maximize any opportunities that came their way by piggybacking on foreign scientific expeditions.[58] They thereby had research access to Chinese Central Asia (Figure 2 and Figure 3). In their efforts to integrate Central Asia into the geo-body of the nation and in accordance with the global system of nation-states, Chinese scientists strove to appropriate foreign scientific imperialism. They thus sought to participate in the transnational scientific discourse. They hoped to gain control or at least to have a say in the production of knowledge and images of Chinese Central Asia—its nature, its history, and its inhabitants.

[57] Japan, the host nation, claimed that the Chinese representatives were not an official delegation and made the procedure difficult for them. The Chinese scientists felt that they were intentionally slighted by the Japanese organizers. Japanese imperialism was an important part of the historical context of this international scientific conference. Xu Minghua, "Zhongyang yanjiuyuan yu Zhongguo kexue yanjiu de zhiduhua," 22 (1993): 233–254. On p. 241.

[58] Grace Shen, *Unearthing the Nation: Modern Geology and Nationalism in Republican China, 1911–1949* (University of Chicago Press, forthcoming), Chapter 4 provides an excellent analysis of such strategies of the Chinese geologists.

Fig. 2 Leaders of the Sino-Swedish Expedition studying a map (c. 1927–28). (Picture courtesy of the Sven Hedin Foundation, Museum of Ethnography, Stockholm). Xu Bingchang, a professor of philosophy at Peking University (left), Sven Hedin (center), and Yuan Fuli, a geologist (right).

Conclusion

In this historical episode, we observe that two closely related trends were at work. On the one hand, there was the expansion and circulation of a global infrastructure and discourse concerning science, law, cultural heritage, nationalism, and internationalism that was exemplified by the international market for Asian antiquities, the museum world which desired fossils, and international congresses of science. It was largely due to this development that antiquities and fossils acquired internationally recognized values—monetary, intellectual, cultural, and scientific. On the other hand, there was the trend to institute legal apparatus and border control over the circulation of these objects. Increasingly, the nation-state insisted on its rightful ownership of such objects. Laws were passed to control their movements. Both trends helped shape the national and international rules of the game in science and were essential to a better understanding of the position, value, and practice of science in a global context.

Fig. 3 Members of the Sino-Swedish Expedition were launching a weather balloon. (Picture courtesy of the Sven Hedin Foundation, Museum of Ethnography, Stockholm) Meteorological fieldwork was an important part of the expedition. Four Chinese meteorology students accompanied the expedition and participated in the meteorological survey of Chinese Central Asia.

It may be said that the circulation of modern science contributed in part to the erection of boundaries to the circulation of certain material objects (e.g. fossils and archaeological remains). The objects now acquired scientific and cultural values that increased their desirability in the eyes of Chinese nationalists and scientists. Consequently, they were not permitted to circulate outside the delimitated space (i.e. national territory) without proper authorization. In this case, it can be argued that the circulation of knowledge actually led to a tighter control of the circulation of material objects. However, a tighter control of material objects did not necessarily mean a blow to the circulation of related knowledge. Archaeology, paleontology, and other field and museum sciences did not crumble after the institutionalization of national ownership of such material objects. Some Western scientists might complain about the new inconveniences and obstacles they now faced; they could no longer carry away scientific objects as they wished. Yet, knowledge production and communication continued. Duplicates, copies, and virtual specimens such as images and descriptions were produced and they traveled across boundaries without difficulty. New forms of collaboration took shape, such as the Sino-Swedish expedition staffed and led by both Chinese and Europeans (principally Sven Hedin). While there may be disagreements as to how successful or unsuccessful the scientific collaboration proved to be, it is hard not to see the event as part of a much broader historical phenomenon.

The Chinese intellectuals were hardly unique in their belief in the power of science, in their commitment to the preservation of antiquities, or in their claim to the ownership of the geo-body of the nation. All of these were components of a global discourse of nation, science, and modernity in which the Chinese intelligentsia anxiously participated. Both Chinese and foreign intellectuals involved in the controversy subscribed to these assumptions—assumptions abundantly reflected in their attempts to explore, study, and represent Chinese Central Asia and in the various institutions devised to preserve antiquities. Antiquities law constituted part of the global system of nation-states, and notably many of the ideas, institutionalization, and practices of antiquities law were first developed in countries that found themselves under the shadow of imperialism. To be sure, it was in these places that the need for such laws was felt most strongly. Still, we cannot ignore the fact that these nations actively participated in and directly contributed to the formation of modern legal regimes that regulated the circulation of goods, currency, and people across national borders. Thus, they helped to shape the rules of international engagement and the international legal order. Equally important, there was an element of cultural internationalism that drew upon transnational and cosmopolitan aspirations, and ultimately established institutions to promote these ideas and practices.

GOING WITH THE FLOW: CHINESE GEOLOGY, INTERNATIONAL SCIENTIFIC MEETINGS AND KNOWLEDGE CIRCULATION

Grace Yen Shen

In recent years, the history of science has become increasingly concerned with the circulation of scientific knowledge. This is no longer the old positivist question of how science spread (or failed to spread) from centers of discovery and innovation to the periphery, but a recognition that communication and exchange are at the heart of how science develops. To take this insight seriously, however, we not only have to explain how science moves, but how specific modes of circulation such as international conferences, journals, and visitors came to dominate modern scientific discourse so much that they were taken for granted.[1]

From the mid-1800s to the first World War international scientific congresses became so common that contemporary sources dubbed the phenomenon a "mania",[2] and as key disciplines began regular tri- or quadrennial meeting cycles, the pulse of international scientific congresses regularly drew interested parties from far and wide to designated host countries. Studies of these early international congresses have emphasized the ways that they fell short of the 'science without borders' ideal they claimed to express, but most of this scholarship has focused on the key players involved in establishing the congress circuits.[3] If we look instead

---

[1] These three modes were used as examples in James A. Secord, "Knowledge in Transit," *Isis* 95, no. 4 (December 2004): 654–672. Secord notes earlier in this talk that the debunking of the positivist view of science has had the ironic consequence of making the movement of science a problem instead of a given. However, it seems important to examine cases in which systems of circulation take on a life of their own and become almost compulsory, moving information around in ways less dependent on content and more dependent on the currents of habit and the economy of motion. International conferences, journals, and visits are each revealing examples of circulations that demand rather than merely mobilize content, and it is worthwhile considering how the myth of universally efficacious science encouraged this trait.

[2] This term and its variants were a commonplace in both popular and specialist scientific journalism from the 1870s to 1910s, and the phenomenon was so well known that it "The Congress Mania" even merited satirical treatment in *Punch* 66 (1874): 135.

[3] See for example, Brigitte Schröder, "Caracteristiques des Relations Scientifiques Internationales, 1870–1914," *Cahiers d'histoire mondiale* 10, no. 1 (1966): 161–177; Paul Forman, "Scientific Internationalism and the Weimar Physicists: The Ideology and Its Manipulation in Germany after World War I," *Isis* 64, no. 2 (1973): 151–180; Anne Rasmussen, "Les Congrès

at the view from the margins, we can trace the ways that the process of establishing national presence could buttress the international scientific project by asserting a viable role for communities previously excluded from intellectual exchange. For newcomers to the international scientific community, it was clear that pilgrimages to the moving metropole of rotating congress venues not only circulated knowledge, they also carved out scientific geographies that could be deciphered and then navigated on one's own terms.[4]

In the pages that follow, I use the example of Chinese geologists to illustrate how the circulation patterns defined by international congresses could both marginalize and stabilize new scientific communities. I begin by considering the newness of the idea of scientific internationalism for Chinese leaders and intellectuals, and tracing their evolving responses. The next section examines several International Geological Congresses of the early twentieth century and demonstrates how Chinese geologists shifted China's strategy of participation so that geological imperatives would no longer threaten their nation's claims to territorial sovereignty. In contrast, the Pan-Pacific Science Congresses covered in the third section were useful in publicizing Chinese science, but they showed Chinese geologists the limits of relying on other nations as hosts. I conclude by suggesting that we pay more attention to the ways that, as specific currents of knowledge circulation became increasingly compulsory, non-circulation took on new meanings and possibilities as well.

Internationalism and Chinese Geology

The late nineteenth and early twentieth centuries were notable for internationalist movements in many arenas, but in the sciences, internationalism took on a particular urgency, for it was central to the accelerating engine

internationaux liés aux Expositions universelles," *Revue d'histoire intellectuelle* 7 (1989): 23–44; and Ronald E. Doel, Dieter Hoffmann, and Nikolai Krementsov, "National States and International Science: A Comparative History of International Science Congresses in Hitler's Germany, Stalin's Russia, and Cold War United States," *Osiris* 20 (2005): 49–76. For a treatment that makes use of the distance between the ideals and practices of international meetings, see Matthew Stanley, *Practical Mystic: Religion, Science, and A. S. Eddington*, illustrated edition (Chicago: University Of Chicago Press, 2007).

4 "Moving metropole" riffs upon Roy Macleod's "moving metropolis" but in this case the metropole is constructed by the act of circulation itself, since congress venues were sometimes relatively marginal outside of meeting dates. See Roy MacLeod, "On Visiting the 'Moving Metropolis': Reflections on the Architecture of Imperial Science," in *Scientific Colonialism: A Cross-Cultural Comparison*, ed. by Nathan Reingold and Marc Rothenberg (Washington: Smithsonian, 1987), 217–49.

(and image) of knowledge-making. Scientific internationalism expressed a belief in shared, objective nature that dictated progress via joint, cumulative efforts, and scientists assembled in disciplinary congresses to discuss issues like the standardization of weights and measures, the integration of worldwide sanitary procedures, and the rationalization of nomenclature and systematics, among many others. They set disciplinary agendas, adjudicated theoretical debates, and facilitated exchange of technical information, often sparking collaborations along the way.[5]

But, as with broader internationalist trends of the age, scientific internationalism existed in constant tension with more particularistic commitments, leading nation-state theorist E. H. Carr to observe, "In no period has there been more talk of cooperation between nations [and]... In few periods, less of the reality."[6] The problem was not necessarily one of insincerity, however. Internationalism simply presupposed the nation as a political form, and even its most unifying successes were generally the product of bureaucratic procedures modeled on the one nation, one vote ideal. European and American participants took decades to grapple with the mismatches of this notion of internationalism with political and scientific realities, but it was even more challenging to outsiders like the Chinese, for whom the very concept of an interlocking community of sovereign nations was still somewhat exotic and new.

When the British petitioned the late Qing government for bilateral diplomatic relations in the early 1800s, for instance, their request was indulgently acknowledged and then summarily rejected. It was an interesting idea, but one that made little sense to a complacent government unaccustomed to thinking of itself as anyone's equal. By the 1860s, when the beleaguered and distraught scholar-official Hu Linyi stood on the bank of the Yangtze River and "vomited blood" at the sight of iron gunboats sailing upstream to enforce treaty rights, Chinese were beginning to see that the new breed of foreigners could not be dealt with using traditional methods of 'barbarian management.'[7] Still, it was not until the next generation of literati-reformers at the turn of the century that this passing of the old transformed into wholehearted engagement of the new. While the Qing court failed to see British pursuit of diplomatic representation as "an

---

[5] See, for example, Simeon E. Baldwin, "The International Congresses and Conferences of the Last Century as Forces Working Toward the Solidarity of the World," *The American Journal of International Law* 1, no. 3 (July 1907): 808–829. The list suggests the breadth of fields involved and the kinds of idealism represented by the internationalist movement.

[6] Edward Hallett Carr, *Nationalism and after* (London: Macmillan, 1967), 29.

[7] 《清史稿 Qing shigao》卷四百六 列传一百九十三 Chapter 406, Biography 193 《胡林翼传 Hu Linyi zhuan》.

invitation...to join the modern world,"[8] the young journalist-reformer Liang Qichao grasped the new geopolitical order of things the first time he saw a world map,[9] and he spent the rest of his life chasing variations of Chinese modernity. The difference, represented both by gunboats and the easy availability of Western-style maps in commercial Shanghai, was that no one was asking China for permission any longer. The "contraction of China from a world to a nation in the world," as the sinologist Joseph Levenson put it, was a consummated fact before many Chinese understood its implications or accepted its reality, and even after the republican revolution of 1911, the practical transition from empire to nation took several decades to take root.[10]

Under these conditions, though Chinese engagement with the international in the early twentieth century took on the themes of cooperation and idealism that colored its Western counterpart, scientific internationalism could not mean what it did in places where science was considered a birthright and nation-states a natural evolution. Internationalism for the Chinese was not an elective good agreed upon and sustained by relative equals; it was a matter of necessity and survival thrust upon them from the outside, and one which was often the rosy face of imperialism.[11]

---

[8] John King Fairbank and Merle Goldman, *China: a New History* (Cambridge, MA: Harvard University Press, 2006), 197.

[9] Xiaobing Tang, *Global Space and the Nationalist Discourse of Modernity: The Historical Thinking of Liang Qichao* (Palo Alto, CA: Stanford University Press, 1996), 13.

[10] Joseph Levenson, "The Genesis of *Confucian China and its Modern Fate*," *The Historian's Workshop: Original Essays by Sixteen Historians*, ed. L. P. Curtis, Jr. (New York: Knopf, 1970), 288. Levenson's ideas, which suggest that China's independent historical trajectory ended with modernity, are understandably unpopular, and a great deal of work over the past four decades has established the myopia of that perspective. However, it is still difficult to argue around the ways that engagement with the West was often perceived as a fundamental crisis by Chinese themselves. One of the primary challenges for contemporary Chinese historians is to escape a teleological view that inherently privileges Western forms (science, democracy, capitalism, etc.) *without* erasing the lived experience of the Chinese actors for whom "the West" and Western knowledge were critical conceptual categories.

[11] Scientific internationalism was hardly neutral and utopian anywhere, but western European and American participation before World War I was based on a common rhetoric of scientific objectivity. For more, see Eckhardt Fuchs, "The Politics of the Republic of Learning: International Scientific Congresses in Europe, the Pacific Rim, and Latin America," in *Across Cultural Borders: Historiography in Global Perspective*, eds. Eckhardt Fuchs and Benedikt Stuchtey (Oxford: Rowman & Littlefield, 2002), 205–244. For a classic study of science and imperialism more broadly, see also Lewis Pyenson, *Cultural imperialism and exact sciences: German expansion overseas, 1900–1930* (New York, Bern, Frankfurt: P. Lang, 1985).

Costs and benefits were, however, difficult to disentangle. The International Settlement in Shanghai, for instance, was a stinging example of China's semi-colonial reality where foreigners governed themselves and were above Chinese law. It was also a place where anti-Qing Chinese nationalists could publish in safety by invoking the protection of the international community.[12] Later on, some would argue, the weak Beiyang government that officially represented China abroad (but was besieged domestically) was mainly kept afloat by the unwillingness of international powers to lose their treaty rights. In both situations, rules of conduct in the international community preyed on Chinese sovereignty while simultaneously offering specific Chinese groups new possibilities for action.[13] In like manner, scientific internationalism often threatened to turn China into a laboratory and Chinese researchers into laborers. And yet, Chinese geologists who tried to turn the tables, preferred to strengthen rather than subvert the system, in hopes that by doing so they could truly bringing about a community of equals ready to embrace their participation.

That said, geology was a particularly tricky field in which to learn the norms of international science. On one hand it was ripe for international mobilization. As early as the 1790s, Alexander von Humboldt drew the geological sciences into an international program of measurement and empiricism. Though other disciplines also organized on an international scale, geology's global awareness was built into its scientific objectives.[14] By the turn of the twentieth century, terrestrial magnetism, geodetic studies of gravity and mountain levels, seismology, volcanology, and soil science were all worldwide projects. These geological investigations did more than standardize and review individual labors. The unity of earth history and a 'manifest destiny' to map and measure the globe inclined geologists towards both communal observation and the systematic integration of local contributions. Some hoped to settle existing disputes and extend known concepts through comparison and correlation. Others sought out geological novelties that would shed light on general processes. All

[12] One of the most famous examples of this was the *Subao* case involving Zhang Binlin and Zou Rong. For an interesting take, see Par Cassel, "Excavating Extraterritoriality: The 'Judicial Sub-Prefect' as a Prototype for the Mixed Court in Shanghai," *Late Imperial China* 24, no. 2 (2003): 156–182.

[13] For several variants on this, see Xiaoqun Xu, *Chinese Professionals and the Republican State: The Rise of Professional Associations in Shanghai, 1912–1937* (Cambridge: Cambridge University Press, 2001).

[14] Stratigraphy, for example, was built upon the comparison and correlation of local series, each of which was but an incomplete record of geohistory.

contrived to overcome the practical difficulty of studying the immensity of the planet by multiplying research capacity, and weaving the particularities of place into a web of pervasive geological meaning.

At the same time, geology was also a decidedly practical and strategic science, and the pull of national interests was perhaps nowhere as explicit or acute. Knowledge of the land meant a great deal of control over its resources and the people who inhabited it. Detailed information about mines and oilfields often translated directly into economic power, but theories about where coal or silver might be were just as important, if territorial expansion or extraction were possibilities. Even without such geopolitical disturbances, the link between geology and material wealth or military advantage encouraged secrecy about maps, specimens, and field data that proved antithetical to internationalist ideals. Even 'international' projects were often instances of one or more powers acting on, rather than with, a less-developed neighbor. Because of the local nature of geological investigation, competing national interests played out as contests for access to specific places, for the object of study could never be transported into a laboratory, purified, and studied in abstraction.[15] Field geology, to invert Bruno Latour's famous formulation, was a science of 'mutable immobiles.'[16]

In this way geology was as inextricably tied to economics, politics, and issues of sovereignty as it was to globalization and cooperation, and it embodied many of the contradictions in scientific internationalism. It was an ideal way for the Chinese to learn both the workings of the new world order in which they found themselves and the pitfalls of internationalism on other nations' terms. Contemporary Chinese historians of geology, however, have seldom moved past the surface of China's early successes in the international scientific arena.[17] They frequently cite foreign recog-

---

[15] Elisabeth Crawford, Terry Shinn, and Sverker Sörlin, "The Nationalization and Denationalization of the Sciences; An introductory essay," in *Denationalizing Science: The Contexts of International Scientific Practice*, eds. Elisabeth Crawford, Terry Shinn, and Sverker Sörlin (Dordrecht, The Netherlands: Kluwer Academic Publisher, 1993), 14–15.

[16] See Bruno Latour, *Science in Action* (Cambridge, MA: Harvard University Press, 1987), 227, for his discussion of "immutable and combinable mobiles".

[17] For some examples, see 李学通 Li Xuetong, "中国地质事业初期若干史实考" ("Zhongguo dizhi shiye chuqi ruogan shishi kao"), 中国科技史杂志, *The Chinese Journal for the History of Science and Technology* 27, no. 1 (2006): 61–74; 霞 奚 Xia Xi, "中国地质事业的奠基人——丁文江" ("Zhongguo dizhi shiye de dianjiren—Ding Wenjiang"), 神州学人 *(Shenzhou xueren)*, no. 3 (1994): 39; 43; 吴凤鸣 Wu Fengming, "中国地质事业的开拓者—章鸿钊" ("Zhongguo dizhi shiye de kaituozhe—Zhang Hongzhao"), 中国科技史料, *China Historical Materials of Science and Technology* 15, no. 1 (1994): 29–40.

nition of Chinese geological work as sterling proof of its quality and significance. Without a scientific track record, international acceptance was indeed hard to come by and therefore precious, but, by marking geological merit in this way, present-day scholars surrender judgments of legitimacy and validity to the international scientific community in a way that Republican Era (1911–1949) geologists themselves did not.

While Chinese historians of geology emphasize the difficulty of gaining international recognition, many Western scholars seem to discount it. James Reardon-Anderson, in his classic study of modern Chinese chemistry, suggests that "Chinese who chose this route [i.e., geological sciences] played at the low easy end of the scientific spectrum," capitalizing on the inclusiveness of natural history without accepting the "challenge" of the "big leagues of international science" in which nations contested universal physical principles. As a descriptive science, geology's internationalism was based on 'filling in the blanks' rather than the immutability of natural laws, and so "Chinese scholars … could quickly enter the game of world science, contribute to the nation at home, and gain recognition from the scientific community abroad, simply by reporting on what they found in their own backyards."[18]

Backyards, unfortunately, were easy to trespass upon, and Reardon-Anderson's account misses the point that participation in international science was as risky as it was rewarding for Chinese geologists. No available account of Chinese geology pries deeply enough into the relationship between international and domestic contexts, settling instead for triumphs or failures in one or the other arena. I argue that such depictions are not only incomplete, they are misleading and uninformative about the ways that science as an internationalizing project disciplined newcomers into supporting roles for established players on the knowledge production chain. By looking at the International Geological Congresses and Pan-Pacific Science Congresses of the early twentieth century, we can follow Chinese geologists through the process of learning scientific norms and examine how they leveled the playing field by investing in international ideals rather than by subverting or resisting them. Along the way, their attention to models and patterns also helps us understand international science as an increasingly coercive system of knowledge flow that linked opportunity to the obligation to circulate.

---

18 James Reardon-Anderson, *The Study of Change: Chemistry in China, 1840–1949*, Studies of the East Asian Institute (Cambridge: Cambridge University Press, 1991), 5.

International Geological Congress

The first International Geological Congress was timed to coincide with the Universal International Exhibition in Paris, 1878. Despite the grandiose pretensions of the Exhibition's name, the scope of the Congress it hosted was rather limited. Of 312 attendees, 227 were members of the Geological Society of France and 8 from the United States, which first proposed the Congress two years earlier.[19] Only three British scientists made the trip across the Channel and, though there were also three German nationals at the Congress, Germany refused to participate formally. The Congress' stated objectives of standardizing geological terminology and mapping conventions were hampered by this lackluster showing from two of geology's powerhouses, and though vocal, most congress participants were not leaders of their field, but simply interested, practicing geologists eager to talk shop. Twenty-two nations were registered, but in actuality this was mostly a tally of citizenship, since attendees generally participated as individual scientists rather than as official delegates with any power to represent or influence their home constituencies.

Little standardization was achieved but the Congress broke ground for future meetings, which were held thereafter at three or four year intervals, barring war and other calamities. By the second congress, Germany was on board, and by the turn of the century most of the established centers of geology were taking part.[20] Over time, the meaning of "international" in the congress' organization shifted as well. In part this shift was a simple matter of preparation. Once the succession of International Geological Congresses was regularized, organizing committees had much more time to distribute circulars, contact foreign organizations individually, and set up themes for discussion. Organizing committees, as rotating hosts, could also be more explicitly national, introducing extensive fieldtrips to showcase the geological interest of their own territory and emphasizing the global import of local geological problems (e.g., glaciers and orogeny in Switzerland).

---

[19] For basics, see Francois Ellenberger, "The First International Geological Congress (1878)", n.d., 2, http://iugs.org/uploads/images/PDF/1st%20IGC.pdf. Debra Adrienne Everett-Lane has a good discussion of competing claims for credit in initiating the first Congress in "International Scientific Congresses, 1878–1913: Community and Conflict in the Pursuit of Knowledge," PhD diss., Columbia University, 2004, 44–54.

[20] Gian Battista Vai, "The Second International Geological Congress, Bologna, 1881," *Episodes* 27, no. 1 (March 2004): 13–20.

While countries with highly developed research communities continued to have large contingents at the congresses (often representing their own academic, commercial, or provincial affiliations rather than national groups), host countries began sending direct invitations to previously unrepresented nations via diplomatic channels to stretch the boundaries of inclusion and point the discipline in uncharted directions. This shift in attitude was reflected in the adoption of multiple official languages (usually French, English and German), and an increase in non-scientific delegates whose presence was symbolic rather than substantive. Whereas internationalism in the first congress had been an attempt (however flawed) to portray geology as the international science that it was, congresses in the early 1900s began to represent geology as the global project that it wanted to be. As the prestige of the congress grew, modernizing nations such as Uruguay, Turkey, and Chile could send representatives to indicate their commitment to progressive ideas, and the congress acknowledged this participation as endorsement of geology's critical importance.[21]

This shift, marked less by explicit directives than by changes in the balance between delegates and presenters, occurred just as Chinese were trying to acclimate to their new position in international affairs. Though records are scanty, one can see the Chinese government struggling with its representation in the International Geological Congress. The Qing court appointed foreign mining engineer Clarence Key as its official delegate to the 1906 Congress in Mexico but did not actually send him to the meeting.[22] Key was useful for establishing contacts and handling correspondence but was probably not seen as a suitable representative for China on the world stage. Indeed, by the next congress in 1910, he was already out of the country, and the court sent Djin Da Min to Stockholm instead.

We know little about Djin, except that he was a Qing dynasty envoy stationed in Berlin. We cannot be sure of his exact government post, since the Qing did not have ambassadors to individual European countries. We also do not know what language he used at the congress, for he was a student and substitute language instructor at Berlin University, but the

---

[21] These shifts can be seen in the *Comptes Rendus* of the various congresses and are supported by Fuchs, "The Politics of the Republic of Learning: International Scientific Congresses in Europe, the Pacific Rim, and Latin America."

[22] Clarence Key was a grandson of Francis Scott Key and was hired to assist the Chinese minister to the United States, Mexico and Peru, Sir Chentung Liang Cheng (Liang Pixu, 梁丕旭) in 1911 because of his skill in Spanish.

technical sessions were officially in French and English only.[23] If anything, we can be more confident about who Djin Da Min was not and what he did not do. He was not a graduate of any Western mining school that we know of, and he did not present a paper at the congress. He did not submit any observations or commentary for publication in the *Compte Rendu*, nor did he take part in any of the geological field trips before, during, or after the congress. No one associated with him took advantage of any of the available ladies' activities, and he did not, in any language, offer an official toast from his imperial court to his hosts, the King and Queen of Sweden. No record or report of his trip exists in any of the standard administrative publications in China. We only know that he was in Stockholm and, as the official delegate from China, was automatically a Vice-Président of the Bureau du Congrès. This nominally put him in the ranks of the world's most accomplished and powerful geologists despite his complete unfamiliarity with the science, and indicated China's attempt at representation by presence, rather than expertise.[24]

At the same time that Djin Da Min was in Stockholm, another government-sponsored Chinese named Weng Wenhao was already at the University of Louvain, busily working on a doctorate in geology, and, based on the honors and accolades he received in Belgium, he might easily have been an unexpectedly impressive Chinese delegate. But, the imperial choice of representative is only puzzling if we assume that attendance at the International Geological Congress was strictly about geological science. If we consider the possibility that the government was primarily concerned with advertising its interest in outside assistance with resource development, it becomes perfectly reasonable that an older, more politically connected representative would be chosen over an untested student of "Western learning." In fact, whether China was invited because of Johan Gunnar Andersson's interest in Chinese iron deposits or the Chinese government made the first move to get the Director of the Swedish Geological Survey to China to study iron, Andersson was soon officially in the service

[23] He also was a student and later a substitute instructor of Chinese at Königlichen Friedrich-Wilhelms-Universität zu Berlin. A more contemporary transliteration of his name is likely Jin Damin, but we have no further details. Thomas Harnisch, *Chinesische Studenten in Deutschland: Geschichte und Wirkung ihrer Studienaufenthalte in den Jahren von 1860 bis 1945* (Hamburg: Institut für Asienkunde, 1999), 108.

[24] Everett-Lane, "International Scientific Congresses," discusses the similar insufficiency of presence in making any contribution on the Congress on pp. 237–8 of her dissertation, where she discusses the limited impact that women's attendance had on early meetings.

of the Chinese government, and it is very likely that the 1910 Congress was the critical contact point.[25]

This tactic seems to have had some benefits for the Qing court, which was under external pressure but secure enough in its claim to power to feel comfortable relying on foreign advisors. However, it was no longer a sensible strategy by the Republican era, when China's new government was struggling to assert control over the declining empire it had inherited and still courting international recognition. Unfortunately, the weak young regime did not have many options. The XIIth International Geological Congress was held in only the second year of the Republic (1913), amidst a vicious struggle for power between Yuan Shikai and the newly formed Guomindang Party. Zhang Hongzhao, who had returned from Japan with a baccalaureate degree in geology from Imperial University of Tokyo in 1911, was desperately trying to convince the government to maintain the small, underfunded Department of Geology he established in Nanjing after the revolution.[26] Zhang did not at the time know any Western languages and Ding Wenjiang, who received a geology degree from Glasgow University in 1911, was an unknown teaching middle school in Shanghai until the end of 1912.[27] Though Zhang and Ding were both campaigning for geology in government circles when the congress actually occurred, the new regime could not have paid for their travel to Canada to attend the congress and they were more effective at home trying to establish the Geology School (September 1913).

Parkin Wong, meanwhile, was an engineer living in Ithaca, NY in 1913. He was a comfortable train ride from Toronto and well-versed in English.[28] Though there was also a mining student at Columbia University listed as a member of the Congress, Wong had political connections and, as China's official delegate, was the only one in attendance at the congress. The time

25 Magnus Fiskesjö and Chen Xingcan, *China before China: The Discovery of China's Prehistory* (Stockholm: Museum of Far Eastern Antiquities, 2004).

26 章鸿钊 Zhang Hongzhao, 六六自述 (*Liuliu zishu*) (武汉 Wuhan: 武汉地质学院出版社 Wuhan dizhi xueyuan chubanshe, 1987), 30.

27 For more on Ding Wenjiang in English, see Charlotte Furth's classic work *Ting Wen-chiang: science and China's new culture* (Cambridge, MA: Harvard University Press, 1970).

28 In fact, there is a well-known story in which, decades later in Hong Kong, Parkin Wong was asked if he understands English, and he is said to have replied "I speak a little. I graduated from Columbia University." See David Faure, *Colonialism and the Hong Kong mentality* (Hong Kong: Centre of Asian Studies, University of Hong Kong, 2003), 26. While there is no record of him ever attending Columbia, he was a popular member of Cornell's Cosmopolitan Club (Class of 1914), and active in the early Chinese Students' Association and Science Society.

for representation by presence had passed, however, for Wong's ceremonial Vice Presidency was no match for Kinosuke Inouye's Presidency of an actual scientific meeting, backed by his Directorship of the Imperial Geological Survey of Japan. It also provided no answer to the attendance of Chutaro Kido from "Mandchourie" (Manchuria), a Japanese geologist who presented material on Manchuria's coal resources based on the fieldwork of the Imperial enterprise, the South Manchuria Railway Company.[29] When one of the more celebrated debates of the congress erupted over estimates of China's much vaunted coal reserves, Wong could offer no response. Instead, Noah Drake, an American from the University of Kansas who had taught at the Beiyang University in Tianjin, and Inouye contested the issue of whether China was coal rich or coal poor, and later commentators lamented that "China had no one at the meeting."[30]

The 1913 Congress' push to catalogue the coal resources of the world followed Sweden's 1910 attempt to estimate the iron resources of the world, and exemplified another aspect of the shift in internationalism of the congress. While standards and conventions were still a major issue, many regions were too unexplored to allow correlation with enough accuracy to settle thornier classificatory problems just yet. When the Swedish organizing committee decided to support the interests of its mining constituency by calling for studies on iron ore, they set a trend of resource compilation which continued at international geological congresses for the next fifty years. This was in many senses the scientific corollary to the organizational push for broader national representation, for it expressed the mission of international geology as inclusive and comprehensive rather than simply coordinated and methodical. Ominously, this meant that if China could not study and survey its own territory, others would be justified in doing so. In his presentation on "The Geological Map of the World," Emmanuel de Margerie of France explained that "A single gap remains towards the east, but surely a very important one: China, with her eighteen provinces." Since "No official institution is at work there," de Margerie suggested that

---

[29] *Compte-rendu de la XIIe session du Congrès géologique international, Canada, 1913* (Imprimerie du gouvernement, 1914), 26. This issue was particularly delicate because of Japan's colonial interests in the region, established after Japan's surprisingly quick victory in the Russo-Japanese War (1904–1905). Neither Russia nor Japan had officially colonized Manchuria, which remained officially part of China. The region was a Japanese "sphere of influence" until 1931, when Japan instigated the Manchurian Incident and established of the puppet state of Manchukuo (for which Japan was booted from the League of Nations).

[30] Noah Fields Drake, "The Coal Resources of China" in *The Coal Resources of the World* (12th International Geological Congress, 1913), I: 129–168.

"critical discussion and comparison of the data . . . should be entrusted to a competent geologist sufficiently familiar with the Far East." Given that in his discussion of the rest of Asia he turned to Britain, Russia, the Netherlands, and Japan as "natural" candidates "because of the political interests and of the territorial possessions which their respective countries possess,"[31] it seems clear that, contra James Reardon-Anderson, Chinese geologists who rose to the task were not playing at the "low easy end of the scientific spectrum."[32] They were in a high-stakes game in which geology and territorial aggression traveled together, and they could no longer avoid entering the fray, because the logic of the map would not permit blank spaces.

If China's own geological capacities at the time were still insufficiently developed to stave off these scientific advances, the realities of oncoming war were effective enough. When, after the Great War ended, a XIIIth International Geological Congress was held in 1922, China was in a much stronger position to speak on geological matters, but few scientists abroad knew it. By 1922, six cohorts of natively trained geology students had graduated from the national Geology School (1913–1916) and the Peking University Department of Geology (1918–). A Geological Survey had been active since 1916, and the geological community was strong enough to support the founding of the Geological Society of China (1922).[33]

With these developments, China's delegation to the Congress could be shaped by the geological community itself, rather than by bureaucrats or convenience. As luck would have it, the XIIIth Congress was organized by Belgium, where the Vice-Director of the Geological Survey, Weng Wenhao, had earned both his undergraduate and doctoral degrees in geology.[34] Weng was therefore a natural choice to send to the Congress, and China's paid membership in the congress was a carefully orchestrated statement of China's arrival on the geological scene. Ding Wenjiang, for example, was not actively involved with the Geological Survey at the time, but he retained his title as Director so that he could represent the Survey on the Congress roster. On the other hand, Amadeus Grabau, who had joined

31 E. De Margerie, "The Geological Map of the World" in *Compte-rendu de la XIIe session du Congrès géologique international, Canada, 1913*, 185.

32 Reardon-Anderson, *The Study of Change*, 5.

33 张九辰 Zhang Jiuchen, 地质学与民国社会: 中央地质调查所研究 (*Dizhixue yu minguo shehui: Zhongyang dizhi diaochasuo yanjiu*) (Jinan: 山东教育出版社 Shandong jiaoyu chubanshe, 2005).

34 李学通 Li Xuetong, 翁文灏: 书生从政 (*Weng Wenhao: Shusheng congzheng*) (兰州 Lanzhou: 兰州大学出版社 Lanzhou daxue chubanshe, 1996).

both the Chinese Geological Survey and Peking University in 1921, was registered under China, but with his former professorship at Columbia University still listed prominently in his credentials.[35] Zhang Hongzhao, who was also Deputy Director of the Survey, was listed in his capacity as the President of the Geological Society of China, though the Society was only founded half a year earlier. Finally, Johan Gunnar Andersson, former Director of the Geological Survey of Sweden and chief organizer of the 1910 Congress, was named as both advisor to the Chinese government and Curator of the Museum of the Geological Survey of China, also founded just a year before.

By spreading the core members of China's geological circle across several institutions, despite their overlap, the Chinese delegation appeared deeper and broader than it would have otherwise, and this hinted to Congress participants that China had not been idle while the West was fighting. More importantly still, out of deference to Weng, the Belgians listed Hango Murakami, Chief Geologist to Japan's South Manchuria Railway Company, under China, rather than separately under Manchuria as in 1913.[36] The Chinese government could not halt Japanese incursions in the northeast provinces, but the geological community could try to put scientific findings in their proper place.

In Chinese sources, Weng's most widely touted achievements were his formal representative functions, especially his role as a Vice-President of the Congress and as president of a technical session on general geology.[37] While the latter was indeed a sign of both his growing reputation and the regard of his former Belgian colleagues, the former, as we know, was an empty honor he shared with both Djin Da Min and Parkin Wong. Weng's most important activities were far less official, and much more mundane. Letters and invitations exchanged after the Congress attest to Weng's ability to make both scientific and social connections at the Congress. After 1922, formal correspondence with the Geological Survey was almost always followed by a personal note to Weng, recalling meetings and conversations or asking for suggestions. In one instance, the Geological Society of

---

35 For more on Amadeus Grabau's move from the United States to China after the first World War, see Allan Mazur, *A Romance in Natural History: The Lives and Works of Amadeus Grabau and Mary Antin* (Syracuse, NY: Garret, 2004).

36 *Comptes rendus de la XIIIe session du Congrès géologique international: en Belgique, 1922. fasc. 1–3* (Imprimerie H. Vaillant-Carmanne, 1926), 18, 30.

37 For a recent example, see 李 Li, "中国地质事业初期若干史实考" ("Zhongguo dizhi shiye chuqi ruogan shishi kao), 68–69. This interpretation follows the views expressed by Weng's contemporaries and in popular media reports of the day.

America even offered to pay his way to the XVIth Congress in Washington, DC, in hopes of continuing exchanges begun in previous meetings.[38]

Though just over five feet tall, Weng was not easy to overlook. At the August 18th meeting of the Congress, Weng dominated the afternoon session. He presented his own paper on the seismological influence of geological structures in China in French and then read two papers in English, by Ding Wenjiang and J. G. Andersson, respectively. In a virtuoso performance, Weng switched easily between the two languages and fielded questions for all three papers, taking up the bulk of the session. Later, both anglophone and francophone correspondents would express relief that they could communicate with him in their native languages, citing linguistic difficulty as one reason for their lack of exchange with certain Japanese researchers and praising his fluency in speech and writing.

Language aside, Weng's paper analyzed earthquakes recorded in Chinese history in terms of geological structures observed through fieldwork in order to locate epicenters outside of the generally recognized seismic belts. Weng grounded China's modern earth sciences in a long tradition of seismic recording, countered arguments that accused historical accounts of gross exaggeration and offered a modern geological explanation for regions of strong seismic activity, while introducing his audience to the work of his Chinese colleagues through a series of prominent citations. By recalibrating historical data on the bases of both earthquake depth and intensity (rather than intensity alone), Weng argued that modern scientific concepts could explain historical materials, but cautioned that Chinese geology was best understood by those invested in its history, not just its present-day features.

When Swiss geologist Emile Argand gave his commanding presentation on "La Tectonique de l'Asie" on the first day of the sectional meetings, Weng's ability to engage Argand's mobilist ideas on the grounds of paleontological and paleoclimatological work in China led to a standing invitation for qualified Chinese students to work with Argand at the Université de Neuchâtel. Far from being threatened by a foreign synthesis of Chinese geology into a global theory, Weng saw an opportunity to bring newer data into the picture and introduce an eminent geologist to recent activity in China. For Weng, Chinese geologists were not only historically

38 Misc. invitations and letter from Geological Society of America to Wong Wen-hao, dated March 25, 1929, Academia Sinica Modern History Institute Archives, File 08–24 5(2).

bound to their homeland, they were also committed to its future and therefore in the best position to engage in ongoing fieldwork prompted by emerging geological questions.

In a toast to his Belgian hosts at the closing dinner of the sectional meetings, Weng expressed the eagerness of the "Chinese Government"—or at least the Chinese Geological Survey—to shoulder the duty of making "its vast domain" known to the world. As Weng saw it, Chinese would contribute to the geological knowledge of the world by studying their own part of it. Though the boundaries of this territory remained undefined, Chinese geologists saw them as representing not just a sovereign right, but an international responsibility that the government and its agencies were soberly growing into. Weng diplomatically thanked his "distinguished colleagues from friendly nations" for the spur to Chinese labors that their "precious competition" provided, and promised greater Chinese efforts, while reminding his audience that China was larger by far than the sum total of all Europe. In this way, Chinese geologists attempted to transform the logic of the map from one that threatened usurpation to one that reserved a place for their own work, however tardy.

To a certain extent, this move was successful. With the existence of China's Geological Survey, and given Weng's strong presence at the 1922 International Geological Congress, Chinese territory was understood to be spoken for, and the Congress no longer delegated investigative responsibilities to Japanese or Anglo-European visitors. Organizing committees for the geological congress requested special reports from the Survey and, if none were received, let the matter drop.[39] Chinese geologists themselves preferred to read papers and send delegates that showed off China's strengths in the more basic aspects of geoscience, such as structural geology and paleontology, but, the more they established their expertise in geology broadly construed, the more they expected to speak for China's natural resources. While non-Chinese had been invited to report on Chinese iron and coal reserves for the 1910 and 1913 resource compilations, subsequent compilations opted to leave China out rather than infringe on the preserve of the Chinese geological community.

---

39 An example of this was the case of soil surveying, for which the Geological Survey received many international requests. The Survey was unequipped to meet these demands without derailing its own research agendas but by 1930, a soil laboratory was established, which began its first national soil survey in 1933. This case illustrates the ways that the international circulation of Chinese Geological Survey members and their language of responsibility staved off threats of foreign work on their behalf while simultaneously shaping the future priorities of local geologists.

Pan-Pacific Science Congress

This strategy of establishing expertise worked in the context of the International Geological Congress, which was a loose organization of professional and disciplinary interest, but was less successful in the Pan-Pacific Science Congresses (PPSC) of the early 1920s, where political interests were far more explicit. The first Pan-Pacific Science Congress was held in Honolulu in 1920, as part of post-World War I scientific interest in finding alternatives to the problematic European metropoles. As one of the founders of the PPSC, Herbert Gregory of Yale and the Bishop Museum in Honolulu, lamented over the dominance of the Atlantic, he complained that "It is difficult for us in Honolulu to realize ... that to know something of the working of the Chinese mind is of greater immediate value than familiarity with the politics of New York City".[40] Ironically, the organizing committee of the second PPSC in Australia neglected to invite either China or the Soviet Union to the 1923 meeting.

China's absence at the first PPSC was a clear result of the hasty organization of that first, rather small conference. The oversight at the 1923 conference, however, stemmed from the PPSC's ambivalent relationship to colonialism. While the push for a "Pacific sense" that motivated the conference was rooted in an American and Australian move for independence from "Atlantic-mindedness," the success of the idea depended as much on the desire of Anglo-European states to buttress scientific investigation in their Pacific colonies as it did on the desire of certain ex-colonies to establish their own scientific dominion. In the balancing of this equation, China was neither a formal colony nor an empire builder, and Japan took it upon itself to represent East Asia on the early organizational committees.[41]

China's only official recognition in the second PPSC of 1923 was a short blurb on "The Areal [sic] Geological Survey in China" submitted by the National Geological Survey of China.[42] Though the Chinese Survey

---

[40] Herbert Gregory, "Remarks" in *Proceedings of the First Pan-Pacific Science Conference: Under the Auspices of the Pan-Pacific Union, Honolulu, Hawaii, August 2 to 20, 1920* (1921): 936–937.

[41] Roy MacLeod and Philip F. Rehbock, "Developing a Sense of the Pacific: The 1923 Pan-Pacific Science Congress in Australia," *Pacific Science* 54, no. 3 (2000): 209–225.

[42] National Geological Survey of China, "The Areal Geological Survey in China" in Australian National Research Council, *Proceedings of the Pan-Pacific science congress, Australia, 1923* Vol. II (By authority: H. J. Green, government printer, 1924), 1319–21. In the same volume, Weng Wenhao and Amadeus Grabau individually contributed "Carboniferous Formations of China", 1033–1043.

was merely introduced with a brief history of its accomplishments and activities plus an overview of map publication, this submission directly contradicted the claims made by the Director of the Japanese Geological Survey at the previous Honolulu meeting that the only mappings of China proper by 1920 were the classic works of Richthofen and Willis, and the more recent surveys by the Tokyo Geographical Society. The article also promised future maps that would cover the whole territory of China, and a figure was provided that unambiguously laid claim to western and northeastern frontier areas, including Manchuria and Tibet, which were objects of Japanese, Russian, and British interest.

Once again, an apology was offered for the incompleteness and slow progress of Chinese geological researches, but, following on the heels of the 1922 International Geological Congress, this was yet another way of staking a claim to geological responsibility for Chinese territory.[43] This move was totally consistent with the ideals put forth by Gregory about scientists and local ties, but the case for expertise and proximity were not enough to fully bring China into the fold at the 1923 Pan-Pacific Science Congress. When a move was made to create a permanent organization for Pacific Science, China was not among the invited countries, largely because it had no national organization, such as a National Research Council, which could fully represent its government and yet maintain scientific autonomy. The PPSC was far broader in scope than just geology, involving all scientific concerns of direct interest or utility to the Pacific Region, so the Geological Survey was both too narrow and too tied to government control to represent China, and China was thus ineligible to deliberate on a constitution for the organization and its future congresses.[44]

The discrepancy between China's scientific participation and its broader representation only increased at the next congress in Tokyo in 1926, when China was able to send eleven representatives (plus four independently funded foreigners affiliated with Chinese institutions). The focus of the Tokyo congress was on geological science, which allowed the Chinese delegation to showcase their greatest strength. Li Siguang presented a paper "On the Age and Distribution of the Paleozoic Coal-bearing Formation

[43] Ding Wenjiang used a similar strategy of framing Chinese interests in terms of China's responsibility to international geology in 丁文江 Ding Wenjiang, "中國地質學者的責任" ("Zhongguo dizhi xuezhe de zeren") 北大地質學會會刊 (*Beida dizhi xuehui huikan*) 4 (April 1930), 1–13.

[44] Fuchs, "The Politics of the Republic of Learning: International Scientific Congresses in Europe, the Pacific Rim, and Latin America," 217–220.

in North China" which used micropaleontological studies of *fusulinidae* to settle a long-standing debate on the existence of Lower Carboniferous formations in northeastern China.[45] This debate, which originated with Richthofen and Willis' early researches and was taken up by various Chinese geologists, was fundamental to estimates of coal resources in China, and its resolution was one of the first instances in which a local geologist successfully deployed advances in basic science to solve the problems of applied geology in China. Weng Wenhao also presented a paper on "Crustal Movements in Eastern China" which directly challenged Bailey Willis' ideas of orogeny based on ocean basin subsidence, suggesting that for the entire Pacific Basin, the Hercynian Period of mountain building was insignificant in comparison with the Zhuluoji.[46] Willis, who was present, gave a paper on "The Metamorphic Process of Orogeny," making this an extremely prominent disagreement, and Weng subsequently invited him to visit China in addition to various Pacific Islands, in order to examine more field data on this problem.

These scientific contributions and Weng's active participation in the administrative meetings on geology easily substantiated the local expertise of Chinese geologists, but expertise alone was not enough. China still had a difficult time gaining official representation on the Pacific Science Council. To a great degree, this was due to Japanese efforts to keep China out of this central organizing body. Official Chinese representative Fen Qin, an astronomer from Peking University, submitted a petition to include the Science Society of China on the Council, but the Japanese representative, geographer Naomasa Yamasaki, pointed out that there were two Chinese governments represented at the Conference since two Chinese representatives were from Canton, which was the seat of the rival Guomingdang government established by Sun Yat-sen in 1924. Thus, he argued, the Science Society of China was not eligible for participation in the Council because it did not represent a unified China. After protracted discussion, during which the two Chinese representatives in question

---

[45] J. S. Lee, "On the Age and Distribution of the Paleozoic Coal-bearing Formation in North China" in *Proceedings of the third Pan-Pacific Science Congress, Tokyo, October 30th–November 11th, 1926*, ed., Gakujutsu Kenkyū Kaigi (Tokyo: National Research Council of Japan, 1929), Vol. II, 1540–1555. *Fusulinidae* are a family of Foraminifera often used to date rocks from the Lower Carboniferous to Upper Permian.

[46] Ibid., Vol. I, 462–485.

denied affiliation with any particular government, the motion to induct the Science Society of China was passed without any further protest.[47]

This type of issue cropped up again and again. After the third PPSC in Java in 1929, Weng Wenhao, China's first representative on the Pacific Science Council, published a summary of his experiences and thoughts concerning the Conference in the Chinese journal, *Science*. Weng detailed the scientific contributions made by the eighteen person Chinese delegation in several different disciplines, especially geology, agricultural science, zoology and botany. However, he devoted most of his attention to suggestions for rethinking China's representation in the conference. In particular he was concerned that Chinese scientists were too impressed by the size of the American, Australian, and Japanese delegations. He argued that China did not have as much experience with modern science and was not in a position to make as many scientific contributions. Rather than try to send thirty or forty delegates to the next conference, Weng hoped that China would field a smaller, more balanced and participatory group. He suggested that China try to have one or two delegates in every major disciplinary division, with at least one scientifically credible paper between them. Moreover, the entire delegation should have representatives from each of the three major academic centers of China (Beijing, Nanjing, and Canton), with greater geographical diversity if possible. China should think of its delegations to international conferences, not as independent actors but as orchestrated units designed to model the scope of national research.[48]

In a similar vein, Weng suggested in both his private report and his published review of the third PPSC, that China bid to host the seventh PPSC. "Given that China has joined the [Pacific Science] Council, how can we continue to stand still and be indifferent, to take without giving?" In fact, he claimed that "there are those who for some time now have been complaining about our failure to take up our turn as host."[49] In fact, there were few outside of China who could even imagine China as capable of hosting a two to three week international conference with over a

[47] *Third Pan-Pacific science congress, Tokyo, 1926: Minutes of final general meeting. Annex, constitution and by-laws of Pacific science association*, 1926 and 翁文灝 Weng Wenhao, "太平洋科學會議之歷史" ("Taipingyang kexue huiyi zhi lishi") Academia Sinica Modern History Institute Archives, File 17–24 1(2).

[48] 翁文灝 Weng Wenhao, "第四次太平洋科學會議紀略" ("Disici taipingyang kexue huiyi jilue") draft, Academia Sinica Modern History Institute Archives, File 17–24 1(2)

[49] 翁文灝 Weng Wenhao, "第四次太平洋科學會議紀略" ("Disici taipingyang kexue huiyi jilue"), 科学 *Science* 14, no. 5 (January 1, 1930): 633.

thousand participants, but once again the language of responsibility and expectation allowed Weng to convey what he thought was the criteria for full participation (and hence recognition) on the world scientific stage.

Weng Wenhao calculated that based on previous meetings, China would have between six and ten years to prepare if its bid for the seventh PPSC was accepted. He reminded the *Science* audience that though China's scientific enterprises were only a few decades old, there were bright spots in the legacy, and that China's territory was temperate and diverse. The climate would be suitable for a conference and various scientific visitors could go to different places to examine the country firsthand. These visitors would also bring their expertise and "seed" the young Chinese scientific community with new ideas, "speeding up … national development and raising interest" in the sciences.[50] Weng also pointed out that geology was one of China's strengths and that, with its "girdle of fire and fringe of islands", geology was the true basis for the structural unity of the Pacific. Unfortunately, while scientific communications did indeed strengthen in geology, Weng's vision relied on an ability to plan far into the future that the Chinese government in Nanjing simply did not possess, and China did not host a major international scientific congress until the 1990s.[51]

International Circulation and Circulation Voids

The scientific value of international scientific meetings was a frequent matter of debate for both participants in early scientific congresses and later analysts who tried to assess the impact of internationalization initiatives. Many participants at early meetings of geologists, chemists, and astronomers noted that the congresses were more social than technical, though this was not necessarily perceived to be a weakness, since the purpose of international meetings was in flux throughout the late 19th and early 20th centuries. For some, exposure to new ideas and discussions of standardization and policy were sufficient reason to organize and attend, but for others different national approaches were reason to stay away, as in the case of early congresses of historians.[52]

---

50 翁 Weng, "第四次太平洋科學會議紀略," 634.

51 This does not count several Communist bloc scientific meetings hosted as showpieces in the 1960s.

52 Karl Dietrich Erdmann et al., "Toward a Global Community of Historians: The International Historical Congresses and the International Committee of Historical Sciences,

More often than not, congresses were understood to be important places for exchange but not necessarily engines of innovation. As August Dillmann commented, in regards to the International Congress of Orientalists:

> Indeed, our intent in these congresses cannot be the direct advancement of science; rather their real significance lies in their international character, in the congregation of men from the widest variety of nations for a common purpose. Science naturally is... international: A light that goes on in one country cannot be hidden; rather, it shines into the other countries as well. Anyone who neglects to seek researchers from abroad is often cheating himself out of the best help. All the nations of Europe have contributed to the scientific wealth that we now enjoy.[53]

Similar remarks were made in the proceedings of other natural and social scientific meetings, though views varied by discipline, period, and nation. What is clear, even for those who were dubious of the value of scientific congresses was that no one wanted to be left out, and despite complaints from some quarters that technical content was too basic or too disconnected, fierce arguments over the hosting of congresses, official languages, and other forms of representation made evident that by the turn of the 20th century, the trend toward international scientific meetings was driven as much by fear of being marginalized as desire to share the cutting-edge. This was no more clear than in the wake of World War I, when the former Central Powers were blocked out of most major congresses and they attempted to organize their own international meetings.[54]

What this points to, and what the experiences of Republican Chinese geologists at the International Geological Congresses and Pan-Pacific Science Congresses helps us understand, is that circulation in the modern period often turned from opportunity to necessity. The purposeful construction of international scientific organizations and the quickening schedule of scientific meetings both regularized the movement of objects,

---

1898–2000," in, 2005, 14, http://ezp-prod1.hul.harvard.edu/login?url=http://search.ebscohost.com/login.aspx?direct=true&db=hia&AN=46295755&site=ehost-live&scope=site.

[53] A. Dillmann, *Verhandlungen des Fünften Internationalen Orientalisten Congresses. Gehalten zu Berlin im September 1881. Ersther Teil. Bericht über die Verhandlungen* (Berlin: 1881), 31, as quoted and translated in Fuchs, "The Politics of the Republic of Learning: International Scientific Congresses in Europe, the Pacific Rim, and Latin America," 211.

[54] Arguably the return to German participation in international scientific meetings after the world wars was a turning point in the purpose and scope of contemporary congresses. Brigitte Schroeder-Gudehus, "Les congres scientifiques et la politique de cooperation internationale des academies des sciences," *Relations Internationales*, no. 62 (1990): 135–148; Forman, "Scientific Internationalism and the Weimar Physicists: The Ideology and Its Manipulation in Germany after World War I."

ideas, and individuals and created voids in which non-participants found it difficult to be heard. Certainly this was not a wholly new phenomenon, for being outside of the 'right' social networks in the early Republic of Letters was similarly marginalizing, but the scale and professionalization of the international scientific congresses made the problem more acute for scientists and scientific communities struggling for legitimacy on the world stage.

On the other hand, the official and explicit nature of the international congresses of the late 19th and early 20th centuries (both in structure and in pledges to support a universal image of science) also invited members of new scientific communities to study the system and self-consciously groom themselves for participation. Though Weng Wenhao's ideas for hosting a future Pan-Pacific Congress never came to pass, his remarks show the ways in which Chinese scientists were thinking about their development within the *system* of international science, and they hint at how geologists based the activities of the Geological Society of China, where they could control the terms of engagement, on their experiences in bigger venues.[55] At the International Geological Congress of 1922, the Chinese geological community strategized its delegates and their listed affiliations in order to display their institutional growth, and seemingly small matters of representation were used to assert claim over Manchuria and other 'Chinese' concerns. Chinese geologists, who had studied abroad and who worked in multiple foreign languages, also learned the sociability of Anglo-European science, turning toasts and dinners to their advantage, and parlaying Congress contacts into future research networks by representing the best of China as individuals instead of as agents of the faltering state.

However, thinking of circulation patterns in terms of necessity and danger does not just explain the initial exclusion of new scientific groups and their eventual moves to participate. As Arjun Appadurai points out, the metaphor of circulation not only allows us to analyze flows and interconnections, it also makes it possible to see how historical actors used global forms to create what he calls 'bumps' or 'obstacles' to cultural flow.[56] For instance, a state may use rule of law—itself a value conveyed through

---

[55] For more on the Geological Society of China and international science, see Grace Yen Shen, *Unearthing the Nation: Modern Geology and National Identity in Republican China* (forthcoming, University of Chicago Press).

[56] Arjun Appadurai, "How Histories Make Geographies: Circulation and Context in a Global Perspective," *Transcultural Studies* 1 (2010): 7.

circulation—to block the flow of information from an undesirable organization. Likewise, in science, the voids produced by the regular flow of people and information through international meetings could be used to create quiet spaces for different kinds of scientific activity.

In the Chinese case, the Geological Society of China used the model of international congresses to attract foreign geologists and paleontologists to China and borrowed the networks of these visitors (and itinerant locals) to project an image of cosmopolitanism upon the local scientific community. But as it did so, through polyglot meetings and its glossy, well-illustrated journal, *The Bulletin of the Geological Society of China*, it deflected attention from more local, less outward-turning activities and created a new 'quiet space' in which Chinese geologists could actually withdraw from international scrutiny. The international face of Chinese geology, made visible by its embrace of global circulation, allowed younger, less confident Chinese to use the Chinese language and non-international forums to escape notice and buy them time to develop their skills.[57] By looking for the voids and eddies created by the intentional and large-scale circulation of knowledge through modern international scientific meetings, it is possible to better understand not only how new scientific communities, such as the Chinese, were able to inject themselves into the global flow, but to see how voluntary pauses in circulation might be just as valuable.

[57] For more on this "quiet space" defined by language and distance from the international, see Grace Shen, "Periodical Space: Language and the Creation of Scientific Community in Republican China," in *Routes of Science*, eds. Benjamin Elman and Jing Tsu (Brill, forthcoming.)

PART FOUR

BUILDING SCIENCE IN MODERN INDIA

HOW MAY WE STUDY SCIENCE AND THE STATE IN POSTCOLONIAL INDIA?

Jahnavi Phalkey

> The field [of South Asian Studies] is in need of a historiography that knows how to distinguish between... secularism as a system of values and secularism as an ideology to buttress the centralising project of a post-colonial nation-state.[1]

This essay is an attempt to trace the journeys of an underlying assumption at work in twentieth century Indian history, especially history of science, that "science" became the unity—and an important part of the secular ideology—behind the centralising project of a post-colonial state in India. This claim is palpable in social scientific writings about India after independence, which forms the bulk of the scholarship. It is also equally palpable in history writing, especially the small body of scholarship in history of science.[2] This palpability is, however, rarely elaborated upon or brought into question. In the coupling of reasons of state with science, while the mechanisms of state power are made apparent—again and again—the idea of "science" and what it meant to those who evoked it, or for that matter its exact relationship to the state, remains unexamined in the history of twentieth century India. Michel Foucault's "*savoir-pouvoir*" has become historical explanation.

It is not my intention, in this essay, to propose an argument about "Science" and its relation to the Indian state. Instead, I want us to notice opportunities in the existing scholarship: Sudipta Kaviraj and his theory of the *decolonised* content of Indian modernity; Pratik Chakrabarti's scientific industrialism; and finally Shiv Visvanathan's scientized technology. From these three, I will argue, we can build or depart towards a

---

[1] Sugata Bose, "Post-Colonial Histories of South Asia: Some Reflections," *Journal of Contemporary History* 38, no. 1 (2003): 135–136.

[2] See for example, the classic, Partha Chatterjee, *Nationalist thought and the colonial world*: a *derivative discourse* (Minneapolis: University of Minnesota Press, 1986). For history of science see, especially, Chapter 6 in David Arnold, *Science, technology, and medicine in Colonial India* (Cambridge: Cambridge University Press, 2000), and Chapter 7 and 8 in Gyan Prakash, *Another reason: science and the imagination of modern India* (Princeton: Princeton University Press, 1999).

better understanding of the ideas about science, development and state formation that were circulating between India, Britain, the United States and the Soviet Union in the middle decades of the 20th century and what forms this debate can take.

Colonial Science

The study of science under colonialism as a field of inquiry is now firmly established within history of science as well as history of colonialism. Indicative are the recent "Focus" sections on colonial science (2005) and on global history (2010) in *Isis*, the journal of the History of Science Society (HSS).[3] As the editor of the section in 2005, Londa Schiebinger, notes, the historiographic essays are complementary to an earlier volume edited by Roy MacLeod for *Osiris* (2000), the sister publication to *Isis*.[4] The previous year had already seen an issue of *Science Technology and Society* edited by Michael A. Osborne addressing the "social history of science, technoscience and imperialism".[5] But this by no means implies that there is scholarly agreement on just what the term 'colonial science' can definitively explain. Mark Harrison's useful genealogy of the term begins by tracing it back to the repeatedly and obligatorily discredited George Basalla.[6] Harrison argues that Basalla has to be understood as a part of the modernisation theory complex where his "universal model for the diffusion of science" makes sense. Colonial science for Basalla was

---

[3] Londa Schiebinger, ed., Focus section *Isis* 96, no. 1 (2005), contains the following essays: Mark Harrison, "Science and the British Empire," Jorge Cañizares-Esguerra, "Iberian Colonial Science," Steven J. Harris, "Jesuit Scientific Activity in the Overseas Missions, 1540–1773," and Michael Osborne, "Science and the French Empire".

[4] Roy MacLeod, ed., "Nature and Empire: Science and the Colonial Enterprise," *Osiris* 15, (2000).

[5] Michael A. Osborne, ed., "The Social History of Science, Technoscience and Imperialism," *Science, Technology and Society* 4, no. 2, (1999). Historians of science, but also scholars engaged in postcolonial science studies are equally committed to understanding colonial science. Similarly suggestive are the following special issues: Itty Abraham and Paula Chakravarty (issue organizers), "The Contradictory Spaces of Postcolonial Technoscience," *Economic and Political Weekly* 41, no. 3 (2006); Maureen McNeil, ed., "Postcolonial Technoscience," *Science as Culture* 14, no. 2 (2005); and Warwick Anderson, ed., "Postcolonial Technoscience," *Social Studies of Science* 32, no. 5/6 (2002). Postcolonial science studies is best characterised as an approach to study scientific practices under colonial rule or practices transformed by colonial rule, and to study periods beginning with colonialism into periods following decolonisation as well.

[6] Mark Harrison, "Science and the British Empire," *Isis* 96, no. 1 (2005): 56–63; George Basalla, "The Spread of Western Science," *Science* 156 (1967): 611–622. The following paragraph is based on Harrison's essay.

characterised by dependence, and in the backlash that followed against modernisation theory and politics of the early Cold War, colonial science came to be characterised as a "tool of empire". Dependency theorists of the 1970s, while continuing to work with Basalla's chronology, framed the question on colonial science through a critique of the colonial enterprise. Colonial science thus seen was an instrument of imperial rule, functioning within the binary of "centre/metropolis and periphery". In the 1980's Roy MacLeod's concept of the "moving metropolis" signalled that historians were willing to allow for dynamism in scientific practice at the periphery.[7] Others began to lay stress on the cultural authority of science in colonial contexts following theoretical questions raised by the works of Michel Foucault and Edward Said.[8] Historians of India continue to engage with extensions of some of these questions, and current scholarship particularly converges upon the relationship between colonialism, science and Indian modernity.

Recent scholarship on the response to, and establishment of, scientific practice in India has now firmly recognized the centrality of "science" in understanding the history of modern India, and its inseparability from the history of imperialism.[9] History of science in India thus framed cannot be understood except as a part of colonial history and all the contingent alliances that shaped this history up until Indian independence when it

[7] Roy MacLeod, "On visiting the 'Moving Metropolis: Reflections on the Architecture of Imperial Science," in *Scientific Colonialism: A Cross-Cultural Comparison*, eds. Nathan Reingold and Marc Rothenberg (Washington DC: Smithsonian Institution Press, 1987), 217–249.

[8] Edward W. Said, *Orientalism* (New York: Penguin, 1978); Michel Foucault, "Governmentality," in *The Foucault Effect: Studies in Governmentality*, eds. Graham Burchell et al. (Chicago: University of Chicago Press, 1991), 87–104.

[9] See Kapil Raj, *Relocating Modern Science: Circulation and the Construction of Scientific Knowledge in South Asia and Europe, 17th and 19th Centuries* (New Delhi: Permanent Black, 2006); Dhruv Raina and S. Irfan Habib, *Domesticating Modern Science: A Social History of Science and Culture in Colonial India* (New Delhi: Tulika Books, 2004); Pratik Chakrabarti, *Western Science in Modern India: Metropolitan Methods, Colonial Practices* (Delhi: Permanent Black, 2004); John Lourdusamy, *Science and National Consciousness in Bengal, 1870–1930* (New Delhi: Orient Longman, 2004); David Arnold, *Science, Technology an Medicine in Colonial India*, The New Cambridge History of India III, 5 (Cambridge: Cambridge University Press, 2000); Gyan Prakash, *Another Reason: Science and the Imagination of Modern India* (Princeton: Princeton University Press, 1999); Zaheer Baber, *The Science of Empire: Scientific Knowledge, Civilisation and Colonial Rule in India* (New Delhi: Oxford University Press, 1998); Deepak Kumar, *Science and the Raj, 1857–1905* (New Delhi: Oxford University Press, 1995); Deepak Kumar, ed., *Science and Empire: Essays in Indian Context* (New Delhi: Anamika Publications, 1991).

became possible to think of India otherwise.[10] The first academic histories of colonial science in India came from Deepak Kumar and his colleagues at the Jawaharlal Nehru University, New Delhi, followed closely by Dhruv Raina and S. Irfan Habib at the National Institute for Science, Technology and Development Studies (NISTADS), again in Delhi. Deepak Kumar's first book (1995), even with its recent revised edition with two extra chapters, remains the authoritative chronicle of British colonial scientific enterprise in India. New entrants to the field have found book topics embedded in the footnotes, a testimony to his extensive archival research in India and Britain. Raina and Habib (2003) have over the years engaged with the question of domestication of modern science in India in an episodic exploration of the history and historiography of science in India.[11] The field of history of science in India then, is relatively young.

Two books that have made a major impact on the understanding of the place of science under colonialism in the history of modern India in the recent years are those by Gyan Prakash (1999) and David Arnold (2000). Both cover the same time period but are remarkably different in their approaches. Gyan Prakash treats science as a cultural discourse in colonial India. He argues that even as the cultural authority of science constituted and informed strategies of control for imperial rule, it also held the promise of "liberty, progress and universal reason" for the Indian nationalist elite and those doubtful of the colonial enterprise. Modern India, he proffers, was shaped within this contradiction. Prakash does not want to look at colonial science in its connection to political ideologies and institutional structures that colonialism put in place, but rather scientific practices themselves as constituting colonialism's political ideology and institutional structure. Scientific practice was not a mere tool of empire,

[10] As a part of nationalist history writing, the Indian National Science Academy (INSA) constituted a board to commission a history of science in India (1959). Leading the committee was the physicist Debendra Mohan Bose, (Jagdish Chandra Bose's nephew), and the book was published in 1970. See D. M. Bose, B. V. Subbarayappa, and S. N. Sen, *A Concise History of Science in India* (Delhi: INSA, 1970). The Academy also began publishing the *Indian Journal for History of Science* beginning 1966. On the other hand, the Needhamian Abdur Rahman's efforts led to the establishment of the National Institute for Science, Technology and Development Studies (NISTADS) in the early 1980's as a laboratory under the Council for Scientific and Industrial Research, dedicated to science policy related research. Rahman was a prolific writer and worked closely with UNESCO, a tradition that continues to date with other academic historians of science in India. See Dhruv Raina, *Images and Contexts: The Historiography of Science and Modernity in India* (New Delhi: Oxford University Press, 2003), especially Chapter 5, "Science, Scientists, and the History of Science in India, 1966–94".

[11] See Kumar, *Science and the Raj*; Raina and Habib, *Domesticating Modern Science*.

he argues, because it was inextricably implicated in the very making and meaning of empire.

David Arnold has argued that Prakash's book does not tell us, "about how science was constituted, disciplined and institutionalised, *even what science was*, in the rapidly changing context of nineteenth and early twentieth century India".[12] Prakash in turn looks at David Arnold's Cambridge history volume as offering no interpretive breakthrough in the field but most certainly "an authoritative interpretation" of existing scholarship on science in colonial India.[13] Having synthesised the wide-ranging scholarship on science, technology and medicine in India, Arnold in fact agrees with Prakash, that science was never a mere tool of empire. British authorities drew upon Indian intellectual traditions in uneven ways and even when they did employ "science" for political purposes, they did not necessarily possess resources or even enough mastery over scientific matters to maintain it as a tool of control. Arnold then identifies three characteristics that can illustrate the history of science in colonial India. First was the continuing presence of variegated traditions of India's own sciences and their legacy for imperial rule. Second, the social and intellectual impact of colonial science and its relationship to the imperial regime in India as well as European science, and finally, the authority of science, technology and medicine as central attributes of India's modernity, drawing upon both Indian and European sources.

Imperialism's scientific authority was fragile because "the cultural voice of science was being trained in Europe when it was first heard in India".[14] But the fragility did not register symmetrically in the colonial setting because the establishment of scientific practice was not separate from the establishment of political authority. Colonial scientific practice was not an autonomous activity when most funding, at least up until the late nineteenth century, came from various institutions that shaped and informed the activities of the colonial state. Governmental agencies continued to function with racial privilege and Indian practitioners rarely acquired leadership positions. It also meant that disciplines considered useful for maintaining colonial rule such as medicine, botany, zoology and

[12] David Arnold, "Review of Gyan Prakash (1999) Another Reason: Science and the Imagination of Modern India," *The Journal of Imperial and Commonwealth History* 28, no. 2 (2000): 163. Emphasis added.

[13] Gyan Prakash, "Review of David Arnold (2000) Science, Technology and Medicine in Colonial India," *Victorian Studies* 45 (Autumn Issue, 2002): 149–151.

[14] Roy MacLeod, "Review of Gyan Prakash, Another Reason: Science and the Imagination of Modern India," *American Historical Review* 105, no. 5 (2000): 1720–1721.

geology were relatively more—even if unevenly—institutionalised, unlike mathematics, physics and chemistry.[15] This uneven development of disciplines had an immense impact on their development after independence and partition.

Historiography of colonial science, we can say, has moved in three phases: in the diffusionist model of the 1950s and 1960s, historians looked at science and technology as tools of empire building and mastery. In the 1980s, historians began to work with less stable concepts of centre and periphery and in the process created space for historians, to claim, in the 1990s and thereafter, that science and technology also contributed to the imagination and reconstruction of free, independent and modern territories outside the European orbit. Within this vibrant new scholarship on the place of science in decolonising territories, scholars have begun to locate the important questions on science and technology informed by Indian history before and after the British Raj. An important book to reset the historiographical agenda for early modern and modern Indian science has been Kapil Raj's recent book (2007).[16] Arguing for circulation as a model to understand the processes of knowledge production, Raj foregrounds a single analytical framework to begin mapping both the colonial encounter and the place of science in it. A combination, therefore, of historiographical approaches that look at the instrumentality, hegemony, authority, utility, ideology and circulation of science are available to examine and assess the place of science in the transformation of India as an independent nation-state.

Decolonisation and After

The historian John Darwin has proposed, "if we could generalize about the causes of imperial failure at the colonial periphery, and weigh them

[15] Even though chemistry was not institutionalised in company science, it became important by 1900 in response to the "blue terror" of synthetic indigo developed in Germany. In 1916, Thomas Holland, earlier with the Indian Geological Survey of India, led the Indian Industrial Commission and asserted the importance of supporting research in chemistry. Plans to institute an Indian Chemical Service in 1920 were opposed strongly by among others Prafulla Chandra Ray, leading the 'School of Indian Chemistry' and founder of Bengal Chemical and Pharmaceutical Works (1899) in Calcutta. See P. C. Ray, *Life and Experiences of a Bengali Chemist* (Calcutta: Chukerverity, Ghattesjee and Co, 1932) and Arnold, "Review of Gyan Prakash," 163–166.

[16] Kapil Raj, *Relocating Modern Science: Circulation and the Construction of Knowledge in South Asia and Europe, 1650–1900* (Houndmills and New York: Palgrave Macmillan, 2007).

against domestic and international constraints upon the behaviour of the colonial powers during and after WWII, we might advance the historiography of decolonization beyond the crude and generalized level at which it now rests".[17] Postcolonial scholarship has already drawn attention to the need for analytical separation between the formal moment of decolonisation and the long-drawn out process of decolonisation itself. This important insight requires fine-tuning in two senses. First, the significance of the formal transformation of the relationship between coloniser and colonised cannot be underestimated. As the colonizing power Britain had to begin to stop thinking like an empire, it opened formal spaces for the nationalist elite in India to no longer frame themselves as subjects of imperial rule.[18] The struggle against subjugation was an omnipresent dimension of colonial rule itself: empires *are* unstable, and colonial administrations constantly have to redefine and reaffirm the modalities of their power through time. However, the historical weight of resistance became more significant as London itself began to lose grip on its hegemonic power. Second, the struggle to define and debate a new sense of what India was, and would become, began long before the formal transfer of power from London to Delhi. The formality of decolonisation lent political credibility to these forces already in the making. From the Indian perspective, a strong case can be made that the formal process of decolonisation began in the 1930s and more decisively in 1939 with the outbreak of WWII. The reader may want to remember that while historians of science have seldom engaged with the question of transformation on the sub-continent, as mentioned earlier in this essay, scholars exploring transformation have also failed to pay close attention to science. However, discussions on science and technology often do show up implicitly in the discussion of modernity.

While most historians and social scientists agree on the importance of science as a metonym for modernity and development within the Nehruvian project, there is far too little social history of the foundational moments of state formation in India. There is, however, a significant and rich body of literature engaged with the connections between the state in independent India and its role as an agent of development that takes the

---

[17] John Darwin, "What was the late colonial state?" *Itinerario*, 3, no. 4 (1999): 73–82.

[18] See Frederick Cooper, *Colonialism in Question: Theory, Knowledge, History* (Berkeley: University of California Press, 2005).

coupling of science and the state for granted.[19] In these studies, science is evoked in the context of state-led technological projects like large dams where science stands in for the violence of modernisation. Ashis Nandy's landmark volume, *Science Hegemony and Violence* is one such important critique of the Nehruvian state.[20] Recent histories and social studies mapping the transformation of the state in independent India have no chapters on scientific practice, institutions or scientific statesmen.[21] None of the above writers elaborate on the sources of the moral and material meanings of science at work in their analyses.

Imperial ambition demanded that decolonisation had to be accomplished in such a manner that it did not imply for Britain a complete abdication of moral, political, cultural and scientific leadership in the post-war era at the same time that it implied for India the gradual affirmation of political autonomy, the redefinition of identity and increasingly the need to reinvent scientific industrialism in the project of shaping a national culture. The political and economic demands of a brutal war against European fascism saw both Indian independence movements and the [British] Indian government (despite being at war) planning and imagining a new India.

---

19 See among others Amita Baviskar, *In the Belly of the River: Tribal Conflicts over Development in the Narmada Valley* (New Delhi: Oxford University Press, 2004); Akhil Gupta, *Postcolonial Developments: Agriculture in the Making of Modern India* (Durham: Duke University Press, 1998); Satyajit Singh, *Taming the Waters: The Political Economy of Large Dams in India* (New Delhi: Oxford University Press, 1997).

20 Ashis Nandy is at the Centre for Studies in Developing Societies, New Delhi (CSDS), an autonomous social science research institution established in 1963. The writings of scholars like Rajni Kothari, D. L. Sheth and Ashis Nandy from the CSDS have become "a point of reference for various attempts from the South to question the global establishment view of democracy". The centre's most widely representative statement on the place of science in independent India is Ashis Nandy, *Science, Hegemony and Violence: A Requiem for Modernity* (New Delhi: Oxford University Press, 1988). In this volume see Shiv Visvanathan, "On the Annals of the Laboratory State", 262–278. For an overview review, see S. Ravi Rajan, "Science, State and Violence: An Indian Critique Reconsidered," *Science as Culture* 14, No. 3, (September 2005): 1–17.

21 See for example Francine R. Frankel, Zoya Hasan, Rajeev Bhargava, and Balveer Arora, *Transforming India: Social and Political Dynamics of Democracy* (New Delhi: Oxford University Press, 2000); Partha Chatterjee, *State and Politics in India* (New Delhi: Oxford University Press, 1998); Sudipta Kaviraj, *Politics in India* (New Delhi: Oxford University Press, 1997). Two notable exceptions are Sunil Khilnani, *The Idea of India* (London: Farrar, Straus and Giroux, 1999), see chapter 2: "Temples of the Future"; and Srirupa Roy, *Beyond Belief: India and the Politics of Postcolonial Nationalism* (Durham: Duke University Press, 2007), see chapter 3: "Indian Darkness: Science, Development and the needs Discourse of the Nation State".

If anti-colonial nationalisms had informed the struggle for self-rule, it was increasingly becoming apparent at the moment of decolonisation that nationalism as ideology could not provide a coherent national identity. Jawaharlal Nehru's *Discovery of India* (1946) written in prison during the last three years of WWII, ends with a deep anxiety about what the idea of India could contain, and for how long. "The discovery of India—what have I discovered? [...] Today she is four hundred million separate individual men and women, each differing from the other, each living in a private universe of thought and feeling."[22] These doubts were not new for Nehru, but he had earlier found enough faith in the idea to argue for the "unity" of India.[23] His shaky belief in this unity was now based upon "invisible threads". "India is a geographical and economic entity," he continued to argue, "a cultural unity amidst diversity, a bundle of contradictions held together by strong but invisible threads."[24] Having led various anti-colonial struggles successfully under the banner of the Indian National Congress, the challenge for the nationalist elite of the new country lay in their ability to articulate and mobilise, even create, precisely these invisible threads of belonging. They would find the answer in the formation of a bureaucratic state.

The political, administrative and territorial entity of India that was created in the aftermath of partition and integration of princely states had never before existed as a single state. Upon decolonisation a majority of Indian citizenry thus begotten, was "unaware of its own national and later constitutional subjectivity—about what it meant to give themselves a republican, democratic and secular political authority".[25] Even in the early nineteenth century, most observers of Indian politics would have located the state as a 'curious marginality' to everyday life in Indian society. Sudipta Kaviraj, an important theorist of the state in India, has argued that pre-colonial states in India were characterised by exteriority; "as long as the rent was extracted, the political regime did not arrogate to itself the right to rearrange productive or more generally ordinary social relations". If British imperial government acted within general rules of

---

[22] Jawaharlal Nehru, *The Discovery of India* (New Delhi: Oxford University Press, 1946, 2003), 562.

[23] Jawaharlal Nehru, *The Unity of India: Collected Writings 1937–40* (London: Lindsay Drummond, 1941).

[24] Nehru, *Discovery of India*, 562.

[25] Sudipta Kaviraj, "The Modern State in India," in *Dynamics of State Formation: India and Europe Compared*, eds. Martin Doornbos and Sudipta Kaviraj (New Delhi: Sage Publications, 1997), 230.

marginality and majesty as a successor to the Mughal state in the early phase of colonial rule, at the height of imperial hegemony it was characterised by reorganisation of social and economic life, encouragement of social reform and a discourse on political morality of imperial authority. It managed to thus generate a debate on the morality of its own political authority, which finally contributed to anti-colonial nationalism. Political mastery though was never really a matter of debate alone and "as the British used barbarism to deal with the 'barbarians', ... they also undercut the very ideals of civilisation and progress that legitimised their power".[26] The practice of colonial rule undid its own foundations.

Partha Chatterjee, one of the most important political theorists of anti-colonial nationalism in India, has argued that ideas of difference and particularity were foregrounded in the shaping of Indian national identity and culture, what he calls nationalism's 'spiritual domain'. But he argues further that 'difference was not a viable criterion in the domain of the material' sphere of national life, which included matters of science, technology and the state.[27] This claim has been criticised first and foremost for rendering 'European modernity' singular, as well as for drawing an unsustainable distinction between the spiritual and material domain of nationalism in India.[28] In my reading, Chatterjee would not have a problem denying variegated interests and uneven effects of colonialism. What he, and other post-colonial scholars, wish to remember, are the dangers of missing the framework of imperialism as mastery. The colonial project may not have been implemented or experienced evenly but it was informed by a shared belief in Britain's right to conquer and rule, in this case India. The expanse of colonial interests—be they mercantile, religious, political, scientific or adventure and therefore necessarily differently configured—were still not mere contagion. In rendering colonialism effortlessly fragmented, there is an inherent danger of neglecting the historically constructed hierarchy between cultures and races, and the violence employed to enforce them.[29]

---

26 Prakash, *Another Reason*, 47.

27 Partha Chatterjee, *The Nation and its Fragments: Colonial and Postcolonial Histories* (Princeton: Princeton University Press, 1994), 9.

28 Among others, see Cooper, *Colonialism in Question*, 140; Prakash, *Another Reason*, 158; and Sugata Bose, "Nation as Mother," in *Nationalism, Democracy and Development: State and Politics in India*, eds. Sugata Bose and Ayesha Jalal (New Delhi: Oxford University Press, 1996), 76–103.

29 The nature of British colonialism in India is a matter of strong debate among the Cambridge School, the Chicago led North American Schools, and the Subaltern School of historiography of the Indian sub-continent. For the debate see among others Partha Chatterjee, *The Nation and its Fragments* (Princeton: Princeton University Press, 1994),

Anti-colonial nationalisms were responses to the foundations, assumptions and workings of this mastery, and to use a very, very, old phrase, using the very tools of the master.

Chatterjee's other claim, of the separation of nationalism's spiritual and material domain, is historically less sustainable. Even if the spiritual domain was made inaccessible to the transformations of European modernity, however understood, the very act of making it so is out of step in proposing that 'public' life does not have the capacity to inform 'private' life. One good question to ask would be if it were even possible to create such a disjunction between public and private. If this divide was affected even rhetorically, a robust history of its meanings for India's various regions and populations is not possible within the scope of this essay. It should suffice to say that this claim is not evenly true for practicing scientists and political leaders among the nationalist elite have I studied. But Chatterjee's work remains important because even if one were to *entirely* deny his theory of the national domain, his characterisation of nationalist reconfiguration of what he calls the material domain continues to be useful. More recently, Chatterjee outlined the strategies of modernisation followed by the nationalist elite:

> Is it possible to accept western modernity without its colonialism? And I think that has been a fundamental strategy of elite nationalism, which is to take the position that there are good things about western modernity and there are bad things about western modernity, but it is possible to accept and adopt and learn the good things of western modernity without accepting colonialism. Right?[30]

Nationalist imagination in India actively sought to reorder the material domain—i.e. the state, and industrial development, but this reorganisation would not proceed by invoking the indigenous, difference or particularity. By the mid-twentieth century, a section of the elite on the sub-continent were convinced of their adoption of the nation-state, science and technology as the good things of European modernity, and this faith was further emboldened when the struggle against colonialism

Chapter 2; David Washbrook, "Progress and Problems: South Asian Economic and Social History, c. 1720–1860," *Modern Asian Studies* 22, no. 1 (1988): 57–96; David Ludden, "Introduction" in *Reading Subaltern Studies: Critical History, Contested Meaning, and the Globalisation of South Asia*, ed. David Ludden (New Delhi: Permanent Black, 2001); and Sugata Bose, "Post-Colonial Histories of South Asia."

30 "Interview: Partha Chatterjee in conversation with Anuradha Dingwaney Needham," *Interventions* 1, no. 3 (1999): 422.

appeared to have finally succeeded. What is clear from Chatterjee's argument is the need to rethink *just what the nationalist elite thought they were doing* when they accepted or rejected ideas or processes introduced by the colonial encounter. Chatterjee's is but one attempt to theorise what anti-colonial nationalism wanted to accomplish. Gyan Prakash offers yet another reading, that Indians received modernity (and science) in translation, and this translation hybridised or led to the undermining of its cultural authority. If Indians could understand and practice science, they could now counter the hegemony of colonial rule on its own terms and lay claim upon modernity.[31] In either case, what Chatterjee and Prakash discuss is the *manner of reception and negotiation* of modernity by the nationalist elite. A more interesting question, to my mind, is to move back from the manner and instead begin again with the pragmatic *content* of modernity—which included science and technology—that the nationalist elite agreed *had* to be appropriated.

By 1945, Nehru was leading the elite nationalist coterie that confronted at least two legitimately authorising power structures towards independence. The first was the bureaucratic and discursive apparatus of the colonial state with its history of repression of those in whose name it was now sovereign. The second was the amorphous Indian National Congress, a 'triumphant national movement' whose history was one of challenging that very state apparatus. In the very first instance, the nationalist leaders had to decide if their goal was to establish "an untrammelled version of the Western state, a purer form of Western modernity than colonialism had permitted". The other option was to abandon that ideal in recognition of its inherently flawed morality in making political subjection feasible, as Mohandas Gandhi continued to insist.[32] Would anti-colonial nationalism civilise the state, or was the administrative state to be transformed into a civilising state?[33] The answer lay in the Congress' own understanding of power: were they going to subject the modern state to its own [nationalist]

31 Prakash, *Another Reason.*

32 Kaviraj, *Politics in India*, 234.

33 For a comprehensive argument on the inseparability of state-formation in Western Europe from its self-understanding of the "civilising process", see Norbert Elias, *The Civilising Process* (Oxford: Basil Blackwell, 1978 [1939]). I propose the idea of a "civilising state" with a sense of irony informed by Elias' understanding of the civilising process and state formation in Western Europe. To begin with, the state makers had to, and in part wanted to, accomplish the affects of Elias' civilising process in India. At the same time though, state makers in independent India had inherited and come to participate in the attributes of a similarly misplaced and indefensible European civilising mission.

civilising impulses, or were they going to subject an already established subject population to a nationalist version of a civilising mission—a *more true* universal modernity?

Mohandas Gandhi claimed that Indian nationalism could make a choice and not accept "western modernity" as something good for the majority of India. The nationalist elite led by among others Subhash Chandra Bose, Jawaharlal Nehru and nationalist scientists like Meghnad Saha argued strongly against Gandhi and accepted economic determinism as a guarantee of political autonomy. "And in that area the choice had already been made—Elsewhere, by History, by 'the spirit of the age' [...] An economy based on cottage and small scale industries was "doomed to failure" because it could only 'fit in with the world framework' as a 'colonial appendage'.[34] The universalised configuration of the modern state (and science) as proposed by History and introduced into India by the colonial project would lead the way to progress and sovereignty.[35]

And what was this thing called universal modernity at the end of WWII when Europe had diverged, contested and sought settlement through violence and fascism? Sudipta Kaviraj raises an important question of how *the content of modernity* was pragmatically determined in the process of state formation.[36] How did the nationalist elite led by Nehru perceive the corpus of: "capitalist industrialisation, the increasing centrality of the state in social order, urbanisation, sociological individuation, secularisation in politics and ethics, creation of a new order of knowledge, vast changes in the organisation of family and intimacy, and finally changes in artistic and literary culture?"[37] The answer lies in how Nehru and the nationalist elite understood the relationships between these various components into which the history of European modernity was rendered "analytically decomposable". The Indian nationalist elite, like European intellectuals of the time, ascribed simultaneity, and symmetrical interdependence to all these processes. The link between them was the overarching logic of

---

34 Partha Chatterjee, *Nationalist Thought and the Colonial World: A derivative discourse?* (New Delhi: Oxford University Press, 1985), 144. Chatterjee is quoting from Nehru's *Discovery of India.*

35 For a controversial but interesting take on "Westernisation" of the political system in the world as "modernisation of the world order" see Bertrand Badie, *The Imported State: The Westernisation of Political Order* [Translated by Claudia Royal] (Stanford: Stanford University Press, 2000).

36 See Sudipta Kaviraj, "An Outline of a Revisionist Theory of Modernity," *Archives of European Sociology* 46, no. 3 (2005): 497–526.

37 Ibid., 508.

increasing rationalisation of public life and bureaucratisation of state practice. This Weberian reading of European modernity proved crucial in their imagination in the first instance of how Europe became modern.[38] It also strengthened the nationalist elite's claims on state formation by packaging modernity as an expressive totality.

In accepting simultaneity as against a "sequential" development of European modernity, Nehru and the nationalist elite explicitly rejected any contradiction between economic growth and democracy.[39] Given that they had accepted economic growth as a guarantee of political autonomy, and political autonomy was a moral assertion against the despotism of the colonial state—suggesting or accepting that prosperity and economic well being historically developed or can develop in the absence of democracy was discomfited. The economic logic of growth, prosperity and progress, and the political logic of democracy and social justice, had to appear congruent. If British failure to develop or even to allow the development of India was ground for discontent, how could the Indian nationalist elite now propose economic growth without political freedom and thus undermine their own legitimacy to hold power? There was of course the added difficulty of reading the emergence of modernity historically and therefore differentially in Britain, Germany, Italy, France and so on—to decide which one was more acceptable, and then arrive at the particular sequence of the corpus of processes therein. Allowing for such a specific reading would perhaps also lead to an introduction of the role of religion in social order. The nationalist elite in India was confronted with the task of state formation, and the state would have to accomplish the establishment of the varied processes that legitimated its own authority to rule over territory and population, and allow for the creation of secular modernity that the state should symbolise. Simultaneous and symmetrical was a pragmatic historical reading that best served their purposes. Informed by this structural reading then, the nationalist elite "deliberately attempted to advance all these processes simultaneously in

[38] Ibid. Kaviraj is referring to the German social scientist Max Weber here.

[39] The goal of economic growth was acceptable to the large interests represented in the Indian National Congress—be it as a historically necessary stage (for the left) or a desirable goal (the liberals and centrists). This agreement began to disaggregate in the mid 1940s with big business and industrial interests becoming concerned (and vocal) about what form of economic growth was desirable. One such expression was the Bombay Plan (1944).

the hope that they would support each other, or fall together in the face of resurgence of tradition".[40]

Colonial rule had not made India productive enough.[41] The nationalist elite saw themselves as legitimate agents who would better perform the task of making India productive. The integrated conclusions found within the National Planning Committee (NPC) of the Indian National Congress (1938) debates were a welcome change for Nehru from the troubling debates in politics. Planning was not to be about politics, but about making a task list for the new state. In this manner, NPC also managed to place debates on the future of India outside politics, in the hands of experts and (therefore) beyond contest. Accountability for development was now outside the purview of exactly those whose future was being discussed. The arguments were well ensconced in economic nationalism of the turn of the century—colonialism had rendered India unproductive and poor. The path to increasing production, economic growth and therefore to progress was through industrialisation but also of increasing rationalisation of public life. The "Note for the Guidance of Sub-Committee's of the National Planning Committee" clearly laid out the priorities on this path and especially marked the "Russian Plan" as the "model of Priorities hitherto found".[42]

> It is essential that much greater attention should be paid to *making scientific and technical research as an integral part of planned economy.* Universities and research institutions should be organised to permit of fundamental and basic applied research on the widest possible basis being undertaken in the country. Further, industrial research should be comprehensively planned

---

[40] Kaviraj, "Outline of a Revisionist Theory of Modernity," 519–520. Kaviraj argues that at the time of independence, "it was simply taken for granted that the symmetrical-functionalist reading of European modernity was correct; indeed, there was no competing hypothesis about how to read that history".

[41] For an argument prioritising political economy as an explanation over modular nationalism in the making of India as a productive space and a discussion of late nineteenth century economic nationalism see Manu Goswami, *Producing India: From Colonial Economy to National Space* (Chicago: The University of Chicago Press, 2004), Chapters 2 and 7.

[42] K. T. Shah, "Appendix III—Note for the Guidance of Sub-Committee's of the National Planning Committee," *Report of the National Planning Committee* (New Delhi: The National Planning Committee of the Indian National Congress, 1938), 65. Emphasis added. The period between 1938 and 1947 saw very systematic expressions of the consensus on scientific industrialism of which Soviet inspired planned development was an important aspect. First was the appointment of a National Planning Committee by the Indian National Congress Ministries in 1938. Soon followed the communist "Peoples Plan" (1943), the Bombay Plan proposed by industrial leaders (1944) and Mohandas Gandhi's "Constructive Program" (revised 1943) based on his ideas of rural development and village republics.

> and linked with the development of industries, and industrial research organisations created, best suited to the needs of individual industries. To put further research actively on a proper basis, highly trained personnel with first-class scientific ability in progressively growing numbers will be needed. To produce these workers, two things require to be done; (1) to build up as rapidly as possible institutions in India which can give the highest type of scientific training of the most varied type and (2) to have men trained abroad to meet the scientific and industrial needs of the country in the very immediate future.[43]

Thus, the nationalist conception of the state *embodied the technological imperative*, and this was most explicitly expressed in the establishment and mandate of the National Planning Committee of the Indian National Congress.[44] The independence of India having been accomplished, Congress nationalist government strategy was to strengthen science-based-industrialisation as an overarching purpose of the state—a talisman to hold up against resurgence of religious and other nationalisms, and to fortify redress of uneven development under imperialism, should the left ever manage to deliver a politically destabilising critique.[45] There was also the question of princely states where government was legitimated by ascribed power. State formation has been characterised with the concentration of power, necessarily through marginalisation and dispossession of other power-wielding or potentially power-holding processes and structures.[46] A centralising impulse thus executed towards the purpose of making India productive, and therefore autonomous and sovereign, could potentially inscribe legitimate political authority of the Indian state

[43] Shah, "Appendix III", 61.

[44] Zachariah, *Developing India*.

[45] For the history of communism in India in the 1940s, see Bhagwan Josh, *Struggle for Hegemony in India: 1920–47*, Volume 2: 1934–41 (New Delhi: Sage Publications: 1992).

[46] Max Weber accounts for the state-formation as successful monopolisation of physical violence over a definite territory, and over the totality of the corresponding population. Norbert Elias accounts for state formation on similar lines but as more than monopolisation of physical violence and as a part of the larger "civilising process" of Western Europe. See Max Weber, *Economy and Society: An Outline of Interpretive Sociology* I. (Berkeley: University of California Press, 1978); Norbert Elias, *State Formation and Civilisation* (Oxford: Basil Blackwell, 1982 [1939]). For more recent accounts see Quentin Skinner, "The State," in *Political Innovation and Conceptual Change*, eds. Terence Ball, James Farr, and Russell L. Hanson (Cambridge: Cambridge University Press, 1989), 90–131; Charles Tilly, *Coercion, Capital and European States, AD 990–1990* (Oxford: Basil Blackwell, 1990). I have found very useful Bourdieu's argument of the genesis of the state as concentration and increasing bureaucratisation of both symbolic and physical power. See Pierre Bourdieu, "Rethinking the State: Genesis and Structure of the Bureaucratic Field," *Sociological Theory* 12, no. 1 (1994): 1–18.

in a territory acquiesced, acquired and annexed from varying sources of power—something the evocation or imagination of a primordial belonging to nation could far from achieve at this moment and moreover, something religion could not be allowed to accomplish—a feeling that some involved in the deliberations held with some degree of conviction. "Science" appeared to suggest, as Yaron Ezrahi has argued, "a cultural strategy for depersonalising authority through the free operation of a voluntary, self-regulating community which evolves universally valid standards".[47] Science-based-industrialisation was made into a secular ideology "to buttress the centralising project of a post-colonial nation-state".[48]

The formal decisions on what a new Indian polity, or at least a new state, would look like were deliberated in the Constituent Assembly of India between 1945 and 1950. The immediate context was no less important in shaping these deliberations than were the larger political and moral concerns with universal modernity that informed anti-colonial nationalism. The immediacy was compounded by the fact that the Constituent Assembly of India also happened to be carrying out the day-to-day functions of government as a provisional parliament.[49] The overwhelming immediate problems were those of maintaining law and order following the violence of partition, establishing the territorial integrity of "India", and transforming an anti-colonial nationalism into a politics of self-government. The period between 1945 and 1950 is most crucial for understanding the imperatives of state formation in India.

The Postcolonial State

The national-state had effectively emerged as an agent of social change in two ways. The Indian National Congress had emerged not only as leaders

---

[47] Yaron Ezrahi, "Science and the Problem of Authority in Democracy," *Transactions of the New York Academy of Sciences* 39, (1980): 44.

[48] Bose, "Post-Colonial Histories of South Asia," 142. In calling for a "dynamic conception of religion and its changing meanings in the pre-colonial ecumene and colonial and postcolonial public sphere", Sugata Bose is convincing when he argues that "The field [of South Asian Studies] is in need of a historiography that knows how to distinguish between religious sensibility and religious bigotry as well as between secularism as a system of values and secularism as an ideology to buttress the centralising project of a post-colonial nation-state".

[49] See Granville Austin, *Working a Democratic Constitution: The Indian Experience* (New Delhi: Oxford University Press, 1999).

of government in independent India but also as a moral alternative to colonial rule, and in that, personified the new state. Their promise to build a nation, the Congress would now have to deliver as a state. No less important was building the consensus on making India productive—the development of India. Given then, the urgency of law and order problems coupled with the fact that the Nehru led nationalist coterie was increasingly facing opposition from both within and without the nationalist movement, the provisional government came to rely heavily upon existing bureaucratic procedure for effective government. Even if independence meant that bureaucratic leadership was now committed to transforming the state, because this was still a largely unreconstructed colonial bureaucracy, "local administration was inseparable from local power" structures. A strong bureaucratic state resting precariously upon local power structures thus began to emerge in the early years of Indian independence as, a guarantor of freedom and citizenship, and as an agent of development.[50] The significance therefore, of nationalism's *explanatory potential* for history of this period has sometimes been overestimated.

With independence, the overarching rationale of anti-colonial nationalism was no longer binding even the Indian National Congress. First and foremost, there were those that no longer identified with Nehru's continuing support for secular politics in the wake of partition. The left-leaning members of the Congress also began to move away from what they perceived as Nehru's liberalism and socialism. As an umbrella organisation, the Congress had allowed for panoply of interests to converge on two broad issues: the issue of Indian independence and an ideology that would bind even those outside the rather diverse ensemble even as it was supposed to undercut the legitimacy of religion: science-based-industrialisation offered a less contentious goal away from politics.[51] What exactly was scientific industrialism? Pratik Chakrabarti has proposed but not sufficiently developed upon the idea of "scientific industrialism" in the short

---

[50] See especially Sudipta Kaviraj, *Politics in India*. See also Francine Frankel, *India's Political Economy 1947–2004: The Gradual Revolution*, 2nd edn (New York: Oxford University Press, 2005), Chapter 3, and Lloyd I. Rudolph and Susanne Hoeber Rudolph, *In Pursuit of Lakshmi: The Political Economy of the Indian State* (Chicago: University of Chicago Press, 1987), Chapter 2.

[51] See among others, Paul R. Brass, *The Politics of India since Independence* (Cambridge: Cambridge University Press, 1990); Rajni Kothari, *Politics in India* (Delhi: Orient Longman, 1970), especially Chapter 5; Stanley A. Kochanek, *The Congress Party of India* (Princeton: Princeton University Press, 1968) and Myron Weiner, *Party Building in a New Nation: The Indian National Congress* (Chicago: University of Chicago Press, 1967).

conclusion to his book.[52] "Orientalist imagination", he argues, particularly of the eighteenth and nineteenth century, had overwhelmingly stressed mysticism and spiritual aspects of the Indian civilisation, and some Europeans and some Indians often used that as an explanation for the lack of "material development" in India to varying degrees, with varying significance, and in different contexts. Beginning late nineteenth century, "scientific industrialism" was co-produced as an ideology around the need to restore or establish material development of the Indian sub-continent through industrialisation, where industrialisation came to mean the development of science based industry. The inspiration came from the various meanings that science came to occupy in its representation of man's triumph over nature intellectually, but also in its consequences for the betterment of living conditions. The ontology of European science promised a new social and economic order, *in relation to the essential materialism that India lacked or lost.*[53]

Science, Chakrabarti claims, was eventually "practiced and adopted in colonial India with a faith in its superior ontology". Indian scientists, especially those working in the 1920s and after, had come to accept the ontology of Western science, perhaps also with a faith in its superiority, although such a claim is bound to be heavily contested through Gandhian thought, subaltern scholars like Partha Chatterjee discussed above, and by post-colonial scholarship in general. Scientific industrialism grew out of the accommodation of this *acceptance* within the larger struggle against subject status and defiance of alien rule. The term is perhaps useful to map the thoughts and beliefs of a select nationalist elite for purposeful organisation of scientific and industrial activity on the national scale.

Shiv Visvanathan, sociologist of science, has provided a useful history of what he calls "scientized technology", a term that has significant overlap with Chakrabarti's "scientific industrialism". Visvanathan traces the roots of "scientized technology" with the establishment of the Indian Association for the Cultivation of Science (IACS) in Calcutta (1876).[54] Mahendra Lal Sircar, founder and promoter of the IACS was convinced that scientific education "must permeate the country before technical education is

---

52 Chakrabarti, *Western Science in Modern India*, 298–300.

53 Ibid., 299.

54 Shiv Visvanathan, *Organising for Science* (New Delhi: Oxford University Press, 1985), 1–132.

even possible".[55] Nationalist press supportive of Sircar's ambitions laid out the message for Bengal's English speaking audiences:

> that between scientific knowledge and its application to the practical pursuit of these industries in India, there is a gulf fixed, and this gulf is due to conditions which mere science is powerless to remove, which are indeed the despair alike of Savant and Statesman. In the vast majority of cases, the application of science to industry or art is not a simple operation which can be performed either by the man of science or man of practice, even by both together, but an extremely difficult operation in which success implies long patient and costly experiment…[56]

At the turn of the century, Sircar would finally argue, "It is through the laboratory that starvation may be eventually turned to plenty".[57] Visvanathan traces the historical trajectory of a more or less coherent "scientized technology" beginning with the *Swadeshi* movement (1905),[58] following through WWI, and the establishment of an Indian Industrial Commission to find ways of reducing Indian dependence on exported goods (1918), compounded with an admiration for the Soviet path to industrialisation (1920s)—all of which discursively culminated in the idea that planned industrial development, based upon goal-oriented scientific research or industrial research, was the most correct path towards progress. The argument was nationalist in spirit but the *employment of scientific knowledge towards practical pursuits* could hardly appear anachronistic—as historians of colonial science have argued, most scientific institutions established under colonial rule were engaged with imperial ambitions of trade, survival in the tropics and territorial control. This perspective on scientific inquiry combined with the critical engagement of Indians with scientific research beginning the late nineteenth century, was increasingly accompanied by the later reductionisms of the editors of journals like *Current Science* and *Science and Culture* in English but also in the vernacular press. By the 1930s, the perceived necessity of material development of India also drew upon varied sources like New Deal America, Soviet planning and industrialisation, reconstruction in Nazi Germany and American

[55] Mahendra Lal Sircar, "Secretary's Address," *IACS Annual Report* (1888): 18.

[56] Excerpt from the *Hindu Patriot*, in *IACS Annual Report* (1891): xiii–xxxiv.

[57] Mahendra Lal Sircar, "Secretary's Address," *IACS Annual Report* (1901): 27.

[58] *Swadeshi* literally translates to "of one's own nation". For a history of the partition of Bengal and the *Swadeshi* movement see: Sumit Sarkar, *The Swadeshi Movement in Bengal, 1903–1908* (Delhi: People's Publishing House, 1973).

philanthropic foundations.[59] Visvanathan's "scientized technology" thus captures the purposeful enrolment of scientific research for material development, and the faith in science-based-industrialisation to advance the modernisation of India. This faith in industrialisation, in the final instance, became the core of a secular ideology—and it could be called "scientific industrialism"—that provided the scaffolding of a free India.[60]

Concluding Thoughts

Just what comprised this unified idea of science evoked time and again in the period immediately before and after the partition of and independence on the Indian sub-continent? And why could science promise a higher truth that could bolster nationalism, and undermine both subnationalism and religion at the same time? Moreover, what role did science as ideology play in the role India sought to create for herself during the Cold War? In this essay, I hope to have suggested (not exhausted, or for that matter comprehensively covered) ideas from the existing scholarship on independence and after that could serve as points of departure: Sudipta Kaviraj and the *decolonised* content of Indian modernity; Pratik Chakrabarti and scientific industrialism; and Shiv Visvanathan and scientized technology. It remains to be seen to what extent the ideas of science and state in India, in conjunction with those in Britain, the United States and the Soviet Union, have complemented each other, travelled together to co-produce and cement the idea of India, and where this falls apart. It is equally important that this departure also serve another purpose: that

[59] Most historians of science have attributed the crystallisation of scientific industrialism in India to the debates generated by and in the pages of *Science and Culture*, launched in 1934 and led by Meghnad Saha and his colleagues at the Indian Association for the Cultivation of Science and University College of Science, Calcutta. See among others Prakash, *Another Reason*; and Chakrabarti, *Western Science in Modern India*. I would suggest that journals like *Current Science* launched in Madras (1932) contributed similarly and equally to the debate but in this case the southern regions of India. Meghnad Saha in Calcutta and Mokshagundam Visvesvaraya in Madras were but two that vociferously argued for depoliticised industrial growth. For an overview on ideas for developing India for the period between 1930 and 1950, see Benjamin Zachariah, *Developing India: An Intellectual and Social History* (New Delhi: Oxford University Press, 2005).

[60] There remained, though, a significantly influential section of the Congress led by Mohandas Gandhi that did not support the necessity of large-scale industrialisation throughout the first half of the twentieth century.

of charting the effects these processes on the actual practice of science in the laboratory and in the field.

In a recent article, the historian Ramachandra Guha noted the neglect of contemporary history writing in and of India. The strict division of labour between social scientists and historians is well reflected also in scholarship on the history of science. At the heart of this essay is a concern to map the historian's opportunity rendered all the more significant by the centrality of science and technology to the self-understanding and reinvention of India's place and position in the current world order.

A WESTERN SCIENTIST IN AN EASTERN CONTEXT: J. B. S. HALDANE'S INVOLVEMENT IN INDIAN SCIENCE

Veena Rao*

> *We ought to have some really good genetics here in about ten years. The trouble is that most geneticists here would like to do the sort of thing you are doing, which needs a lot of apparatus, highly purified amino acids and so on. Whereas plenty of things can be done here that can't be done in London or Long Island.*
>
> J. B. S. Haldane[1]

The article that follows concerns the 'Indian period' of J. B. S. Haldane, a British scientist of international renown who chose to move to India and work there for the last seven years of his life. Helen Spurway, his wife and a well-known geneticist in her own right, accompanied him. Their Indian period was unique in many ways and offers a fascinating example of cultural exchange between East and West. Haldane gave a thrust to the development of animal and human genetics in India, to science education and statistical studies in biology. The course of his life in India is an interesting example of the circulation of knowledge in the early years of post-colonialism. Passage from India to the West was something of a rite for aspiring young Indian men. For one thing there was the prestige associated with the English language and Western culture; for another, material benefits awaited someone who acquired an English education—or more, managed successfully to pass the Indian Civil Service examination.

* My sincere thanks to the following: the Centre for Cellular and Molecular Biology, Hyderabad, India, for access to its Haldane Collections; Profs. P. P. Majumder and K. B. Sinha, Indian Statistical Institute, Kolkata for providing me with the facilities to work at the Prasanta Chandra Mahalanobis Memorial Museum and Archives, ISI library, Kolkata, India; the Department of Photography, ISI, Kolkata, India for granting permission to use the photographs; the Archives of University College London, U. K.; the Archives of the National Library of Scotland, Edinburgh, U. K. I am grateful to Prof. Gordon McQuat for inviting me to speak at the meeting on "Circulating Knowledge, East and West" in July 2010, where this work was presented. The research carried out in the course of preparing this article was part of my project "J. B. S. Haldane's Indian Period" which was supported by the Indian National Science Academy, New Delhi, India.

[1] Letter dated 17th September 1958 from J. B. S. Haldane to M. Demerec (MS 20536, f. 92–105, National Library of Scotland, Edinburgh, U. K.).

But never before in modern times had a Western scientist of his calibre chosen to move to India—not to speak of becoming a citizen. Haldane had to grapple with many obstacles in his attempt to inculcate a Western attitude to doing science, specifically biological science, in an Indian setting; and, at the same time, to foster (to the extent possible) a distinctly Indian science that took advantage of what was locally available and was free from the dominance of Western science. The story also involves a westerner who rejected British imperialism to the extent of wanting to 'go native'... and in the process landed himself in a culture whose mores retained annoying traces of its colonial past.

Leaving Britain for India

J. B. S. Haldane was a brilliant scientist who remains unrivalled for the vast range of his contributions to science.[2] He was one of the founders of the mathematical approach to evolution known as population genetics. It would not be wrong to consider him as the father of biochemical genetics. A gifted mathematician, aristocrat, socialist, atheist, materialist and superb populariser of science, Haldane enjoyed his eccentricities and was recognised by his peers as a genius. His personality was such that it continues to stimulate and interest scholars today, almost half a century after his death in December 1964. He first held a position at the Indian Statistical Institute, Calcutta (now Kolkata) and later at Bhubaneshwar, Orissa, where he died. It is striking that he left Great Britain when Western science was making spectacular progress and landmark discoveries were being made, especially in biology. At the time the state of science in India was not such as to offer him the intellectual environment, technological support or library facilities that he was used to in Britain. Yet he emigrated from Britain to India. The reasons were many and complex. The reasons behind the move included a long standing fascination with India, disenchantment with post-war Britain, an appreciation of what P. C. Mahalanobis was attempting to do through the application of statistical methods at the Indian Statistical Institute, Jawaharlal Nehru's

[2] For more information see Ronald Clark, *The Life and Work of J. B. S. Haldane* (Oxford: Oxford University Press, 1968); K. R. Dronamraju, *Haldane. The Life and Work of J. B. S. Haldane with special reference to India* (Aberdeen: Aberdeen University Press, 1985); K. R. Dronamraju, "J. B. S. Haldane's Last Years: His Life and Work in India (1957–1964)," *Genetics* 185, no. 1 (2010): 5–10; and Partha P. Majumder, "Haldane's contributions to Biological Research in India," *Resonance* (December 1998): 32–35.

emphasis on planning and scientific temper as two pillars on which to build a modern India, and, quite likely, a spirit of contrariness.[3]

Haldane was a Marxist and an open supporter of the Communist Party by the latter part of the 1930s though he did not formally join the party till 1942. He had visited the Soviet Union and admired the science being done there. The Lysenko affair changed things.[4] Trofim Lysenko was a practical plant breeder with a strong belief in the ability of nurture to influence hereditary traits and little time for the principles of Mendelian genetics. He soon became a favourite of Stalin and used his influence to destroy the careers of many. One of those who suffered for opposing Lysenko was Nikolai Vavilov, an internationally known botanist and a friend of Haldane. Vavilov had hosted Haldane during his visit to the Soviet Union. Initially Haldane did not take a strong stand against Lysenko. But by 1949 he found it difficult to remain silent and began distancing himself from the Communist Party. This was also a time when Haldane was feeling uncomfortable with his old comrades and let down at his work place, University College London, which, he felt, was not giving him his due. He wrote to D. W. Logan at the Senate House:

> I am leaving London prematurely because promises made to me over the last twenty years both by the administrative authorities of the university and of this college, have been systematically broken..... Further, in the past, the university systematically behaved as if it found me an embarrassment, which may well have been the case.[5]

In addition, he disapproved of the politics of Western Europe. The Anglo-French invasion of Suez in 1956 acted as an impetus for him publicly to voice his anger at the British government.

A semi-farcical incident compounded his disenchantment with Britain. His wife Helen Spurway had got herself arrested for allegedly stamping on a dog's tail; she was said to have been rude to the police; and she refused to apologise. University College London (UCL), where both of them were working, asked for her resignation. She promptly provided it and called Britain a police state. William C. Clarke the American student who was with Helen Spurway during this incident stated, "She pleaded not guilty

---

[3] Vidyanand Nanjundiah, "J. B. S. Haldane: His Life and Science," *Current Science* 63, no. 9–10 (1992): 10–25.

[4] Zhores A. Medvedev, *The Rise and Fall of T. D. Lysenko*, trans. Michael Lerner (New York: Columbia University Press, 1969).

[5] Letter dated 23rd July 1957 from J. B. S. Haldane to D. W. Logan, Senate House, London (Box 22, Special Collections, UCL Archives, London).

to any of the charges but was found guilty. I also pleaded not guilty to the charge of obstructing the police and was given an absolute discharge."[6] According to John Maynard Smith, well known evolutionary biologist and Haldane's student, Helen Spurway had quarrelled with everyone in the department at UCL except himself and also had fallen in love with India.[7] Be that as it may, she had her own reasons for moving. Haldane was due to retire soon and he wished for a place where both of them could carry out research and teach young scholars.

Haldane was leaning towards the new India with its hopes of ushering in socialism. Apart from Mahalonobis, there were others for whom he had high regard: the astrophysicist and socialist Meghnad Saha, the naturalist Salim Ali, the philosopher Sarvepalli Radhakrishnan, the person who worked in the U.K. for the cause of Indian independence, V. K. Krishna Menon, and above all the new prime minister of India, Jawaharlal Nehru. Delivering a lecture at a memorial function for his friend Meghnad Saha, he confided: "When about 1953 I decided that I would if possible spend my old age in India, one of my reasons was that I hoped to see more of Meghnad Saha. I have met very few men in my life for whose company I formed so instant a liking."[8]

Haldane had visited free India thrice, in 1952, 1954 and 1956, on each occasion with his wife.[9] Helen Spurway fell in love with India; for Haldane his old enchantment for the country was renewed during the visits.[10] The diverse flora and fauna and unusual marriage customs they encountered equally suggested themes for exciting research work in population genetics that could be carried out with minimal technical equipment. As Haldane put it in a letter to his friend and fellow evolutionist Ernst Mayr, "perhaps nowhere in the world are there more plants and animals, which seem to shout questions at one or more brilliant young men and women whose capacities for research are not used."[11]

After three years of achievement and happiness during which he also experienced disappointments and frustrations with the establishment

---

[6] William C. Clarke. A statement dated 20th December 1956 (543 F. 329, Prashanta Chandra Mahalanobis Memorial Museum and Archives, ISI Library, Kolkata, India).

[7] Letter dated 8th September 2003 from John Maynard Smith to Veena Rao.

[8] J. B. S. Haldane, "The genetic effects of quanta and particles of high energy," *Science and Culture* 24 (1958): 16–22.

[9] Letter dated September 1960 J. B. S. Haldane to the Home Secretary, West Bengal (Acc.9589, National Library of Scotland, Edinburgh U. K.).

[10] J. B. S. Haldane, Autobiographical Notes, Haldane Collections, Centre for Cellular and Molecular Biology, Hyderabad, India.

[11] Letter dated 27th April 1961 J. B. S. Haldane to Ernst Mayr. (No. HUGFP 74.7, BOX 7, 761; Pusey Library, Harvard University Archives, Cambridge, MA, USA).

somewhat similar to what he had in Great Britain, Haldane desired to be an Indian citizen; and he became one in the year 1961.[12] This could well have been another way of rejecting British Imperialism. He explained to his friend V. K. Krishna Menon, by then the Defence Minister:

> You will perhaps appreciate the reasons for my desire to become a citizen of the Indian Republic rather than a subject of Queen Elizabeth II. They do not include a blind admiration of everything Indian but my admiration of some of them is sufficient to make me feel it worthwhile to criticize some others. In England at present, I doubt if this is worthwhile.[13]

Also of importance was the foreign policy of India, which they preferred to that of Britain. Haldane was clear in his mind regarding this matter. While applying for Indian Citizenship, he mentioned to the Home Secretary of West Bengal, "We also prefer the declared policy of the government of India in internal affairs to that of the either British parties."[14]

Mahalanobis's Invitation

The Indian Statistical Institute in Calcutta (now Kolkata) had been established in 1931 for, among other reasons, according statistics its rightful place as a distinct subject by itself. It offered Haldane a position as a full time research professor. The Institute was headed by his friend P. C. Mahalanobis, a statistician of international repute (and, like Haldane, a Fellow of the Royal Society). It had become known as a scientific research centre equal to any in the West for producing world class statisticians. Mahalanobis played a prominent role in the planning that was an integral part of Prime Minister Jawaharlal Nehru's vision of a modern, industrialised India. He was instrumental in implementing the National Sample Survey, deemed an essential statistical input for scientific planning, and was involved in the National Planning Commission of India. He was the honorary statistical advisor to the government of India and was close to Nehru, who in turn was known to Haldane through his sister Naomi Mitchison from the days when India was still under British rule.[15] Nehru

[12] Haldane and his wife's British passports were reaching the date of expiry by the end of 1960.

[13] Letter dated 12th September 1960 from J. B. S. Haldane to V. K. Krishna Menon (Acc. 9589, National Library of Scotland, Edinburgh, U. K.).

[14] Letter dated September 1960 J. B. S. Haldane to the Home Secretary, West Bengal (Acc.9589, National Library of Scotland, Edinburgh, U. K.).

[15] Naomi Mitchison, *You May Well Ask* (London: Victor Golloncz LTD, 1979), 198. Naomi Mitchison, a friend of Nehru, had arranged a meeting of Nehru and Stafford Cripps at

and Mahalanobis were in agreement on the importance of science and science education for social transformation of a country.

As General Secretary of the Indian Science Congress Association, in 1946 Mahalanobis took the initiative of inviting foreign scientists to the annually held Indian Science Congress. Haldane was one among them. The invitation included a proposal to deliver a course of lectures at the Indian Statistical Institute. Haldane initially accepted with great enthusiasm:

> May I say once more how much I hope, that it may be possible for me to come to India. I have a number of ideas which I believe Indian workers might care to follow up, and I am sure that I should learn as much from them as they would from me.[16]

However, he had also been invited to attend a conference at Princeton University, USA, and the dates overlapped. He expressed a preference to visit India if travel grants could be arranged. Unfortunately the grants were not forthcoming in time and he went to the Princeton conference.[17] Eventually Mahalanobis made a more substantial offer and the Haldanes finally took up positions at the Indian Statistical Institute (ISI) in 1957.[18]

The ISI Experience

Their stay began well. Haldane's hopes appeared to be justified. He managed to gather around him a group of bright young men who were on the way to making a name for themselves in their respective fields; his pride in them was that of a father. Some years after he and his wife had made India his home he expressed this in one of many similar letters that he wrote to his friends in the West. He wrote to Dr. Pembry, "My 4 years here (Calcutta) have been among the most fruitful of my life. I have discovered 4 young men of (Bengal, Andhra Pradesh, Kerala, Maharashtra) amazing ability."[19] He was also pleased with the continuing

a dinner party in her house in 1934. J. B. S. Haldane and Julian Huxley were the other guests.

[16] Letter dated 25th July 1946 from J. B. S. Haldane to P. C. Mahalanobis (Box 35, Special Collections, UCL Archives, London).

[17] Veena Rao and Nanjundiah, Vidyanand, "J. B. S. Haldane, Ernst Mayr, and the Beanbag Genetics-Dispute," *Journal of the History of Biology* 44 (2011): 233–281, on 242.

[18] Letter dated 20th July 1957 Haldane J. B. S. to P. C. Mahalanobis (543. B. 54, 36. Prashanta Chandra Mahalanobis Memorial Museum and Archives, ISI Library, Kolkata, India).

[19] Letter dated 21st December 1961 J. B. S. Haldane to Pembry (Acc.9589, National Library of Scotland, Edinburgh, U. K.). The letter from Haldane to his friend Dr. Pembry shows that even after the ISI resignation and problems with his new post with the CSIR,

success of the *Journal of Genetics* which he (as Editor) had brought from the U. K. The journal, whose roots went back to the origin of the field itself, had become well known due to its high-quality publications. Haldane's sharp book reviews made it even more readable.[20] He often helped scientists by pin-pointing some vital fact in a paper that had been missed by the author or by putting authors in touch with scientists elsewhere in order to benefit their research. Haldane himself published some important papers from India in the journal, especially along with S. D. Jayakar, his colleague.[21] Through an unforeseen turn of events that Haldane might have relished, his final publication from the Indian period appeared 32 years after his death.[22]

Yet things were not all that smooth below the surface; the relationship between Haldane and the autocratic Mahalanobis was gradually souring. This was not surprising. As the traditional saying goes, 'two kings cannot co-exist under one roof.' Haldane began to accumulate reasons for feeling slighted. He claimed that certain promises given him had not been kept; for one thing, Mahalonobis had done nothing to clarify the element of vagueness that remained attached to Haldane's official status and responsibilities. He finally spoke up on 27th 1960, July, "I have now been here for three years. For some purposes, I am treated as head of a department, for others as a subordinate."[23] He thought that the ISI environment was becoming formal and less conducive than before for free scientific research. Haldane was appalled to learn that a worker of the institute was sacked because he was found smoking in front of his senior. He commented, "It appears that some persons in the Indian Statistical Institute wish to model it on the court of a British Viceroy rather than on a scientific laboratory. If they succeed they will certainly make original work there impossible, and this may well be their aim."[24]

Haldane was still optimistic about the future as he was confident about the scientific potential of the Indian youth and was deeply attached to some of them.

[20] See J. B. S. Haldane, "A Parade of Ignorance," *Journal of Genetics* 58, no. 1(1962): 138–142.

[21] J. B. S. Haldane and S. D. Jayakar, "An enumeration of some human relationships," *Journal of Genetics* 58 (1962): 81, and idem., "Polymorphism due to selection of varying direction," *Journal of Genetics* 58 (1963): 237.

[22] J. B. S. Haldane, "The negative heritability of neonatal jaundice," *Annals of Human Genetics* 60, no. 1 (1996): 3–5. It had been mislaid after being submitted by Lionel Penrose.

[23] Letter dated 27th July 1960 from, J. B. S. Haldane to Joint Director, ISI (543.61, 225. Prasanta Chandra-Mahalanobis Memorial Museum and Archives. ISI Library, Kolkata, India).

[24] Ronald Clark, *The Life and Work of J. B. S. Haldane* (London: Oxford University Press, 1984), 231.

Matters came to a head in February 1961 during an official visit by Alexei Kosygin, the future Prime Minister of the USSR. In honour of the visit the Institute had asked all departments to put their work for exhibition. Haldane and his group duly did so. Haldane was also taking care of the exhibits of geology unit in their absence. Mahalanobis, who had been out of Calcutta, returned just before Kosygin's visit, and interfered with the way the exhibits had been arranged. This sparked off a heated argument. Haldane was not willing to change the arrangements because he believed that doing so would interfere with ongoing research work. That was something he was unwilling to do for anyone, however important the person may have been. Both of them refused to step back; Haldane announced his resignation from the institute then and there. His wife too followed suit.[25]

After ISI

Around this time S. Radhakrishnan, Vice-President of India, spoke with Haldane about the possibility of his working in a proposed national laboratory for biological science.[26] This was followed up by the director of the Council of Scientific and Industrial Research (CSIR), M. S. Thacker, who suggested that Haldane should start such a unit in Calcutta under the aegis of CSIR. There was also an offer of a professorship in genetics from Panjab University in Chandigarh, a modern city designed by Le Corbusier.[27] But by then Haldane had committed himself to Thacker. He wrote to his friend, the ornithologist, Salim Ali, on 6th March 1961:

> I have finally have had to resign my post here [ISI] as conditions for teaching and research have deteriorated. However I have been offered a better job, so far only verbally...... It is in fact the leadership of the proposed institute of biology. Thacker offered it and at first I refused. It was renewed and I have accepted.[28]

---

[25] Letter dated 24th February 1961 J. B. S. Haldane to P. C. Mahalanobis (543. F. 50, 282. Prasanta Chandra Mahalanobis Memorial Museum and Archives, ISI Library, Kolkata, India).

[26] Ronald Clark, *The Life and Work of J. B. S. Haldane* (London: Oxford University Press, 1984), 233.

[27] Letter dated May 31st 1961 Vice-Chancellor of Panjab University to J. B. S. Haldane (Acc.9589, National Library of Scotland, Edinburgh, U. K.).

[28] Letter dated 6th March 1961 J. B. S. Haldane to Salim Ali (Acc. 9589, National Library of Scotland, Edinburgh, U. K.).

Soon Haldane started a Genetics and Biometry Research Unit, initially at his own home; the Unit was crammed amongst the bedroom, bathroom and balcony of the house. Thacker had said that a laboratory, buildings and grounds would follow shortly. The promise did not materialise even after months. Space was needed to manage the *Journal of Genetics*, Suresh Jayakar as the editor had to be paid, and the house was stacked with files and papers. There was no room for a laboratory nor could the CSIR give Haldane a definite assurance as to when it would be forthcoming. On top of this, Haldane got the unwelcome news that having become a government employee, his freedoms were restricted. He was banned from taking part in politics, participating in anti-nuclear bomb campaigns, and approaching any foreign organisation for help to travel abroad.

Such a state of affairs proved intolerable. It culminated in his resigning from his position with a parting shot. He said 'CSIR' stood for 'Conspiracy for the Suppression of Independent Research'.[29] Haldane was a public figure and the issue of his resignation was raised in the Indian parliament. When his explanation was sought Thacker tried to brazen it out; he claimed that all of Haldane's requirements had been met. Not the person to take such a situation lying down, Haldane called the Press and stated, "It is the intolerable conditions imposed by the bureaucrats, and not the low salaries or the lack of equipment, which cause so many Indians to take up posts abroad. I hope that my resignation may do something to draw public attention to these conditions."[30] The wide publicity that was created gave rise to the rumour that Haldane was leaving India. But, around the same time, on June 15th, 1962, Haldane was writing to his friend Howard Florey, the man who took the lead in successfully exploiting the discovery of penicillin, regarding this matter:

> Please don't think I am fed up with this country...... But young men are grand and so are a lot of plants and animals. Also I can make a more effective stink here in cases like this than I could in England.[31]

For a while, after the misadventure with the CSIR, Haldane gave as his 'present address' simply 'c/o Journal of Genetics, 18/1 Barrackpore Trunk Road, P. O. Belghoria, 24 Parganas, W. Bengal, India,' to some publications

---

[29] Haldane J. B. S. 22nd March 1962. (MS 20543 F. 132–142, National Library of Scotland, Edinburgh, U. K.).

[30] Ronald Clark, *The Life and Work of J. B. S. Haldane* (London: Oxford University Press, 1984), 241.

[31] Letter dated 15th June 1962 J. B. S. Haldane to Howard Florey (Acc. 9589. National Library of Scotland, Edinburgh, U. K.).

for want of a definite official address.[32] Even as Haldane and his associates were feeling frustrated with the inefficiency of CSIR, in April 1962, deliverance came from an unexpected source. Biju Patnaik, Chief Minister of the neighbouring state of Orissa, was an unconventional politician and a forward thinking man with a strong commitment to the industrial development of his state. He put before Haldane the prospect of heading a brand-new laboratory in Bhubaneshwar, Orissa. The offer was taken up and Haldane appeared to be happy in his new surroundings.[33] He picked up active research and soon, both by himself and together with his group of young researchers began shooting off scientific papers in true Haldanian style.[34] But the sudden discovery of cancer put paid to all long-term plans; Haldane succumbed to carcinoma of the colon in December 1964.

Scientific Contributions Before the Move to India

Even among the great biologists of the 20th century, Haldane's originality and prescience make him an extraordinary figure. He was one of a small number who showed that the Darwinian theory of evolution, natural selection, was consistent with Mendelian genetics.[35] Together with Briggs he developed a comprehensive mathematical theory of enzyme action.[36] By making it plausible that metabolism and physiology were consequences of the activities of genes, he helped found the field of biochemical genetics. He put forward the 'one gene one enzyme' concept well before Beadle and Tatum.[37] He contributed to developing human genetics as a distinct discipline. Nearly five decades after his death, many of his theories are

[32] K. R. Dronamraju and J. B. S. Haldane, "Inheritance of Hairy Pinnae," *American Journal of Human Genetics*. 14, no. 1 (1962): 102–3.

[33] Rao and Nanjundiah, "J. B. S. Haldane, Ernst Mayr, and the Beanbag Genetics Dispute," 252.

[34] J. B. S. Haldane, "A Defense of Beanbag Genetics," *Perspectives in Biology and Medicine* 7 (1964): 343–359; J. B. S. Haldane and S. D. Jayakar, "The solution of some equations occurring in population genetics," *Journal of Genetics* 58 (1963): 291.

[35] Rao and Nanjundiah, "J. B. S. Haldane, Ernst Mayr, and the Beanbag Genetics Dispute," 241.

[36] G. E. Briggs and J. B. S. Haldane, "A Note on the Kinetics of Enzyme Action," *Biochemical Journal* 29 (1925): 338–339. Along with Briggs, Haldane gave the correct interpretation of the kinetics of the simplest enzyme reaction; he also wrote an influential book on enzymes.

[37] "His experience as a biochemist in Cambridge had had him to accept the one-gene-one-enzyme hypothesis, originally formulated by Garrod, long before Beadle and Tatum, in their work on *Neurospora* in 1941, raised it to the level of a principle in experimental genetics." See F. A. E. Crew, "Haldane as a Geneticist," in *Haldane and Modern Biology*, ed. K. R. Dronamraju. (Baltimore: The Johns Hopkins Press, 1968), 18.

still bearing fruit. The ideas of *in vitro* fertilization (IVF) or ectogenesis,[38] transgenics,[39] genetic counselling (pre-natal diagnosis),[40] and disease as a factor in evolution[41] all originated with Haldane. He discovered genetic linkage in a vertebrate—the mouse—and much later in life, estimated its extent in man; developed a formula for estimating the physical separation of genes along chromosomes; proposed a rule relating to the sterility of inter-species hybrids; and estimated the rate of mutation in humans.[42] As early as 1941 he foresaw the necessity for a gene to consist of a complementary pair of 'left-right' molecules, each half capable of acting as a template for the synthesis of its complement; this was demonstrated by Meselson and Stahl in 1958 and has become famous as the semi-conservative replication of DNA.[43] Independently of A. I. Oparin (perhaps, because Oparin's publication was in Russian language) he put forward a chemical model for the origin of life based on the 'hot dilute soup' constitution of primitive oceans.[44] Haldane was no ivory tower scientist. He was a superb communicator of science for whom 'popular' writing and publishing technical papers were equally important. He stated: "I think, however, that the public has a right to know what is going on inside the laboratories, for some of which it pays."[45] Towards this end he wrote articles and scientific essays in a wide array of journals and newspapers, gave radio talks, published books and gave many lectures.

---

38 J. B. S. Haldane, *Daedalus or Science and the Future* (London: Kegan Paul, Trench, Trubner & Co, 1930), 65–67.

39 Ibid., 69.

40 J. B. S. Haldane, "The Future of Biology," in his *Possible Worlds* (London: Evergreen Books, 1940), 139–140; D. J. Weatherall, "Daedalus, Haldane and Medical Science," *Haldane's Daedalus Revisited*, ed. Krishna R. Dronamraju (Oxford: Oxford University Press, 1995), 103–123.

41 J. B. S. Haldane, "Disease and evolution," *La Ricerca Scientifica Supplemento* 19 (1949): 2–11.

42 Sewall Wright, "Contributions to Genetics," in *Haldane and Modern Biology*, ed. K. R. Dronamraju (Baltimore: The Johns Hopkins Press, 1968) 1–9.

43 J. B. S. Haldane, "How a Gene Reproduces Itself," in *On Being the Right Size and other essays*, ed. John Maynard Smith (Oxford: Oxford University Press, 1985) 178–81. Haldane was an unusual scientist whose interest spanned diverse fields of scientific research and he could hold his own in all of them. But he was not an experimenter. He thought out new ideas and new modes of investigation into important problems which he himself failed to follow up for a host of reasons; later many became landmark discoveries in the annals of biology.

44 J. B. S. Haldane, "The Origin of Life," in *Science and Life* (London: Pemberton Publishing Co Ltd, 1968), 1–11. Haldane came back to this subject in 1964. See Haldane, "Data needed for the blueprint of the first organism," in *The Origins of Prebiological Systems and Their Molecular Matrices*, ed. S. W. Fox (New York: Academic Press, 1964), 11.

45 Haldane, *Possible Worlds*, 7.

Fig. 1 Haldane, attired in traditional Indian dress, celebrating his birthday at "Amrapali", Mahalanobis's residence at the ISI campus. Note the sacred ash on his forehead. Standing behind him is Mahalanobis. The date is 5th November 1957. Photo courtesy: Indian Statistical Institute, Kolkata, India.

Contributions to Developing Science in India

Biological research in general: Haldane did his utmost to give a push to the development of modern biology, especially human genetics, in India. He had expressed a wish to contribute his expertise for the development of science in India to Nehru after his very first visit in 1952 and followed it up with a list of practical suggestions for doable and interesting research in biology, physiology and animal genetics. Haldane perceived serious omissions in the fields of animal genetics and human physiology, two areas of study where India offered special advantages. Domestic animals were among the country's great natural resources and so their genetics was a worthy object of careful investigation: significant social and economic gains were to be had from looking into the genetics of Indian breeds of poultry for egg yield and disease resistance; and in the case of cattle, for increasing their milk yield. As with all his suggestions for scientific activity in India, Haldane's clincher was that none of these studies would need expensive apparatus or special buildings. He went on to stress that biological research conducted using non-violent methods (as a breeding program would be) was peculiarly appropriate to India. He said he would be glad to assist in any way that he could; he felt obliged to do so to help a country whose government was doing as much for world peace as the Indian government.[46] But he added:

> However, I fully realise that the time has ceased when an Englishman man can claim any right to advise Indians. If such a view is taken, I can make no complaint. If it is not, perhaps I may be of some service to India.[47]

Nehru tried to utilise the expertise and experience of Haldane whenever possible and more than once helped in removing bureaucratic hurdles in his path.

Haldane was largely responsible for planning and implementing a unique integrated course for the Bachelor's and Master's degrees in statistics at ISI. Mahalanobis acknowledged Haldane's help when he inaugurated the courses. He said, "As a matter of fact, the integrated teaching programme which we have, could not have been put through, I am almost certain, without his very massive support."[48] Perhaps for the first time, quantitative biological research gained momentum during Haldane's time in India. To his mind quantitative thinking and statistical analysis were indispensible for excellence in research work.[49] He bemoaned the fact that in Indian universities the syllabus for biology students excluded mathematics and statistics and thereby closed avenues for doing important research of value in developing agriculture or husbandry. Haldane on his part had made all his research projects require statistical methods at least

46 Letter dated 15th February 1952 from J. B. S. Haldane to Jawaharlal Nehru (Box 35, Special Collections, UCL Archives, London).

47 Letter dated 15th February 1952 from J. B. S. Haldane to Jawaharlal Nehru (Box 35, Special Collections, UCL Archives, London).

48 P. C. Mahalanobis, "Farewell to J. B. S. Haldane," *J. B. S. Haldane, a tribute* (Calcutta: Indian Statitical Institute, 1992), 49.

49 His precept was "An ounce of algebra is worth of a ton of verbal argument." See Maynard Smith, "Professor J. B. S. Haldane, FRS," *Nature* 206 (1965), 239.

at one stage. He encouraged all his students to develop statistical skills which would contribute to the growth of India.[50] He travelled extensively as head of a University Grants Commission sub-committee to assess universities and research laboratories, critically evaluating them and making suggestions for rectifying their drawbacks where he could.[51] Because of this he had a good feeling for the state of science in India. He put to use old friendships with prominent figures such as Jawaharlal Nehru, S. Radhakrishnan and V. K. Krishna Menon in his efforts to establish a strong base for distinct Indian science.[52]

The first Drosophila laboratory: One of the highlights of Haldane's Indian period was in helping set up a laboratory to carry out modern cytogenetics. In this instance Haldane was truly instrumental in circulating knowledge across two continents. Haldane and his wife had been invited by the Indian Science Congress and the Indian Statistical Institute, Calcutta, in 1952. While paying a visit to the Zoology Department of Calcutta University Haldane ran into S. P. Ray Chaudhuri. The two had met before when Haldane had gone to examine Ray Chaudhuri when the latter was a student in Edinburgh. Ray Chaudhuri had worked on the genetics of the fruit fly Drosophila (which had become the cornerstone of modern genetics) under Prof. H. J. Muller in Edinburgh.[53] Haldane was unhappy to see that the young man was not carrying out research in the area of his expertise—Drosophila genetics—solely because of the lack of an air conditioned laboratory in which to rear the flies. He made a personal appeal to Nehru for help; the laboratory was set up in record time. Drosophila stock cultures too arrived from Muller's laboratory through Haldane's intervention. The birth of this Drosophila laboratory later gave rise to others in India, as scientists trained by Ray Chaudhuri moved on to Benares Hindu University and Mysore University to establish similar research laboratories.[54]

The incident made Haldane ask researchers to find an alternative to the imported Drosophila, an insect that was adapted to local conditions,

---

[50] J. B. S. Haldane, "Let us develop a Statistical Conscience," *J. B. S. Haldane, a tribute* (Calcutta: Indian Statitical Institute, 1992), 25–32.

[51] Besides providing grants, the University Grants Commission is responsible for overseeing standards of teaching and research in universities.

[52] Nehru helped Haldane in establishing the first Drosophila laboratory in India and invited him to be part of UGC. S. Radhakrishnan and Krishna Menon often resolved many bureaucratic problems.

[53] H. J. Muller had to leave Soviet Union in 1937 because of the nature of the political situation. With about 250 strains of Drosophila he moved to the Institute of Animal Genetics in Edinburgh, U. K.

[54] "First Drosophila Laboratory in India," *J. B. S. Haldane Commemoration Volume* (Calcutta: Proceedings of the Zoological Society of Bengal, 1993).

in this case high temperature and humidity, besides possessing other desirable qualities such as a short generation time that would make it useful for genetical studies in the laboratory. He said as much in a letter to Nehru when he wrote, "A man who undertakes the task of finding an easily bred Indian insect suitable for genetical teaching will also be serving the country."[55]

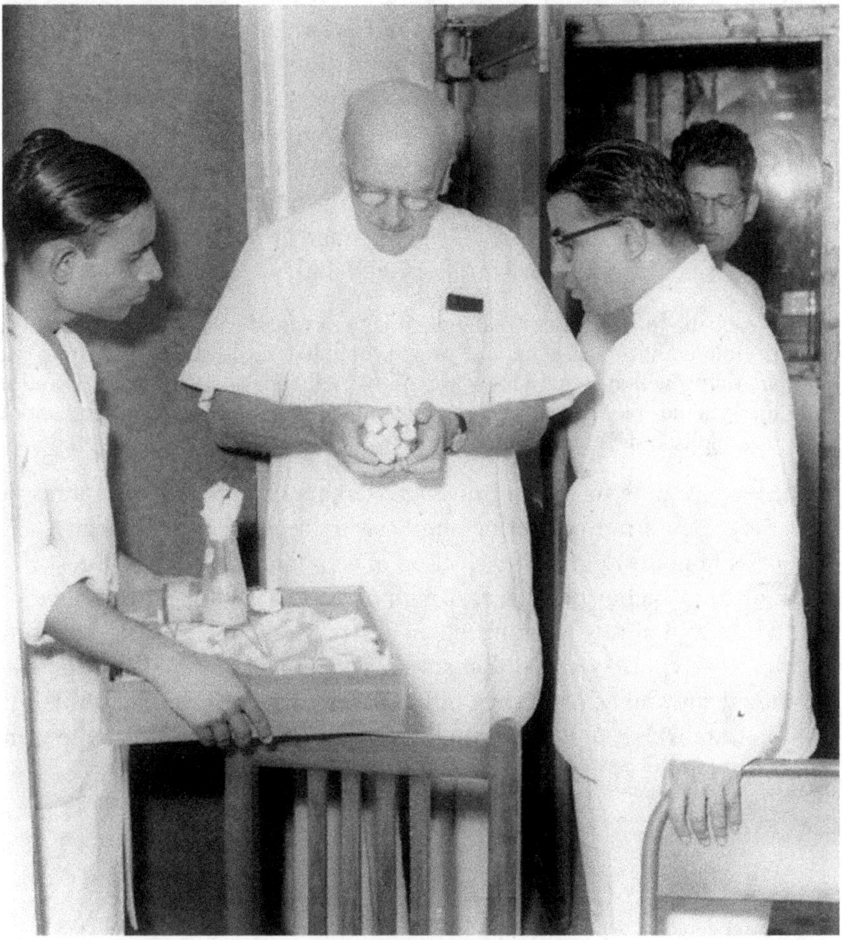

Fig. 2 Haldane, in his laboratory with students, at the ISI, Kolkata, India. Photo courtesy: Indian Statistical Institute, Kolkata, India.

[55] Letter dated 15th February 1952 from J. B. S. Haldane to Jawaharlal Nehru (Box 35, Special Collections, UCL Archives, London).

Working Towards a Distinct, Indigenous Science

Haldane tried to motivate young scholars by assuring them that good research work up to the level of international standards was possible in India in spite of all its drawbacks. He wanted to woo them away from restricting themselves to imitative Western science, an attitude of mind that he thought was a relic of colonial hangover. He advised the UGC (University Grants Commission) board to get rid of its obsession for foreign degrees, because a super-specialised Indian scientist would be unable to work effectively in an Indian laboratory. Relying on foreign degrees is dangerous in agricultural biology, he said, as one would not be able to relate with the local flora. He tried to instil confidence in young researchers to take up projects which were possible only in India and which would add to its development.[56] Haldane shared his thoughts with his friend, the well known geneticist Miloslav Demerec, then at the famous Cold Spring Harbor Research Laboratory in the United States:

> We ought to have some really good genetics here in about ten years. The trouble is that most geneticists here would like to do the sort of thing you are doing, which needs a lot of apparatus, highly purified amino acids and so on. Whereas plenty of things can be done here that can't be done in London or Long Island.[57]

In his suggestions to UGC (University Grants Commission) he stressed that they should not lay undue emphasis on a post-graduate degree or a first class honours degree for selection of candidates for doctoral-research work or for a higher pay scale. A high class obtained in an examination could be a sign of rote learning which destroys original thinking, so essential for research. He warned that such a system would merely put a brake on the advancement of science in India. He gave the example of his student in Cambridge, A. J. P. Martin, a second-class degree holder who went on to receive the Nobel Prize.[58]

---

[56] Letter dated 1st September 1959 from J. B. S. Haldane to UGC committee (MS 20537, f. 116–128, National Library of Scotland, Edinburgh U. K.).

[57] Letter dated 17th September 1958 from J. B. S. Haldane to M. Demerec (MS 20536, f. 92–105, National Library of Scotland, Edinburgh, U. K.).

[58] Martin shared the 1952 Nobel Prize in Chemistry with R. L. M. Synge for developing partition chromatography. He wanted to study chemical engineering in Cambridge but Haldane persuaded him to become a biochemist. See http://www.nobelprize.org/nobel_prizes/chemistry/laureates/1952/martin.html.

He advised teachers at the universities that their aim in science teaching should be to encourage original thinking and not rote learning which may result in a first-class marks but detrimental for research career.[59] Haldane felt that bright young scientists were suppressed by their seniors and were not given freedom in their research, causing many of them to go abroad. Writing to his friend and Defence Minister V. K. Krishna Menon, Haldane commented "... some of my best colleagues are likely to take posts abroad not because they are better paid... but because they have freedom of research."[60] He felt that the hierarchy prevailing in universities and research institutions was a relic of imperialism and would stunt the growth of science. At times he felt helpless about the situation and requested his friends (for example Krishna Menon), to lend a hand to clean up education, scientific and research departments; he felt that the young researchers he met were far abler than the average post graduate student in London and equal to any of those that he had supervised in Cambridge.[61] He also felt that there should be adequate employment for deserving scholars. He remarked in a letter to vice president S. Radhakrishnan, "It is quite useless to give one or two lectures which are described as inspiring and it is even worse to give courses to people who know that they will not lead to jobs at all, let alone safe jobs."[62] At conferences Haldane was disturbed to see seniors hogging the limelight by utilising the research work done by their juniors. He felt the science conferences were an organised conspiracy against originality in Indian science and was an example of the repression of young scholars by their seniors.[63] He stressed the need to encourage young scientists engaged in active research to participate in conferences rather than 'than old fossils' like himself and followed what he preached, going to the extent of financing their travel.[64] When he was criticised in ISI for this unusual habit, he replied to Mahalanobis

---

[59] Letter dated 3rd February 1961 from J. B. S. Haldane to UGC (MS 20541, f. 1–10, National Library of Scotland, Edinburgh, U. K.).

[60] Letter dated 5th March 1962 J. B. S. Haldane to Krishna Menon (MS 20543, f. 57–67 National Library of Scotland, Edinburgh, U. K.)

[61] J. B. S. Haldane. "Suggestions to UGC Board." (MS 20538, f. 46–56; 26–35, National Library of Scotland, Edinburgh, U. K.).

[62] Letter dated 30th December 1960 from J. B. S. Haldane to S. Radhakrishnan (Acc 9589, National Library of Scotland, Edinburgh, U. K.).

[63] J. B. S. Haldane, "The Scandal of the Science Congress," in *Science and Indian Culture* (Calcutta: New Age Publishers, 1965), 21–25.

[64] Meera Khan, "J. B. S. Haldane as an Informal Teacher: Some Personal Reminiscences," *J. B. S. Haldane: A Tribute* (Calcutta: Indian Statistical Institute, 1992), 99–109.

that as he had no children, he considered his students whose work he admired as he would his own children, if he had had any.[65]

Haldane also elaborated some important reasons for low scientific efficiency in India. At most universities which he visited, he was discouraged from seeing research work or local plants and animals whereas he was forced to meet vice-chancellors or make after dinner speeches or listen to them. He commented that this attitude had been borrowed from the British administrators of India and is unworthy of citizens of an independent country.[66] Haldane was reinforcing the earlier comments during a lecture given in Ceylon (now Sri Lanka) for the Association for the Advancement of Science in November 1960. He bemoaned the fact that science was in a bad way in the former colonies of western European nations because a foreign government would not encourage the independence of mind so essential for scientific research. Biological research in India, he felt, could reach high standards if it was research better carried out in India than in any western countries, even with their expensive apparatus. He suggested to the people of Ceylon that one should work on animals and plants of one's own country, especially those of economic importance which would help the growth of their country. According to Haldane such research was also possible in India. True to this belief, he and his wife had initiated research work on silk moths, rice, coconut yields and animal genetics.

He initiated and fostered international scientific collaborations between scientists from abroad and his students so that there was a free flow of knowledge. He was known to finance many scientific projects personally, rather than go through a lot of delay and paper work while applying for funds. Haldane received the International Feltrinelli prize, a prestigious award given by the Italian Academia Dei Lincei, in 1961.[67] By then he had become a citizen of India. On hearing the good news Haldane rejoiced:

> Such a great sum would be of help to a biologist anywhere. In India it may actually influence the whole development of the biological sciences ... I am now a citizen of Indian Republic. Will you permit me to attend in what I

65 Letter from J. B. S. Haldane to P. C. Mahalanobis. (543 F. 135, 197, Prasanta Chandra Mahalanobis Memorial Museum and Archives, ISI Library, Kolkata, India).

66 Letter dated 1st September 1959 J. B. S. Haldane to UGC (MS 20537, f. 116–128, National Library of Scotland, Edinburgh U. K.).

67 Meant to recognise distinguished achievement in the humanities or natural sciences. The awardees have included Thomas Mann, Igor Stravinsky, W. H. Auden, Georges Braque (before Haldane) and Henry Moore, Rita Levi-Montalcini and Joan Mirò (after Haldane). No Indian names are apparent. See http://www.lincei.it/premi/assegnati_feltrinelli.php?PHPSESSID=foe8178fbbdc12616354b93a6efabe45.

> regard as the dignified costume of this country?.. I hope that I may not be the last Indian to receive this prize.[68]

An international collaboration in population genetics, implemented in the malarial areas of coastal Andhra Pradesh, was personally financed at the Indian end entirely with Feltrinelli Prize money. He had grand plans for many projects in human genetics. His collaborator for one such, a W. H. O. project on resistance to malaria, was the Italian biologist Marcello Siniscalco. Siniscalco was working on genetically determined abnormalities of haemoglobin, which are supposed to provide protection from malaria.[69] This was also an idea put forward by Haldane in 1949 which gained acceptance later. Haldane felt that India could offer an excellent field to investigate the genetic basis for malarial resistance as there were many areas rampant with malaria. Haldane hoped for a long term collaboration with Siniscalco in this field which, he felt, would make an important contribution to research in human genetics in India. He financed the entire travel of one of his students, P. Meera Khan (to be recognised later as a renowned cytogeneticist), to Sardinia and Naples for training under Siniscalco. In 1963 Haldane forwarded all the money that he earned by his talks on BBC and lectures in Talahassee, Florida, in the United States, towards this project. This is an extraordinary example of a western scientist contributing in every way possible to build a strong foundation for science in India. Unfortunately he did not live long enough to see this project through.[70]

Haldane tried to bring over many eminent scientists to India for a reasonable period, so that it would benefit the researchers in India by giving them exposure to new ideas and fresh insights into their own research work. Amongst them were Dr. Pamela Robinson from UCL, a Paleontologist, and T. Dobzhansky, an American evolutionary biologist.[71] The invigorating atmosphere thus created influenced the minds of many young scholars. By inviting his friend and old colleague, Pamela Robinson, from University College London, Haldane helped in establishing a palaeontology unit at ISI, Calcutta. Years later, the research work of this unit yielded

68 Letter dated 20th May 1961 from J. B. S. Haldane to Arangio-Ruiz (MS 20541, f. 216–225, National Library of Scotland, Edinburgh U. K.).

69 For example, abnormal haemoglobins, thalassemia in heterozygotes and glucose-6-phosphate dehydrogenase deficiency of red blood cells.

70 Khan. "J. B. S. Haldane as an Informal Teacher," 99–109.

71 T. Dobzhansky talked to the scholars at the ISI, Calcutta about his well known work on Drosophila.

a skeleton of Jurassic mammal which was aptly named *Kotatherium haldanei*.[72] Haldane had frequently laid emphasis on 'Evolutionary Studies' in India because of the presence of rich Jurassic beds.

Learning about the large percentage of inbreeding amongst people of Andhra Pradesh in India, Haldane initiated a novel large-scale study of genetic impact and frequency of marriages between biologically related persons based on the unique marriage customs in South India. The custom encouraged marriage between uncle and nieces or first cousins. This was found very important as progeny from such marriages have a high frequency of genetic abnormalities and it would help if detected early. As this was a valuable statistical data, the government of India undertook a survey on the incidence of inbreeding in all states and union territories as part of 1961 'Census,' as advised by Haldane.[73]

Mentoring

As already mentioned, Haldane's efforts to improve the functioning of universities and, in particular, to improve the organisation of scientific research, were many and impressive. But according to Haldane himself his most noteworthy contribution in India was to motivate and nurture several bright young scholars; many of them received international acclaim later in their scientific careers. His feelings came out in a statement made on being elected to the US National Academy of Sciences in 1964. He asserted, "Since I have done little independent work in the last 7 years, I venture to hope that, my election is in part a recognition of the research done by my colleagues in India."[74]

Supporting Indian scientists and working towards building a strong base in modern biological research in India was always uppermost in his mind. Within the short period of seven years Haldane's mentoring led at least five of his junior colleagues to make a mark in the world of science. They were T. A. Davies, K. R. Dronamraju, P. Meera Khan, S. D. Jayakar and A. K. Ray. Davies was spotted by Haldane when he was working at a coconut research centre in Kayangulam, Kerala. He became renowned as an

---

72 T. K. Roy Chowdhury, "J. B. S. Haldane and Palaentology." *J. B. S. Haldane: a tribute* (Calcutta: Indian Statistical Institute, 1992), 157–159.

73 A. K. Roy Choudhury, "Incidence of Inbreeding in Different States of India," *Demography India* 5, no. 1&2 (1976): 108.

74 Khan, "J. B. S. Haldane as an Informal Teacher," 99.

authority on the biology of the coconut. His expertise ranged from practical matters related to coconut propagation to an extraordinary study on the handedness of the leaf whorls, in which he showed that right- or left-handedness had more to do with the latitude than genes.[75] Dronamraju and Meera Khan were initiated into human genetics by Haldane. They took advantage of the prevalence of consanguinity—a feature of marriage systems in South India—to study the patterns of inheritance of various genetic traits. Dronamraju went on to write extensively on Haldane and Meera Khan became known for his work on somatic cell genetics, genetic disorders and cancer at the University of Leiden. Based on publications, Jayakar would seem to have engaged Haldane's efforts more directly than the others. The two collaborated extensively on a variety of problems in mathematical population genetics; Jayakar established himself as one of the leading workers in the field later at the University of Pavia.[76] At the time when he came into contact with Haldane in Calcutta A. K. Ray was with the famous anthropologist Nirmal Kumar Bose.[77] Ray was to make his mark in anthropological genetics at the University of Toronto. Haldane induced him to examine anthropological issues from the viewpoint of genetics. Haldane's Bhubaneswar (Orissa) period led to a posthumous publication with Ray, who had been invited by Haldane to join him there.[78] Combining Ray's field observations with Haldane's algebraic and statistical analyses, the work followed up a fortuitous observation that Ray had made while he was engaged on a study of colour vision: there were some people with abnormally short fourth toes. According to Ray, 'while [he] was observing the bare feet', Haldane also 'joined [him]...with a pocket calculator, counting feverishly'.[79] It can be seen that the scientific contributions of each of the five are characteristic of different but overlapping interests of Haldane.

The efforts at disseminating scientific findings to the general public continued in India. His articles in the daily news papers, radio talks and

---

[75] And in fact may have had a geophysical origin. Davis's work has re-attracted interest lately; see P. V. Minorsky and N. B. Bronstein, "Natural Experiments Indicate That Geomagnetic Variations Cause Spatial and Temporal Variations in Coconut Palm Asymmetry," *Plant Physiology* 142, no. 1 (Sept., 2006): 40–44.

[76] After Haldane's death Helen Spurway and Jayakar took over the running of *Journal of Genetics*.

[77] Also associate and chronicler of Mahatma Gandhi.

[78] A. K. Ray and J. B. S. Haldane, "The genetics of a common Indian digital abnormality," *Proceedings of the National Academy of Sciences of the United States* 53 (1965): 1050–1053.

[79] A. K. Ray, "Remembering J. B. S. Haldane," *J. B. S. Haldane: a tribute* (Calcutta: Indian Statistical Institute, 1992), 139–145.

public lectures inspired and evoked an enthusiastic response from the public, who flooded him with scientific queries. His regular pieces in the daily newspaper 'The Hindu' attracted a wide readership and were later compiled into a book.[80] His lecture "The Unity and Diversity of Life" delivered in memory of Sardar Vallabhbhai Patel in December 1957 on All India Radio is still remembered by old-timers.[81]

Summing Up

In the seven short years that he spent in India J. B. S. Haldane taught, lectured, tried to shake up people and institutions, clashed with officialdom and, all the while, carried out research with an intensity that was characteristic. He constantly drew attention to the superb circumstances available in India for doing science: a wealth of tradition, customs, practices and modes of thought no less than the wealth of flora and fauna. He pointed out that biology offered an endless variety of fundamental problems to work on; one just had to look around. But the 'looking around' had to include careful observation, formulating a clear hypothesis, constructing a plausible model and, as far as possible, quantitative analysis. Peerless communicator that he was, many workers learnt the art of presenting scientific data from him. He invited scientists from abroad to collaborate on research in human genetics in India because of the unusual circumstances afforded by its marriage customs. A commitment to India and the cause of Indian science remained steadfast through all the vicissitudes of his experience with Indian ways. Resplendent in South Indian dress, at international science conferences he projected India as his country.

How should one view Haldane's engagement with India and more so with Indian science? The answer is mixed. It is important to note that the India that Haldane knew and fell in love with does not exist. Socialism appears rarely in political discourse, the country's foreign policy is that of any large nation state that aspires to power and prestige, tolerance for diversity is lessening, non-violence is apparent mostly by its absence and the dominant value system is one that emphasises industrialisation and the generation of private capital more than the reduction of disparities. To begin with, therefore, the broader philosophical reasons that attracted

---

[80] Haldane, *Science and Indian Culture.*

[81] J. B. S. Haldane, *The Unity and Diversity of Life. Sardar Vallabhbhai Patel Lectures* (Delhi: The Publications Division, 1957).

him to India have largely disappeared. That apart, for Haldane 'science' meant both the natural sciences and the humanities; logic and philosophy were particular interests. He would have been horrified at the treatment being meted out to teaching and research in the humanities.[82]

But if one restricts oneself to the natural sciences, he appears to have come at the right time. First-person accounts make it obvious that he ignited the spark of scientific inquiry amongst the young men and women who came under his spell. Some of them carried forward his legacy successfully, though much of it was outside India. He catalysed research in Drosophila genetics and more so in human genetics; both continue to be actively pursued areas in the country today. His exhortation as far as possible to exploit local circumstances in choosing a problem for study is being picked up (after a gap) by students of plant and animal behaviour, the environment and ecology. Research on infectious diseases and human genetics too, especially the latter, shows similar tendencies though to a lesser extent. However, much of what goes on in the better funded and well equipped centres of biological research continues to be laboratory-based and involves probing plant, animal or microbial systems more in line with Western science. As much as anything else the work appears to be motivated by the aim of attaining the prestige attached to Western science, and, at least in the matter of the number of publications, with science in China.[83] In this context the notion that "plenty of things can be done here that can't be done in London or Long Island," in the epigram to this chapter, appears strange. The possibility of doing interesting science without spending a great deal of money has been more or less given up in these centres but retains proponents and practitioners among smaller, 'people's' science movements.[84] Otherwise, 'taking science to the people' remains a restricted activity; newspapers tend to depend on reports from western news agencies.[85] Haldane thought the seeds of modern biology should take root in the colleges and universities of India. However, by

[82] For example, see 'The real classical languages debate' by S. Pollock; *The Hindu*, 27 November 2008 (http://www.hindu.com/2008/11/27/stories/2008112753100900.htm).

[83] To that extent the practice of modern science in India has become even more delinked from Indian culture than it was during Haldane's stay.

[84] The Kerala Sastra Sahitya Parishat an example of a popular and successful movement. Given the wider trend in India, it stands out in using a motto that Haldane would have liked, 'Science for Social Revolution'.

[85] For example, see http://www.jncasr.ac.in/etumain.php/Science-Outreach-Program/11/1/4/ and http://hindu.com/fline/fl2721/stories/20101022272111400.htm.

and large, the universities have been allowed to run down.[86] The quality of science administration is widely recognised as dismal and one can only imagine how much more Haldane would have been provoked by today's officialdom.

On the positive side, a rapid increase in opportunities for higher education has motivated a larger number of bright young men and women to study science than ever before. Curiously most of Haldane's students were compelled to seek positions abroad after Haldane's demise. One reason could be that though brilliant, none of them had a doctoral degree. The Indian system, as Haldane time and again had criticised, laid emphasis on such degrees and not on one's original thinking capacity. The fact that all those students flourished very well abroad, proves this point. Haldane's role in circulating knowledge cannot be summed up better than in his own words. He wrote, "I regard the help which I have been able to give to young men of great ability who would have had little chance without me, as a service to science as important as, and perhaps more difficult than, the first estimation of a human mutation rate or the discovery that oxygen at seven atmospheres' pressure, has a taste."[87]

[86] Instead of grappling with the difficult problem of improving them—and in some cases restoring their past status—, the response has been to establish a set of new universities. For instance, see *Frontline* (vol. 26 (14), July 04–17 2009; http://www.hindu.com/fline/fl2614/stories/20090717261401000.htm).

[87] Khan, "J. B. S. Haldane as an Informal Teacher," 99.

PART FIVE

CONCLUSION

TRANSLATION AS METHOD: IMPLICATIONS FOR HISTORY OF SCIENCE

Sundar Sarukkai*

Two significant problems in the history of science are related to the transmission and circulation of ideas and knowledge across different cultures. The first one, as to whether civilizations such as China and India 'had' science, has a slightly longer history. The second question concerns the possibility of transmission of seminal scientific concepts from China and India to Europe before the origin of 'modern science'. The idea that modern science has multi-cultural origins is an argument related to the latter.[1] In this paper, I focus on the use of translation and the theoretical assumptions in such translational activities inherent in any question that asks whether non-European cultures 'had' science. I also use the same matrix to enquire into the relation between translation and transmission. I hope to show that the framework of translation studies is extremely relevant to the discussion on transmission and circulation of knowledge, thereby suggesting yet another reason to consider the theoretical and practical aspects of translation as serious methodological issues in the history of science.

Ideas travel across cultures that are both spatially and temporally distinct. When ideas move from one culture to another they are encountered primarily through translation. Let me first clarify what I mean by translation. I follow Jakobson in his typology of three types of translation: intralingual, interlingual and intersemiotic.[2] Intralingual translations are 'translations' of a word in one language into word(s) of the same language. Finding synonyms for a word is an act of intralingual translation.

---

* This is a revised version of a paper presented at the Conference on "Circulation and Transmission of Knowledge: East-West Encounters", University of King's College, Halifax, Canada, July 21–23, 2010. I thank Gordon McOuat for his interest and support, and also the Editors of this volume, particularly Bernard Lightman, for their thoughtful comments on earlier drafts. I thank Jobin Mathew for supplying me the material on Malayalam words in science textbooks, Anu Joy for discussions on translation of scientific terms and M. D. Srinivas for clarifying some doubts about the Kerala mathematicians.

[1] See Arun Bala, *The Dialogue of Civilizations in the Birth of Modern Science* (New York: Palgrave Macmillan, 2006).

[2] R. Jakobson, *On Linguistic Aspects of Translation* in *On Translation*, ed. R. A. Brower (Cambridge, MA: Harvard University Press, 1959), 232–39.

Interlingual translations correspond to translations from one language to another. Intersemiotic translations are those translations in which words in natural language get 'replaced' by other semiotic expressions. Common examples include symbolization of a word such as replacing 'mass' with 'm' or using a graph to indicate a process. Both intralingual and intersemiotic are very important modes of translation activities within the everyday practice of science.[3] In the global transmission of science, interlingual translation also becomes important since many seminal texts are translated from one language to another.

When ideas travel across cultures, there is, similarly, a process of interlingual translation when these ideas are translated from one language into another. When ideas travel across time in the 'same' culture, these ideas get modified (for example, they acquire new meanings) through the processes of intralingual translation.

Following a similar typology, we can characterize the movement of ideas and concepts into two types: intracultural and intercultural transmission and circulation. In the former, concepts move through the 'same' culture. A paradigmatic example of this could be scientific concepts where there is movement of concepts across different theories but this movement is within the same 'culture' which would involve some common theoretical and experimental structures. The development of the concepts of atoms, mass, space and motion, for example, illustrate how the 'same' concept changes over time within the scientific community. A similar example of a non-scientific term is the changing meaning of 'mistress', which in earlier times had positive value but over time begins to develop negative connotations.

Intercultural transmission and circulation correspond to movement of ideas, concepts and linguistic terms across cultures.[4] Typically they involve interlingual translation but are, often, more than that. The movement of mathematical ideas from India to Persia or the transmission of ideas from Europe into China are fundamentally concerned with translation from one language to another. In this sense, any global history of science has to—in principle—draw upon translation studies in order to understand the complexities of the translation process and how these influence the transmission and circulation of ideas across cultures. What

[3] See Sundar Sarukkai, "Translation and Science," *Meta* XLVI (2001): 646–663.

[4] I do not mean to suggest that ideas, concepts and linguistic terms are necessarily distinct but only want to specify these as separate categories in the discussion that follows.

happens to ideas when they are translated and placed within the linguistic and cultural space of another society? Do they get modified? Do they transmute? Do they acquire new connotations? If so, how?

In this paper, I will argue that it is primarily through certain processes of translation that an original idea is appropriated into another culture and through this act the original idea might take on new characteristics. In general, it will be difficult to find an original idea in the same form after translation for the simple reason that ideas and concepts in a culture are supported by various metaphysical presuppositions and belief structures.[5] Very rarely does an idea get transmitted in its structural entirety. How then does one make sense of an idea from a culture without understanding the other concepts which support it? Very often, when ideas get transmitted, they are selectively transmitted. Consider, for example, the transmission of the ideas related to numbers, decimal system and zero from India to Arabia. At one level, there is a functional transmission of these concepts in the sense that the complete worldview of numbers and zero in Indian philosophical and mathematical systems was not borrowed in their entirety by the Arabs. In fact, these concepts get absorbed into the metaphysics and the larger supporting conceptual infrastructure of numbers in the other tradition. Another way to understand this is to say that ideas are often part of a 'theory' and when ideas are transmitted it is not necessary that the whole supporting theory gets transmitted. In contemporary times, non-allopathic systems of medicine offer a good example of this process. For example, Ayurvedic medicines and concepts of health and healing get transmitted to different cultures but rarely (if ever!) do we see the whole theory of the body (that is so essential to Ayurveda) also getting transmitted in the new culture. Similarly, when concepts of mass and energy are translated into Indian languages, very often the complete theoretical presuppositions behind these concepts are often ignored.

This phenomenon leads to a methodological problem in the context of analysing transmission and circulation of ideas. If we remove an idea or concept and view it in isolation then it makes the comparative enterprise far more difficult. I believe that the way out of this difficulty is to take the concerns of translation seriously. In fact, to make sense of an isolated

[5] Even if we ignore the larger cultural issues surrounding translation, even a simple linguistic translation modifies the original in various ways. This view is almost a given in contemporary translation studies. For example, see Andrew Benjamin, *Translation and the Nature of Philosophy* (London: Routledge, 1989); Lawrence Venuti, *The Scandals of Translation* (New York: Routledge, 1998).

concept from another culture it is best to use translation as a method (an idea which I borrow from the way science creates meanings for many of its concepts) which is not a search for equivalences but an active generation of new meaning for these concepts. Therefore, when we evaluate concepts across cultures we cannot be looking for equivalences but only for the potential to bear possible meanings, what I refer to in this paper as 'meaning-bearing capacity'. Translation as a particular method is the most optimum way towards generating these new possible meanings for concepts. In this sense, I consider translation and its role in meaning-making as important methodological issues for history of science.

Translating Scientific Terms

Let me begin with translation of scientific terms from one culture to another. Almost every language has had a difficult history of translating science and scientific terms.[6] Wright points out that Westerners, when confronted with the nature of Chinese language, doubted whether this language 'was even capable of dealing with science.'[7] When eventual translation of scientific texts (from English to Chinese) had been accomplished, there were a number of techniques developed to deal with scientific terms. Wright isolates the following strategies: 1. Not translating the term at all; 2. Transliteration of the sounds into Chinese characters chosen for their phonetic values; 3. The use of existing terms (such as the attempt to retain alchemical terms); 4. The formation of a new term by juxtaposing two or more existing characters (descriptive translation such as oxygen = nourish + vapour/gas, hydrogen = light gas); 5. The resuscitation of an archaic character; 6. The creation of a new Chinese character; 7. The use of Japanese loanwords.[8] We should note that these strategies were those followed in the early days of translating science and one can see similar strategies at work even in present day translations of scientific terms in different languages.

---

[6] See the many articles in the special issue on "Science in Translation," *The Translator* 17 (2011): 179–464.

[7] David Wright, "The Translation of Modern Western Science in Nineteenth-Century China, 1840–1895," *Isis*, 89 (1998): 653–673; 658. See also Martha Cheung, "From 'Theory' to 'Discourse': The Making of a Translation Anthology," *Bulletin of SOAS* 66 (2003): 390–401.

[8] Wright, "The Translation of Modern Western Science in Nineteenth-Century China, 1840–1895," 667–671.

In the case of early Arab translation of scientific terms, the problem, in comparison to the Chinese case, was quite different primarily because of the presence of a contiguous tradition of science and philosophy in the Arabic world. The translation of Hellenistic science into Arabic was accomplished by several translators and with the help of many methods.[9] According to Endress, there were several ways by which new terms in Arabic were created in early Islamic philosophy. One was functional: a good example of this was the 'adoption of loan words', which were words taken from the original language and re-expressed in the translated language with little or no modification. Another method was 'loan-translation' where there is a literal translation of the semantic content of the original word. One example of an 'ad hoc transliteration' was the new *falsafa* (from the Greek *philosophia*) instead of the Arabic *hikma* meaning 'wise saying'/wisdom'.[10] Endress notes that loan-translations as well as loan-words, 'function as shells for the concepts they are appointed to represent.' Another method is what Endress calls 'Paradigmatical' where 'indigenous Arabic words were applied to technical concepts by analogy, extension.'[11]

Given the importance of translation in making sense of science in the historical context, it is surprising that historians of science, as Elshakry also points out, have largely ignored the problem of translation.[12] Even when engaging with translation, they have not, in general, taken into account the theoretical insights from the discipline of translation studies. A primary reason for this indifference is the commonly-held belief that translation is a passive activity. Philosophy and translation studies have for long disputed this belief since a translated text is a new text and the space of meanings created by a translated text is different from that of the original text.[13] Elshakry cites the example of Shumayyil's translation (from yet another translation) to indicate that the 'languages between which scientific translation takes place may themselves be shifting and, indeed,

[9] Gerhard Endress, "The Language of Demonstration: Translating Science and the Formation of Terminology in Arabic Philosophy and Science," *Early Science and Medicine* 7 (2002): 231–254; 234.

[10] Ibid., 235.

[11] Ibid., 236.

[12] Marwa Elshakry, "Knowledge in Motion: The Cultural Politics of Modern Science Translations in Arabic," *Isis*, 99 (2008): 701–730; 702.

[13] See Andrew Benjamin, *Translation and the Nature of Philosophy* (London: Routledge, 1989); Lawrence Venuti, *The Scandals of Translation*, (New York: Routledge, 1998).

altered by the process of translation itself.'[14] In early Arabic translation of scientific terms, there was a process of 'domesticating' foreign terms. Transliteration of terms often indicates a difficulty in finding appropriate terms in a translated language. Such acts of transliteration, as Elshakry points out, was also politically sensitive since there was resistance to adding new foreign words to a language since, among other things, 'excessive borrowing' was seen 'as a sign of political subservience.'[15] Thus, in the early period of the 20th century, there was a focussed attempt to discover or modify Arabic terms for scientific ones. Elshakry concludes by suggesting that the transliteration of scientific terms into Arabic 'was operating under the mistaken impression that the language of science was universal' and that later translation work overturned this belief. The belief about the universality of scientific language was also open to criticism since there were political overtones in such claims.

In the case of India, almost all Indian languages encounter similar problems of translation of scientific terms. As an illustrative example, I will consider one language, Malayalam, a language spoken in the state of Kerala. Similar strategies of translation of scientific terms can be found in other Indian language science textbooks.

Here are some examples of scientific concepts in Malayalam used in the Department of Education school science textbooks (10th standard, 2004) for Malayalam medium in Kerala. For some concepts, new words in Malayalam are coined. For example, 'temperature' is translated as *thāpanila* (thāpam—heat; nila—'level' which can be seen also as a measure); 'melting' as *dravikaranam* (karanam—process and *dravikaranam* as process of becoming liquid); 'latent heat' as *leenathāpam* (leena—latent); 'electric current' as *vaidyutha pravāha theevratha* (electric flow intensity). In these cases specific scientific terms are replaced with words from Malayalam. So temperature is translated as level of heat and electric current as the intensity of flow of electricity. The semantic ambiguity arises in using a Malayalam word for heat which has its own connotations in its cultural uses.

Interestingly, there are various other terms which are not translated at all. 'Intrinsic semiconductor' is translated as *intrinsic ardhachālakam*; the word 'intrinsic' now becomes a 'part' of Malayalam. Similarly for

---

[14] Elshakry, "Knowledge in Motion: The Cultural Politics of Modern Science Translations in Arabic," 704–5.

[15] Ibid., 719.

extrinsic in 'extrinsic semiconductor'. The translation of n-type semiconductor retains 'n-type' in Malayalam. Other scientific terms that are retained as English words include the following: 'Loudspeaker', 'Mass defect', 'Binding energy', 'Radioactivity', 'Chain reaction', 'Nuclear fusion', 'Electron', 'Electrode', 'Ion', 'Electrolyte', 'Anode', 'Cathode', 'Electrolysis', 'Atomic', 'Mole', 'Nucleus', 'Electronegativity', 'Leaching', 'Calcination', 'Hardening', 'Annealing', 'Oxidation' (although for de-oxidation they use 'nir-oxi-karanam' but they don't use oxi-karanam for oxi-dation where *karanam* is process). There are many more such examples.

Why do translators adopt such complex use of words and translation from other languages? This process is not new as illustrated in the extensive work done on loanwords, loan translation/calque, loan rendering and so on. But is there something special in these types of translation that becomes the mark of translation of scientific terms? Does it say anything about science as much as it says something about translation?[16] Elshakry points to the politics of naming when growing Arab nationalism resisted the direct borrowing of non-Arabic words as part of the scientific lexicon. Here I want to point to the epistemological presuppositions hidden in such practices.

Transliteration is preferred in certain cases when the translators feel that there is no appropriate word in their language for the original word. But what is an appropriate word? While there are many dimensions to this question, for the purposes of my argument I want to highlight just one aspect, namely, the analogy between the transliteration of scientific terms and the translation of proper names. When a novel is translated into another language, the standard practice is to retain the names of the characters. So Hamlet will continue to be referred to as Hamlet in a translated version of that text. Proper names are not translated but only transliterated. Why are proper names not usually translated? In the case of proper names, one believes that there is an entity 'independent' of the name, one that functions perhaps as a 'rigid designator' of that name.[17] And since this entity is not defined by its meanings as much as by its existence, there is really no need to translate the proper name since the

---

[16] One could perhaps see this against the background of the politics of English in India and the relation between science and English. For a flavor of the larger debate about English in India, see Probal Dasgupta, *The Otherness of English: India's Auntie Tongue Syndrome* (New Delhi: Sage, 1993).

[17] Such a philosophical position is very influential in some traditions of philosophy of language. Russell's work on proper names as well as Kripke's use of rigid designators to describe proper names are well known in the literature.

reference remains the same across linguistic utterances of different languages. So when scientific terms are transliterated they are essentially functioning as proper names: radioactivity and Hamlet are both transliterated and not translated.

Many times transliteration is preferred by translators because of the perceived difficulty in finding a term in the translated language which matches the original term. Most often this difficulty reflects certain uneasiness in translating *concepts*. Sometimes transliteration becomes an easy way out when confronted with the problems of translating concepts. There are two problems with transliteration that are of relevance to this paper. One, it doesn't let the concept take on new connotations in the translated language. Second, it masks the dynamics inherent in the processes of transmission and circulation. Transliteration keeps 'alien' concepts as alien and is often used to validate strict boundaries for concepts. The debate on the problems of translating concepts such as 'science', 'logic' or 'rationality', as well as the problems of translating some ancient concepts in terms of modern scientific ones, illustrate this constant tension between transliteration and translation. Moreover, the tendency to transliterate sometimes inhibits meaningful dialogue that is necessary to understand multicultural transmission of ideas since transliteration essentially suggests that the translated language does not have the resources to handle the original concept, whether they be larger concepts such as science, logic and rationality, or scientific concepts such as mass, matter, and limit. For history of science, this tension is of some significance since it may be easier to claim that concepts such as science and rationality, or concepts related to calculus, were not available in other civilizations.[18] The point I wish to reiterate here is not that these concepts were available to different cultures but only that these issues of translation must become an integral part of a methodology of history of science.

---

[18] Almost all seminal thinkers from Europe have shared the belief that Asiatic cultures in general did not 'possess' science, logic or reason. These thinkers range from Hegel to Gadamer and include Locke and others. These were also claims that were essential to the colonial discourse. See Hans-Georg Gadamer, *The Beginning of Knowledge*, trans. R. Coltman (New York: Continuum, 2001). For the European response to claims of logic and philosophy in India, see Jonardon Ganeri, ed., *Indian logic: A Reader* (Surrey: Curzon Press, 2001). For the European response to claims of science and technology in India, see Michael Ada, *Machines as the Measure of Men* (Ithaca: Cornell University Press, 1989), and Claude Alvares, *Decolonising History*, (New York: The Apex Press, 1991). For a larger discussion on the encounter of civilizations see Wilhelm Halbfass, *India and Europe: An Essay in Understanding* (Albany: SUNY, 1988).

No doubt, the questions as to whether Indians had science or technology or possessed concepts in calculus before its development in Europe or had developed the modified heliocentric model (Tycho Brahe model which postdates a similar model described earlier by the Kerala astronomers) go beyond the issue of transliteration and translation although they may be catalysed by the activity of translation. To illustrate the problems of translation in such claims, consider the recent debate on the contribution of the Kerala astronomy school and in particular the claims of their having anticipated calculus through the creation of ideas such as infinite series and limits.

As evidence one can see the sixteenth century (c. 1530) text in Malayalam called the *Ganita-Yukti-Bhāṣā* by Jyeṣṭhadeva.[19] This text introduces seminal ideas of calculus before the origin of similar ideas in Europe. There have been claims that conceptual ideas that led to modern calculus were first taken from this text and transmitted to Europe. There have been counter arguments to this theory of transmission but here I only want to look at the problems of translation of certain terms like limit and infinite series. If we translate a term in this text as limit or infinitesimal then the associated suggestion is that the idea of limit (in the context of infinite series and calculus) was first developed in the Indian schools before it developed in Europe. So how then do we decide to translate a term in this text as 'limit'? Are such translations done with a view to claiming primacy of certain ideas?

In Chapter 6 of the *Ganita-Yukti-Bhāṣā*, there is a significant discussion on infinite series and error corrections in the context of calculating the circumference. Basically, this leads to an infinite series expansion for π. Now, if we look at the original text this expansion is written in natural language and not in a 'modern mathematical' form. In the English translation of the text alluded to above, the editors have also given a modern symbolic rewriting of the chapters. The symbolic rendering of the sixth chapter exhibits clearly—for those who are familiar with series in mathematics—what kind of a series is being discussed, what kind of conceptual ideas of calculus are possibly present in this text and so on. Without this rendering into the symbolic domain it is quite difficult to perceive links to calculus since the conceptual world in the natural language text is so different. For

[19] Jyesthadeva, *Ganita-Yukti-Bhāṣā, Vol. 1: Mathematics*, trans. and ed. K. V. Sarma, Explanatory notes by K. Ramasubramanian, M. D. Srinivas and M. S. Sriram (Delhi: Hindustan Book Agency, 2008).

example, this text gives methods to calculate Rsine values. The Sanskrit term for these values is clearly translatable to Rsine. In Malayalam it is referred to as jya whereas in Sanskrit both jya and jiva are used.

The notion of limit is a little more complicated. Although there have been claims that this text describes the idea of limits the situation is perhaps a little more complex.[20] If we ask whether there is a specific word standing for 'limit' which is used by the Kerala mathematicians, then the answer is probably no since the idea of limit is also linked to the ideas of infinity and the infinitesimal. Even in the history of calculus in Europe, the idea of limit comes much after the introduction of the infinitesimal.

However, we can discover the overlap with certain seminal themes in calculus even in texts before the *Ganita-Yukti-Bhāṣā* such as the notion of instantaneous velocity of the planets (as described by Bhāskarācārya, c. 1150) in contrast to the 'true daily motion'.[21] The word used for 'instantaneous velocity' is tātkālikagati. The first seven chapters of the *Yukti-Bhāṣā* (comprising the *Ganita-Yukti-Bhāṣā* mentioned above) primarily deals with various topics in mathematics. Ramasubramanian and Srinivas describe some of the seminal ideas in these chapters which are related to calculus: "... detailed demonstrations of the results of Mādhava such as the infinite series for π, the arc-tangent, sine and the cosine functions, the estimation of correction terms and their use in the generation of faster convergent series. Demonstrations are also provided for the classical results of Aryabhata (c. 499) on kuṭṭākāra (linear indeterminate equations), of Brahmagupta (c. 628) on the diagonals and the area of a cyclic quadrilateral, and of Bhāskara II (c. 1150) on the surface area and volume of a sphere."[22] Moreover, this text also deals with the "estimation of the end-correction terms and the transformation of the π-series to achieve faster convergence and the derivation of the infinite series for Rsine and Rcosine due to Mādhava."[23] Later work on more elaborate calculation of the instantaneous velocity of planets only reinforces the conceptual world of calculus that is explicitly present in this tradition (and one might add, results that were obtained before similar ideas arose in Europe).

[20] C. K. Raju, "Computers, Mathematics Education, and the Alternative Epistemology of the Calculus in theYuktibhāṣā," *Philosophy East and West* 51 (2001): 325–362.

[21] Jyeshtadeva, *Ganita-Yukti-Bhāṣā, Vol. 1: Mathematics*, 224–227.

[22] K. Ramasubramanian and M. D. Srinivas, "Development of Calculus", in *Studies in the History of Indian Mathematics*, ed. C. S. Seshadri (Delhi: Hindusthan Book Agency, 2010), 237.

[23] Ibid., 237.

The mathematical texts were primarily in Sanskrit (most commonly in 'poetic' form) although the *Yukti-Bhāṣā* was written in prose form in Malayalam. Translators and interpreters of these texts have to match concepts—for example, instantaneous velocity equated with tātkālikagati and kuṭṭākāra with linear indeterminate equations. The words used for summations, series and approximations are more literal and the matching seems clearer. Now, how do we show that the concepts of 'instantaneous velocity' and 'linear indeterminate' equations were indeed part of these mathematicians' vocabulary? The best way to illustrate this equivalence is through translation. What Ramasubramanian and Srinivas do is rewrite the original Malayalam prose text in symbolic form of contemporary mathematics and through this rewriting (translation) the structure of the concepts in the original text becomes clear. Thus, translation actually allows us to discover the structure of the concepts which were described in other languages.

Such a method has been used in understanding some scientific implications of Newton's work. Chandrasekhar's monumental rewriting of Newton's *Principia* is a wonderful example of how this method of translation can be used for the purposes of history of science.[24] In this rewriting by Chandrasekhar, we can clearly note the significance of the concepts inherent, but not made explicit, in Newton's book. It is clear from this symbolic rewriting what the conceptual world of the *Principia* was. Symbolic rewriting of the text is nothing but a translation of the text from one language to another. Thus, an act of translation actually serves to exhibit the common conceptual world of two seemingly different discourses such as the *Principia* and modern classical physics.

My point is that to really understand what concepts and ideas were transmitted and circulated we have to first know what concepts and ideas were available in these cultures. Translation is perhaps the only way to do this, particularly when cultures have different languages. But to say that translation is the only way is to say little unless we specify what theoretical methods or practical strategies one uses in these translations. The debates on priority of concepts across different cultures (which is also one way of looking at the Needham question) are often not sensitive to the theoretical issues surrounding the activity of translation. Therefore, this sometimes leads to rigid claims of the availability and priority of concepts.

---

[24] S. Chandrasekhar, *Newton's Principia for the Common Reader* (Oxford: Clarendon Press, 1995).

Translations are actually ways to create new meanings for concepts. Questions of translation have often been caught up in the question of whether a given set of meanings of a term in a source language maps onto similar set of meanings in the translated or the target language. This is a static view of translation. The active view of translation begins by asking how meaning gets accrued to terms. How do terms get the set of meanings that they end up possessing at a given moment? How is it that we meaningfully grapple with concepts that seem to be so different when compared to the ones we are used to? The examples of *Yukti-Bhāṣā* and *Principia* suggest that translation is a *method* to generate and exhibit meanings for terms as well as a method to compare concepts from different cultures which may look 'alien' at the first instance but through acts of translation exhibit their interconnectedness. We can extend this to claim that meanings of concepts are also partly created through the activity of translation. This is particularly true of science and, in my view, is necessary for the creation of meaning of scientific concepts. The next section is a brief description of the mechanism of this process in science. My point in describing it here is to alert us to the role of complex methods of translation that are needed in any comparative work in the history and philosophy of science.

Example of Translation as Method

Interestingly, it is in science that we see a fertile use of translation as a method for creating meaning of scientific concepts. Scientific concepts are very different from our ordinary concepts and in this sense can be called alien concepts.[25] In fact, there is one mark of alien-ness of scientific concepts and that is the creation of concepts that challenge commonsense. So how does science create a coherent discourse even though there is a constant creation of such alien concepts? One simple answer is this: through giving these alien concepts meaning without worrying in the first instance whether these meanings are correct or not. It is the generation of possible meanings of alien concepts that brings these concepts into a common discursive world. And this meaning-generation is largely catalyzed by translation. So translation becomes a method to find the boundaries of concepts and to discover the range of meanings that concepts can

25 R. M. Dancy, "Alien Concepts," *Synthese* 56 (1983): 283–300.

bear. So an essential aspect of scientific (particularly theoretical) activity is to explore the *meaning-bearing capacity* of new concepts, examples of which I discuss below. A similar approach to meaning of concepts, including concepts like 'science', might be useful in enriching the debates in history of science regarding transmission and circulation of knowledge across cultures.

Scientific concepts expand their semantic domain through the 'use' of strategies of translation. As we saw earlier, translation includes intralingual, interlingual and intersemiotic translation. In what follows, I will restrict myself to pointing out how the activity of translation actually increases the semantic content of scientific concepts.

Symbolization is an act of translation and the effectiveness of this process to compare seemingly different concepts was described in the last section. Here, I will briefly discuss one aspect of the act of symbolization in the sciences and its relation to the creation of meaning. First of all, symbolization is primarily an act of intersemiotic translation. Replacing 'time' by 't' is firstly an act of translation where a word in one language is translated into a 'word' in another 'language'. The fundamental difference between interlingual translation (where, for example, 'time' is translated to 'kāla' in Sanskrit) and intersemiotic one is based on meaning. Kāla has a semantic domain associated with it whereas 't' is devoid of meaning. The power of symbolic manipulation lies in this capacity to strip meaning from words and then, after subjecting the symbol to manipulations, put meaning back. This act is a classic description of the activity of translation whereby a word gets translated into another language and then the 'expressive capacities' of that language dictate the flow of ideas associated with the translated word. It is also indicative of a special strategy of retranslation in that there is a constant to and fro translation between the source language and the target language.[26]

Scientific discourse implicitly uses this strategy of translation through the process of symbolization. And there is a reason for this discourse to do this so effectively because this act of semiotic translation generates new meanings in ways not possible otherwise. That is, the shift into the symbolic mode is an important mode through which scientific concepts generate surplus of meaning. Consider a common example. When mass is translated into 'm' it is placed within a new language dealing with other

[26] See Sundar Sarukkai, *Translating the World: Science and Language* (Lanham: University Press of America, 2002).

symbols and operators. Once mass is put into the symbolic domain then one can *do* things with this 'word' in this new language. One can take the square root of 'm' although it doesn't make sense to take the square root of 'mass' as a word. If we look at the history of mass, one can see how every new theory of mass created new meanings of mass primarily based on interpretation of results of some symbolic manipulations.[27] So the translation into symbols allows us to do things to words which we cannot do otherwise. And the major reason for this efficacy is that symbols are not laden with meaning when these operations are performed. So scientific theorizing is a special act of translation where questions of meaning are constantly deferred. Translation is 'free' in the best sense of the word.

Now consider the problems of translating 'mass' into Indian languages. If we translate it into a word that is already available then what kind of mass is that referring to? It cannot be referring to the qualities of Newtonian and Einsteinian masses. But note that this is exactly the 'problem' when we translate mass to m. What is the meaning associated with 'm'—which kind of mass is it referring to? This question does not arise in this case since we have not decided the semantic space of 'm'. In the same way, translation of mass into a Malayalam word should be seen as a step in the formation of new meanings for mass in that language. However, the irony is that in almost all scientific contexts, including school science textbooks, mass is not translated into Malayalam but only transliterated. A counter argument to my position would be to point to the use of Sanskrit words in Malayalam. These words have often been transliterated and have become part of the vocabulary of Malayalam (and many other Indian languages also). But there are two important points of difference in absorbing Sanskrit words as against technical scientific words: one is that the languages themselves have high usage of Sanskrit words and significant portions of the languages are derived from Sanskrit. This is different from transliteration of words from English, Latin or other European languages. Second, the transliterated technical words have cultural currency too—for example, the word dravyam (from Sanskrit but now part of Malayalam, standing for 'matter') has many other connotations in the cultural space and is not restricted to the science textbooks alone. This makes transliterated words from a language like Sanskrit part of a larger process of meaning-making.

[27] Max Jammer, *Concepts of Mass in Classical and Modern Physics* (Cambridge, MA: Harvard University Press, 1961).

Scientific concepts cannot create the rich narratives about nature without the specific strategy of translation described above.[28] As much as science depends on theoretical structures (which are more than the use of symbols alone), it also needs a particular act of translation to create semantic plurality which is the most fundamental mark of scientific theorizing. In other words, translation and retranslation must be seen as part of scientific methodology and are indispensable to it. Given this emphasis on symbolization it might seem that I am primarily concerned with the mathematical and the theoretical sciences.[29] What about scientific theorizing that does not involve mathematics? First of all, there is a strong foundation of mathematics even in those areas of science which do not manifest it explicitly—for example, through dependence on prior fundamental theories. Secondly, the way language is used in these non-mathematical sciences (like some parts of biology or chemistry) illustrates a constant attempt of a move towards a 'technical' language. Nominalization is one well-known strategy that marks scientific writing even when symbolic systems are not used.[30]

So what is the consequence of understanding translation as an important method in the creation of scientific discourse? As mentioned earlier, there are contentious issues when we discover words in other cultures (both spatially and temporally displaced from the contemporary one) which we claim are translational equivalents of concepts used in contemporary discourse. Examples of terms in calculus, astronomy and indeed even larger concepts like 'science' and 'logic' have raised questions of priority when these terms have been matched with words already present in the translated language. The objection is primarily based on the claim that the meanings associated with these concepts today are not the same as terms used as translational equivalents from a different culture and

---

[28] See Sundar Sarukkai, *Translating the World.*

[29] I proceed with the supposition that the theoretical sciences are intrinsically dependent on mathematics, where mathematics should be understood broadly as a semiotic system with rules for manipulation of symbols. Galileo's influential claim that mathematics is the 'language' of nature leads to the position that other natural languages *cannot* be equivalent to it. This places mathematics in a unique domain: as a language which cannot in principle be translated. See Sundar Sarukkai, "Mathematics, Language and Translation," *Meta* 46 (2001): 664–674.

[30] See Herbert Gerstberger, "Transitions to Nominalization in Newton's 'Opticks'," *Proceedings of the Second International Conference on Story in Science Teaching*, 2008, http://sci-ed.org/Story-08-Proc.htm (accessed January 1, 2012); M. A. K. Halliday and C. M. I. M. Matthiessen, *Construing Experience through Meaning. A Language Based Approach to Cognition* (London: Cassel, 1999).

different era.[31] But this argument misses the fundamental point about how we make sense of very different concepts. The semantic domain of such concepts is actually discovered through translation. *Through translation the capacity of a concept to bear new meaning is tested.* It is an inclusive experimental strategy and not an exclusionist one. This experimental method of translation is a model that is recommended for dealing with translation of Indian science/mathematics texts. By allowing 'logic' to be translated as 'anumāna' or some words in Indian mathematics to be translated into modern mathematical concepts, we are only increasing the semantic space associated with these concepts. There are always choices made about which set of meanings is preferred but this judgement cannot occur before these acts of translation takes place.

Alien Concepts and the Ambiguity of Translation

I shall use the phrase 'alien concepts' to indicate concepts that are foreign to the conceptual structure that informs cognitive capacities of different communities. Dancy suggests that alien concepts are those that are 'essentially different' for a particular community.[32] When a modern mathematician encounters Indian mathematics, it is highly probable that even she might have trouble in making sense of the conceptual world of, say, the *Yukti-Bhāṣā*, just as much as these ancient mathematicians might have had with modern concepts in calculus. We are often confronted with completely different conceptual structures of some cultures and this leads to difficulties in understanding the behaviour of the members of these cultures. Interestingly, paradigm examples of alien concepts are mathematical and scientific concepts. I believe that scientific concepts are paradigmatically alien in that they challenge our normal cognitive capacities of sense-making. Nevertheless, we are able to negotiate with these concepts. It will be my contention that meanings are created in such concepts through the activity of translation and retranslation.[33] In this process, we

[31] For an example of the complexities in translating even a 'simple' concept, see Harsh Narain, "Finding an English equivalent for 'Guna'," *Philosophy East and West* 11 (1961): 45–51.

[32] R. M. Dancy, "Alien Concepts," *Synthese* 56 (1983): 283–300.

[33] Normally retranslation is used to refer to translated works which are again translated into the same language. Here I am using retranslation in a broader sense. It refers to the act of translating the translated text back into the source language, which is sometimes called as 'back translation'. The most effective example of this process is the translation of verbal text into mathematics which is then translated back into the original language

can see how translation begins to function like a method to create meaning in alien concepts.

Translation is arguably the dominant process by which one understands alien concepts. In fact, it is the first possible means of entry into the semantic space of such concepts. The enduring discussion on whether a culture 'had' certain concepts in its conceptual framework is often catalysed by a criticism of how these concepts have been translated. Very good examples of this process are endemic to translation of scientific terms and indeed the translation of the concept 'science' itself. In this section, I will begin with a brief discussion on the relationship between alien concepts and translation which then leads to the analysis of scientific terms as alien concepts.

Dancy makes an interesting point about alien concepts in the context of ostension.[34] The problems in understanding concepts across theories have been much discussed as the problem of incommensurability. The charge of incommensurability is not necessarily based on interlingual translation as concepts in succeeding theories (in the same 'language') are sometimes seen to be incommensurable. The larger question here is whether one can make any sense of concepts that are very different from the ones we hold? Dancy points out that many, including Davidson, answer in the negative.[35] How then can we make 'sense' of concepts from ancient Indian and Chinese societies or even contemporary scientific ones? The claim that I am making here is that specific modes of re-translations are ways to make the alien more familiar. In terms of meaning, this claim implies that meanings of concepts are actually created through engaging with the alien in *ambiguous* ways.

In the context of translating concepts across diverse cultures, I would suggest that translation functions like ostension. Merely from a translated word we cannot know what element this word is referring to in the original concept. When a translator translates logic as anumāna, she is doing the following. She is 'pointing' to the concept 'logic' and uttering the word 'anumāna' just as a native utters 'Gavagai' on seeing a rabbit. The word 'anumāna' points to 'logic' just as, following Quine's example, gavagai points to the rabbit. The ambiguity present in ostension is present in translation also and it is exactly of the same kind: *we cannot be*

after which comparison is made between the two source language texts. This is the way by which scientific theory creates new narratives.

34 Dancy, "Alien Concepts."

35 Ibid., 285.

sure what it is in the semantic space of 'anumāna' that points to what of 'logic'. The richer task of translation comes from the way science engages in re-translations in order to increase the semantic content of concepts. It is the ambiguity of translation that is necessary for the expansion of this semantic space.

One can extend this analogy between translation and ostension to understand what happens in symbolization, which is often the first step to developing theory in physics. When a symbol comes to stand for a term, say 'm' for 'mass', this is an act of translation. Like ostension, one can point to 'm' and utter 'mass'. A person might not understand what in 'm' really refers to 'mass'. Symbolization is an extension of the ostension problem except that in this case it is *maximally ambiguous*. There is really little that constrains what 'm' could be and this allows its use in symbolic manipulations that are allowed under that particular theory. At various points, the expressions based on 'm' are then retranslated back into the language associated with 'mass' and new meanings are thus added to 'mass'. This process is well exemplified in the long history of mass from Newtonian to the quantum formulation of mass.[36] Even fundamental principles like energy-mass equivalence follow a similar process. Re-translation really functions as a scientific method which allows the creation of rich semantic spaces associated with scientific concepts. In fact, scientific texts use multiple semiotic systems like figures, diagrams, graphs, pictures and symbols primarily to create a rich surplus of meaning.[37]

It is also through translation and retranslation that concepts get new meaning. Boundaries of concepts are expanded through retranslations like in the case of mass. Alien concepts cannot be grasped in full and ambiguity characterises the relation between the different linguistic descriptions for a concept. This ambiguity is not a problem; on the contrary, this is the mode by which the boundaries of concepts get redefined. In the case of concepts of calculus discussed by the Kerala mathematicians one would have to say that in principle it is not a question of exact matching of concepts of calculus in their tradition with the European ones but it is

36 See Jammer, *Concepts of Mass in Classical and Modern Physics*, on how the concepts of mass get reformulated over centuries.

37 See Sundar Sarukkai, *Translating the World*, for a detailed discussion of this process. For the nature of scientific texts, see J. Lemke, "Multiplying Meaning" in *Reading Science*, eds. J. R. Martin and R. Veel, (London: Routledge, 1988), and Timothy Lenoir, ed., *Inscribing Science: Scientific Texts and the Materiality of Communication* (Stanford: Stanford University Press, 1998).

more a question of how retranslations add new possible meaning to these concepts in *both* these traditions.

Similarly, the meaning of 'logic' gets expanded once there is the engagement with its translated counterpart, 'anumāna'. Claiming that anumāna is not strictly like logic, as understood by the Greeks, is not a useful strategy. Instead, once logic is translated as anumāna, the next step is to explore how the meaning of logic changes in negotiation with the meanings of anumāna.[38] This observation also leads us to conclude that translation of concepts as if they were proper names (using the foreign words like names) is antithetical to this process of meaning making of concepts. I would rather see Malayalam words for 'mass', 'intrinsic', 'extrinsic', 'nuclear fission' etc., in science textbooks for they will suggest new meanings of these terms which will further catalyse the growth of these ideas within the community of Malayalam speakers leading to creative work in science. To the argument that transliterated words have often become part of another language and hence there is really no problem in these transliterations, I would only suggest that transliteration of scientific terms show far more obduracy in becoming part of the semantic world of a verbal language. They retain their exclusivity by various discursive means. The argument in this paper about the creative uses of translation in matters of interest to history of science is not against the presumed hegemony and universality of science nor is it to primarily support local language writing in science. It is purely a matter of following a practice that has had a great success in scientific theorizing. Needless argument about whether the Kerala mathematicians 'really' had knowledge of 'calculus' misses the point about the relation between translation and meaning. A dynamic and open view of translation and re-translation should inform our understanding of the circulation of knowledge across cultures.

[38] For more on the nature of Indian logic, see B. K. Matilal, *Logic, Language and Reality: Indian Philosophy and Contemporary Issues* (Delhi: Motilal Banarsidass, 1985); J. N. Mohanty, *Reason and Tradition in Indian Thought: An Essay on the Nature of Indian Philosophical Thinking* (Oxford: Clarendon Press, 1992); Sundar Sarukkai, *Indian Philosophy and Philosophy of Science* (Delhi: CSC/Motilal Banarsidass, 2005).

INDEX